Plant Genetics

Plant Genetics

Editor: Ben Davies

www.callistoreference.com

Callisto Reference,
118-35 Queens Blvd., Suite 400,
Forest Hills, NY 11375, USA

Visit us on the World Wide Web at:
www.callistoreference.com

ISBN: 978-1-63239-966-3 (Hardback)

Cataloging-in-Publication Data

Plant genetics / edited by Ben Davies.
        p. cm.
Includes bibliographical references and index.
ISBN 978-1-63239-966-3
1. Plant genetics. I. Davies, Ben.
QK981 .P53 2018
581.35--dc23

Table of Contents

Permissions

List of Contributors

Index

Preface

Every book is initially just a concept; it takes months of research and hard work to give it the final shape in which the readers receive it. In its early stages, this book also went through rigorous reviewing. The notable contributions made by experts from across the globe were first molded into patterned chapters and then arranged in a sensibly sequential manner to bring out the best results.

Plant genetics and plant breeding are rapidly expanding areas. The two main ways of genetically modifying plants are gene gun method and agrobacterium method. New scientific advances in the field of genetics are exploring new frontiers in plant breeding. Crops are genetically modified for various reasons such as to increase yield, develop resistance towards pathogens, etc. Different approaches, evaluations, methodologies and advanced studies on plant genetics have been included in this book. It aims to serve as a resource guide for students and experts alike and contribute to the growth of the discipline.

It has been my immense pleasure to be a part of this project and to contribute my years of learning in such a meaningful form. I would like to take this opportunity to thank all the people who have been associated with the completion of this book at any step.

Editor

Genetic Divergence among Camu-Camu Plant Populations Based on the Initial Characteristics of the Plants

Bardales-Lozano Ricardo Manuel[1], Edvan Alves Chagas[2], Oscar Smiderle[2], Abanto-Rodriguez Carlos[1], Pollyana Cardoso Chagas[3], Adamor Barbosa Mota Filho[3], Olisson Mesquita Souza[3] & Antonio Carlos Centeno Cordeiro[2]

[1] Ret Bionorte (Multi-institutional Programme of Amazon), Brazil

[2] Empresa Brasileira de Pesquisa Agropecuária, Embrapa, Brazil

[3] Universidade Federal de Roraima (UFRR), Brazil

Correspondence: Bardales-Lozano Ricardo Manuel, Ret Bionorte (Multi-institutional Programme of Amazon), Brazil. E-mail: rbardaleslozano@yahoo.es

Abstract

The objective in the present work was to evaluate the genetic diversity among 15 indigenous populations of camu-camu plants, identifying important characteristics in the evaluation of genetic divergence, based on the initial characteristics of the seedlings. Seeds extracted from fruits deriving from fifteen indigenous populations of camu-camu were collected. The experimental design was entirely random, with fifteen treatments (populations), and fifteen repetitions (each sub-sample), considering 30 seeds per subsample as an experimental unit. At 40 days after sowing the following were evaluated: the percentage of emergence, the index of emergence velocity, the average time of germination, the height of the seedling and the number of leaves. The data obtained was submitted to variance analysis, and the averages were grouped by the Scott and Knott (1974) test. The genetic diversity was studied according to the Tocher grouping method, based on the Mahalanobis distance (D^2_{ii}) and canonical variables. The fifteen populations are divergent among themselves and the Rio Branco Estirão do Veado, Rio Branco Onofre and Igarapé Agua Boa populations are indicated to have hybridization with other populations due to the high divergence, as well as the rates of emergence and vigor of the seedlings. The height of the seedlings, percentage and speed of emergence, are those that most indicate genetic divergence. The measuring techniques of genetic divergence, canonical variables Mahalanobis distances are useful and corroborating in the evaluation of genetic divergence of the camu-camu plant.

Keywords: Amazonia, canonical variables, genetic variability, multivariated analysis, *Myrciaria dubia*

1. Introduction

The camu-camu (*Myrciaria dubia* (kunth) McVaugh), is an Amazonian fruit species of the Mirtaceae family, that stands out for its elevated level of vitamin C, which can reach from 3 to 8 g per 100 g of pulp, exceeding values presented by the majority of plants cultivated in Brazil (Bardales et al., 2014; Chagas et al., 2015), in addition to containing diverse antioxidant and nutritional composites (Zanata & Mercadante, 2007; Chirinos et al., 2010; Akter, Oh, Eun, & Ahmed, 2011; Imán, Pinedo, & Melchor, 2011).

In Amazonia, the potential for camu-camu is in its use in the preparation of foods like juices, sweets and ferments (Rodrigues et al., 2004; Teixeira et al., 2004; Chirinos et al., 2010; Akter, Oh, Eun, & Ahmed, 2011). It also constitutes a raw material for the cosmetic, chemical, pharmacological industries, food preservation and production of aerated beverages (Correa, 2000; Yuyama, 2011). Thus, the production and the utilization of the fruit appear to be viable alternatives in regional development, as a means of aggregating value from the natural resources available in the region (Welter et al., 2011; Chagas et al., 2015).

The principal function of genetic improvement of the camu-camu is selecting the genotypes which maximize yield from the first phases of their development (Pinedo, Linares, Mendoza, & Anguiz, 2004; Yuyama & Valente, 2011). Genetic divergence is one of the most important parameters evaluated by improvers of the plant in the initial phase of a genetic improvement programme, because, adequately explored, it may accelerate the genetic progress of particular characteristics (Negreiros et al., 2008). In relation to the germination and emergence of seeds of the camu-camu plant, the beginning and the end occur in an irregular fashion, with this period possibly being from 15 to 120 days, making it difficult to form and produce shoots, as they are not uniform (Pinedo, Linares, Mendoza, & Anguiz, 2004; Bardales et al., 2014).

The genetic diversity may be investigated precociously by the physiological quality of the seeds, utilizing vigour tests (Dias & Marcos Filho, 1995, Pinedo, Linares, Mendoza, & Anguiz, 2004). This way, the average time for germination, emergence and uniformity can facilitate the production of shoots on the commercial scale in a more efficient manner (Negreiros et al., 2008).

In the prediction of genetic divergence, various multivariated methods can be applied such as analysis of principle components, canonical variables and the agglomerated methods. The choice of the most appropriate method should be done in relation to the level of precision desired, the ease of analysis and the form in which the data was obtained (Cruz, Carvalho, & Vencovsky, 2004). These multivariated techniques to estimate the genetic divergence among populations have been utilized in various works and with diverse species, such as the coconut (Ribeiro, Soares, & Ramalho, 1999), açaí plant (Oliveira, Ferreira, & Santos, 2007), passion fruit plant (Negreiros et al., 2008), castanheira-do-gurgueia (Ribeiro, Souza, & Lopes, 2012) and pupunha plant (Negreiros, Bergo, Miqueloni, & Lunz, 2013).

In this context, the objective of the present work was to evaluate the genetic diversity among populations of the camu-camu plant, separating the more important characteristics in genetic divergence, based on the initial characteristics of the plants.

2. Methods

The work was done at the Fruticulture Sector of Embrapa Roraima with seeds extracted from fruits derived from fifteen indigenous populations of the camu-camu (*M. dubia*) of the State of Roraima (Figure 1).

At the moment of collection, the fruits presented stage seven (green redish) and eight (red-wine coloured) maturation, according to Inga et al. (2001). In each collection location, the average size of the samples (n) was 15, and for each subsample, 60 fruits were collected, conditioned in polypropylene sacks, maintained in expanded polystyrene boxes, with ice, and carried to the laboratory. Each population received a code with the abbreviation of the name of the river or creek of origin (Table 1).

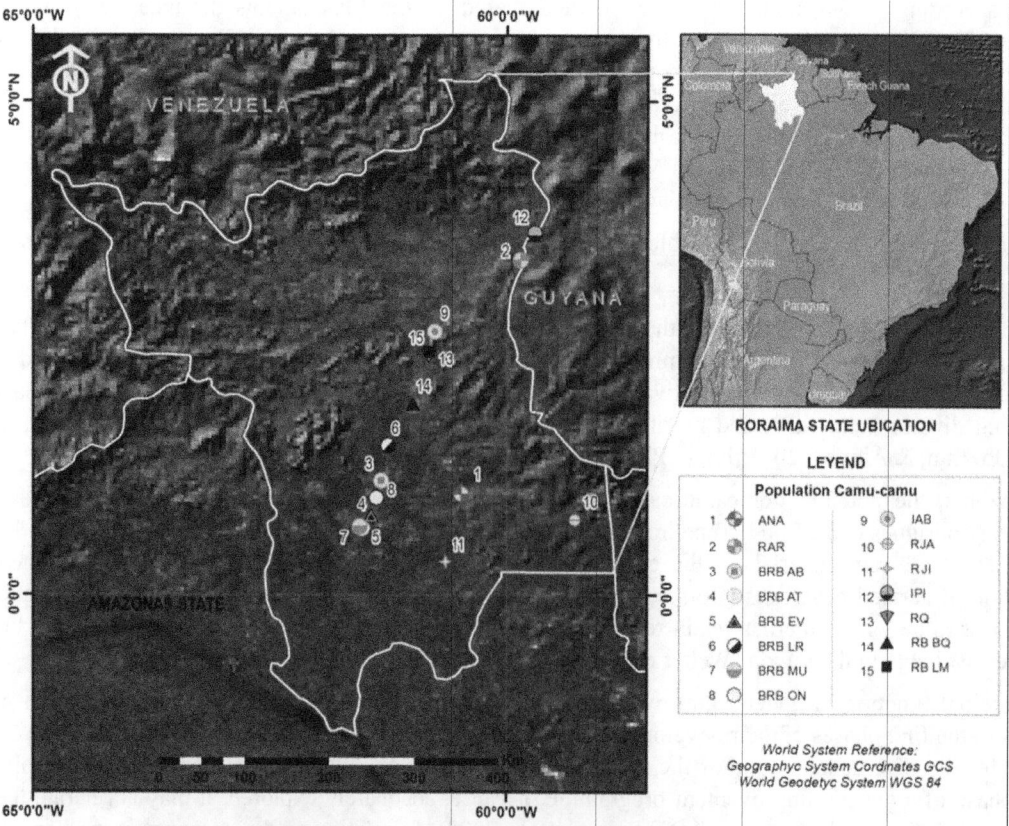

Figure 1. Location of sampling points of 15 camu-camu populations in the State of Roraima

The seeds were manually separated from the fruit and the residual pulp, by friction in a fine grade sieve. After removal, the seeds were washed in running water and treated with sodium hypochlorite solution at 10% for five minutes. The seeds were not submitted to artificial drying. Thereafter, the seeds were stored for 15 days in cold chamber at 10 °C in polypropylene sacks, and the relative humidity of the air was maintained at 50%. Subsequently, they were planted in seedbeds containing sand substrate and sawdust in 1:1 proportions. The seedbed was placed in a nebulization chamber, with irrigation intervals at four times per day, for a period of 10 minutes. The seeds were distributed at a distance of 2 cm between rows and 1 cm between them in line, and at 1.5 cm of depth. The evaluation criteria adopted was counting as the epicotyl appeared 1.5 cm above the surface of the substrate, in emergence stage.

Table 1. Natural camu-camu populations prospected in the State of Roraima in different locations, municipalities and Hydrographical Regions

Population	Location	Municipality	Region Hydrographic
RAR	Rio Arraia	Bonfim	Alto Rio Branco
IPI	Rio Tacutu- Igarapé (Ig.) Pirara	Normandia	Alto Rio Branco
RB LM	Rio Branco- Lago da Morena	Cantá	Alto Rio Branco
IAB	Rio Mucajaí- Ig. Água Boa	Mucajaí	Alto Rio Branco
RQ	Rio Quitauaú	Cantá	Alto Rio Branco
RB BQ	Rio Branco- Bem Querer	Caracaraí	Médio Rio Branco
RAN	Rio Anauá	Rorainópolis	Baixo Rio Branco
BRB AB	Rio Branco- Ig. Água Boa	Caracaraí	Baixo Rio Branco
BRB AT	Rio Branco- Ig. Açaí Tuba	Caracaraí	Baixo Rio Branco
BRB EV	Rio Branco- Ig. Estirão do Veado	Caracaraí	Baixo Rio Branco
BRB LR	Rio Branco- Lago do Rei	Caracaraí	Baixo Rio Branco
BRB UM	Rio Branco- Lago Muçum	Caracaraí	Baixo Rio Branco
BRB ON	Rio Branco- Ig. Onofre	Caracaraí	Baixo Rio Branco
RJI	Rio Jauaperí	Rorainópolis	Sub-Bacia Rio Negro
RJA	Rio Jatapu	Caroebe/Entre Rios	Sub-Bacia Rio Amazonas

The experimental delineation was entirely random, with fifteen treatments (indigenous populations) and fifteen repetitions (subsamples), considering thirty seeds per subsample as an experimental unit, totalling 450 seeds per treatment.

At 40 days after planting, the experiment was finalized as two populations had already presented 100% emergence. The percentage of emergence was evaluated, the speed emergence of seedling (SES, index) (Maguire, 1962), the average time of emergence (Yuyama, Mendes, & Valente, 2011), the height of the plant shoots (cm) and the number of leaves emitted. The emergence percentage data were transformed in square root of arcsine "$x/100$" and the SES in square root "$x + 0.5$" (Gotelli & Ellison, 2011).

$$arcosine\sqrt{x/100} \tag{1}$$

$$\sqrt{x + 0.5} \tag{2}$$

The data from the rest of the variables was not transformed. The SES was established from the emergence test, with daily evaluations being done upon the emergence of the first plants up until the 40th day.

The data was submitted to variance analysis in order to verify the existence of genetic variability among the populations, being that their averages were grouped in accordance with the Scott and Knott (1974) test, at 5% probability. Thereafter, multivariated analyses were used, applying the grouping and canonical variable techniques with the assistance of INFOGEN software, version 2013 (Balzarini & Di Rienzo, 2013).

In the grouping technique, Mahalanobis generalized distance was utilized (D2ii) (Mahalanobis, 1936) as a dissimilarity measure. In the group delimitations, the Tocher optimization method was used, adopting the criteria that the average of the measurements of genetic divergence within each group aught to be less than the average distances between groups (Cruz, Regazzi, & Carneiro, 2004).

Additionally, the relative contributions of the characteristics to genetic divergence was quantified by Mahalanobis generalized distances, utilizing the criteria proposed by Singh (1981), analyzed with the assistance of GENES software, version 2005 (Cruz, 2008).

3. Results

By univariated variance analysis, there were significant differences between the population averages ($p < 0.01$), through testing of F, for all the evaluated characteristics (Table 2), indicating at least divergence among the populations.

Based on the grouping of averages, it was verified that only two populations (RJI and RAN) presented percentages of emergence (EP) below 50%, while the population average of emergence had been 73.8%. The BRB EV and IAB populations obtained 100% EP, indicating that these materials are promissory for future work in improvement. The populations which presented the highest indices of emergence velocity, plant shoot height and number of leaves were those that obtained values above 80% emergence, with the stand out populations being the BRB EV, RB LM and IAB. The shortest emergence times were registered in the RB LM and BRB ON populations with values less than 33 days (Table 2).

Table 2. Average values for emergence of plants (EP, %), speed emergence of seedling (SES, index), average time of emergence (ATE, days), height of plants (HP, cm) and number of leaves (NL) obtained for 15 indigenous populations of the camu-camu plant

Treatments	Initial Characteristics									
	EP (%)		SES (index)		ATE (days)		HP (cm)		NL	
BRB AB	56[*]	c	0.83	b	33..52	b	15..03	b	15..67	c
BRB AT	60	c	1.28	b	33..55	b	15..90	a	18..55	a
BRB EV	100	a	2.09	a	33..30	b	15..39	a	16..89	b
BRB LR	87	b	1.35	b	34.64	d	11.36	c	16.21	b
BRB UM	69	c	1.10	b	34.90	d	9.67	d	15.40	c
BRB ON	84	b	1.45	b	32.91	b	15.60	a	17.20	b
IAB	100	a	2.08	a	34.39	d	15.48	a	18.60	a
IPI	75	b	1.38	b	33.07	b	12.70	b	14.00	c
RAN	49	c	0.92	b	33.20	b	13.02	b	16.40	b
RAR	80	b	1.36	b	33.35	b	8.85	d	13.33	c
RB BQ	77	b	1.32	b	33.80	c	9.79	d	15.00	c
RB LM	82	b	2.06	a	32.02	a	10.44	c	17.17	b
RJA	64	c	1.18	b	33.09	b	10.67	c	14.44	c
RJI	45	c	1.04	b	33.32	b	12.83	b	17.17	b
RQ	78	b	1.47	b	33.41	b	9.97	d	14.62	c
F (population)	5.18	**	7.23	**	9.30	**	39.78	**	6.99	**
Overall average	73.80		1.39		33.50		12.45		16.04	
CV (%)	29.07		11.28		3.35		11.61		13.14	

Note. [*]: Averages with the same letter, in column, belong to the same Scott-Knott grouping at 5% probability; **: Significant in F testing at 1% probability; CV = variation coefficient %.

With a basis in the relative magnitude of D2ii values, the formation of seven distinct groups was verified using the Tocher grouping method, being that the largest concentration of populations was in the first group (Table 3).

Table 3. Grouping of 15 indigenous populations of the camu-camu plant by the Tocher optimization method, based on the Mahalanobis generalized distance (D2ii)

Group	Population				
< 1 >	BRB MU	RB BQ	BRB LR	RQ	RJA
< 2 >	RAN	RJI			
< 3 >	BRB EV	BRB ON	IAB		
< 4 >	BRB AB	BRB AT			
< 5 >	IPI				
< 6 >	RB LM				
< 7 >	RAR				

Note. Greatest distance between the minimums: 3.43.

Group 1 was composed of five populations that were grouped according to those that presented the lowest average values in the plant height characteristic (< 11.5 cm). Of these, the BRB LR and BRB MU populations presented the most time for emergence (> 34.6 days). Group 2 was composed of two populations that presented emergence percentages of less than 50%.

Group 3 was composed of populations that presented the highest values in SES, height of plant shoots and number of leaves, the same obtained values higher than 80% of emergence, and therefore, is indicated for future work in improvement such as crossing, with an eye to obtaining progenies with high heterosis in emergence and plant vigour. Group 4 was made up of two populations, grouped according to the following characteristics: percentage of emergence (≤ 60%) and ATE (33.5 days).

The populations that remained isolated, those are the RB LM and RAR populations, which presented considerable divergence, with percentage of emergence of 80 and 82%, ATE of 32.02 and 33.35 days, and plant shoot height of 10.40 and 8.85 cm, respectively (Table 2). These results were similar to those of the formation of some groups through canonical variance analysis (Figure 2), where a possible structuring of the group was observed, similar to the Tocher method (Table 3), which resulted in two auto-values or canonical variables, of which the first two constituted 86.07% of original variance of the data (CV1 = 68.31%; CV2 = 17.76%).

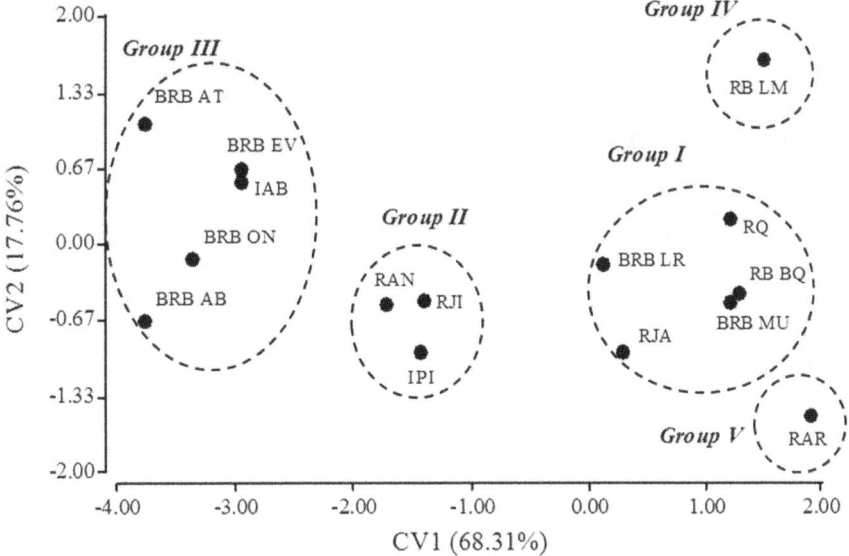

Figure 2. Graphic dispersal by scores of 15 indigenous camu-camu populations and respective groupings, in relation to the representative axes of the canonical variables (CV1 and CV2)

4. Discussion

According to Tekrony and Egli (1991), the vigour of plants, observed in the field by the ability of the seed to emerge and grow rapidly and vigorously, is a factor which can influence the productivity of cultures. According to Pinedo, Linares, Mendoza, and Anguiz, (2004), part of genetic improvement of the camu-camu is related to the selection of genotypes based on their rapidity, percentage of emergence and vigour of plant shoots, among other agronomic characteristics.

According to Cruz, Regazzi, and Carneiro (2004), when the first two canonical variables are above 80% of total variation, their utilization is satisfactory in the study of genetic divergence by way of the evaluation of the graphic dispersal of the scores in relation to the canonical variables (CV1 and CV2). This constant structure, through the formation of group coincidence in the utilization of complementary methods for morpho-agronomic characteristics, generates greater confidence in the results (Sudré et al., 2005; Oliveira, Ferreira, & Santos, 2007; Negreiros, Bergo, Miqueloni, & Lunz, 2013). Thus, the BRB AB and BRB AT populations that had to be classified as one group (Group 4), with the Tocher method, were classified by canonical variables within group III, together with the BRB EV, BRB ON and IAB populations. However, the IPI population, isolated in the Tocher method, was classified, by canonical variables, within group II, together with the RAN and RJI populations (Figure 2).

In group III, the populations of the lower Branco river BRB AB, AT, EV and ON, are interconnected by the same hydrographical region, which probably is related to the high values registered in the initial plant characteristics, representing a region of great potential for obtaining promissory genotypes for future improvement works.

The plant shoot height variable was the most important in distinguishing the populations in the first canonical variable (CV1) (68.31%), followed by the percentage of emergence and the average time of emergence, being that the relative contribution of these three characteristics to the genetic diversity among the 15 populations is confirmed, with a basis in the criteria proposed by Singh (1981) (Table 4). Thus contribution was verified in descending order as follows: plant shoot height, percentage of emergence, average time of emergence and SES. The number of leaves presented the lowest estimate of genetic diversity between the populations (S.j), not being important for the evaluation of genetic divergence between populations.

Table 4. Estimates of relative contribution of each characteristic (S.j) to the genetic divergence between the camu-camu plant populations, based on the partition of the total D^2_{ii}

Variable	S.j	Value (%)
Plant height (cm)	972.35	75.99
Plant shoot emergence (%)	131.11	10.25
Average time for emergence (days)	86.18	6.73
speed emergence of seedling(index)	70.1	5.48
Number of leaves	19.9	1.56

Note. S.j = S is the average relative importance for each variable; j = for the study of genetic diversity.

The percentage of emergence and plant shoot height characteristics contributed to 86.24% of divergence, which could be the initial agronomic parameters to be considered in the selection of genotypes in future works in genetic improvement in the species, bearing the values observed in mind.

5. Conclusions

There is genetic divergence among the fifteen populations and the BRB EV, BRB ON and IAB populations can be indicated for hybridization with other populations due to high genetic divergence, rate of emergence and vigour of the plant shoots.

The height of the plant shoots, percentage of emergence and the ATE are the characteristics of highest contribution to genetic divergence identified in the camu-camu plant.

The canonical variables and the Mahalanobis distance are useful and complementary in the evaluation of genetic divergence of the camu-camu plant.

References

Akter, M. S., Oh, S., Eun, J. B., & Ahmed, M. (2011). Nutritional compositions and health promoting phytochemicals of camu-camu (*Myrciaria dubia*) fruit: A review. *Food Research International, 44*(7), 1728-173. http://dx.doi.org/10.1016/j.foodres.2011.03.045

Bacelar-Lima, C. G. (2009). *Estudos da biologia reprodutiva, morfologia e polinização aplicadas à produção de frutos de Camu-camu (Myrciaria dubia (H.B.K.) McVaugh) adaptadas à terra firme da Amazônia Central/Brasil* (Doctoral dissertation, p. 121). Instituto Nacional de Pesquisas da Amazônia-INPA, Universidade Federal Da Amazônia-UFAM.

Balzarini, M. G., & Di Rienzo, J. A. (2013). *InfoGen versión Software estadístico para el análisis de datos genéticos*. FCA, Universidad Nacional de Córdoba, Argentina. Retrieved from http://www.info-gen.com.ar

Bardales, A. E. M., Pisco, E. G. C., Flores, A. J. F., Mashacuri, N. R., Ruiz, M. C., Correa, S. A. I., & Gómez, J. C. C. (2014). Semillas e plántulas de *Myrciaria dubia* 'camu-camu': Biometría, germinación y crecimiento inicial. *Scientia Agropecuária, 5*(1), 85-92. http://dx.doi.org/10.17268/sci.agropecu.2014.02.03

Chagas, E. A., Lozano, R. M. B., Chagas, C, P., Bacelar-Lima, C. G., Garcia, M. I. R., Oliveira, J. V., ... Araújo, M. C. R. (2015). Variabilidade intraespecífica de frutos de camu-camu em populações nativas na Amazônia Setentrional. *Crop Breeding and Applied Biotechnology, 15*(4), 265-271. http://dx.doi.org/10.1590/1984-70332015v15n4a44

Chirinos, R. Galarza, J., Betalleluz-Pallardel, I., Pedreschi, R., & Campos, D. (2010). Antioxidant compounds and antioxidant capacity of Peruvian camu camu (*Myrciaria dubia* (H.B.K.) McVaugh) fruit at different maturity stages. *Food Chemistry, 120*(4), 1019-1024. http://dx.doi.org/10.1016/j.foodchem.2009.11.041

Correa, S. A. I. (2000). *Cultivo de camu-camu Myrciaria dubia H.B.K. en la Región de Loreto* (p. 32). Iquitos: INIA.

Cruz, C. D., Regazzi, A. J., & Carneiro, P. C. S. (2004). *Modelos biométricos aplicados ao melhoramento genético* (Vol. 1, pp. 171-201). Viçosa: Editora UFV.

Cruz, C. D., Carvalho, S. P., & Vencovsky, R. (2004). Estudo sobre divergência genética. I. Fatores que afetam a predição do comportamento de híbridos. *Revista Ceres, 41*, 178-182. Retrieved from http://www.ceres.ufv.br/ojs/index.php/ceres/article/viewFile/2068/115

Cruz, C. D. (2008). *Programa Genes: Aplicativo computacional em genética estatística*. Versão para Windows. Viçosa: Editora UFV.

Dias, D. C. F. S., & Marcos Filho, J. (1995) Testes de vigor baseados na permeabilidade das membranas celulares: I. Condutividade elétrica. *Informativo Abrates, 5*, 26-33. Retrieved from http://www.scielo.br/scielo.php?script=sci_nlinks&ref=000071&pid=S0101-31222011000100001400005&lng=en

Gotelli, N. J., & Ellison, A. M. (2011). *Princípios de estatística em ecologia*. Porto Alegre: Armed Editora S.A.

Imán, S., Pinedo, S., & Melchor, M. M. (2011). Caracterización morfológica y evaluación de la colección nacional de germoplasma de camu camu *Myrciaria dubia* (H.B.K) McVaugh, del INIA Loreto-Perú. *Scientia Agropecuaria, 2*, 189-201. http://dx.doi.org/10.17268/sci.agropecu.2011.04.01

Inga, H., Pinedo, M., Delgado, C., Linares, C., & Mejía, K. (2001). Fenologia reprodutiva de *Myrciaria dubia* McVaugh (H.B.K.) camu camu, *Folia Amazónica, 12*(1-2), 99-106. Retrieved from http://www.iiap.org.pe/Upload/Publicacion/PUBL697.pdf

Maguire, J. D. (1962). Speed of germination: And in selection and evaluation for seedling emergence and vigor. *Crop Science, 2*(3), 176-177. http://dx.doi.org/10.2135/cropsci1962.0011183X000200020033x

Mahalanobis, P. C. (1936). On the generalized distance in statistics. *Proceedings of the National Institute of Science of India, 2*, 49-55.

Negreiros, J. R. S., Bergo, C. L., Miqueloni, D. P., & Lunz, A. M. P. (2013). Divergência genética entre progênies de pupunheira quanto a caracteres de palmito. *Pesquisa Agropecuária Brasileira, 48*(5), 496-503. http://dx.doi.org/10.1590/S0100-204X2013000500005

Negreiros, J. R. S., Alexandre, R. S., Álvares, V. S., Bruckner, C. H., & Cruz, C. D. (2008). Divergência genética entre progênies de maracujazeiro-amarelo com base em características das plântulas. *Revista Brasileira de Fruticultura, 30*(1), 197-201. http://dx.doi.org/10.1590/S0100-29452008000100036

Oliveira, M. do S. P. de, Ferreira, D. F., & Santos, J. B. dos. (2007). Divergência genética entre acessos de açaizeiro fundamentada em descritores morfoagronômicos. *Pesquisa Agropecuária Brasileira, 42*, 501-506. http://dx.doi.org/10.1590/S0100-204X2007000400007

Pinedo, M., Linares, C., Mendoza, H., & Anguiz, R. (2004). *Plan de mejoramiento genético de camu camu* (1st ed., p. 54). Iquitos-Perú, IIAP.

Ribeiro, F. E., Soares, A. R., & Ramalho, M. A. P. (1999). Divergência genética entre populações de coqueiro-gigante-do-Brasil. *Pesquisa Agropecuária Brasileira, 34*(9), 1615-1622. http://dx.doi.org/10.1590/S0100-204X1999000900012

Ribeiro, F. S. de C., Souza, V. A. B. de, & Lopes, A. C. de A. (2012). Diversidade genética em castanheira-do-gurgueia (*Dipterix lacunifera* Ducke) com base em características físicas e químiconutricionais do fruto. *Revista Brasileira de Fruticultura, 34*, 190-199. http://dx.doi.org/10.1590/S0100-29452012000100026

Rodrigues, R. B., Menezes, H. C., Cabral, L. M. C., Dornier, M., Rios, G. M., & Rynes, M. (2004). Evaluation of reverse osmosis and osmotic evaporation to concentrate camu-camu juice (*Myrciaria dubia*). *Journal of Food Engineering, 63*(1), 97-102. http://dx.doi.org/10.1016/j.jfoodeng.2003.07.009

Scott, A., & Knott, M. (1974). Cluster-analysis method for grouping means in analysis of variance. *Biometrics, 30*(3), 507-512. http://dx.doi.org/10.2307/2529204

Singh, D. (1981). The relative importance of characters affecting genetic divergence. *The Indian Journal of Genetic and Plant Breeding, 41*(1), 237-245.

Sudré, C. P., Rodrigues, R., Riva, E. M., Karasawa, M., & Amaral Júnior, A. T. do. (2005). Divergência genética entre acessos de pimenta e pimentão utilizando técnicas multivariadas. *Horticultura Brasileira, 23*, 22-27. http://dx.doi.org/10.1590/S0102-05362005000100005

Teixeira, S. A., Chaves, S. L., & Yuyama, K. (2004). Esterases no exame da estrutura populacional de Camu-camu (*Myrciaria dubia* (Kunth) McVaugh-Myrtaceae). *Acta Amazônica, 34*(1), 89-96. http://dx.doi.org/10.1590/S0044-59672004000100011

Tekrony, M. D., & Egli, D. B. (n.d.). Relationship of seed vigor to crop yield: A review. *Crop Science, 31*(3), 816-822. http://dx.doi.org/10.2135/cropsci1991.0011183X003100030054x

Welter, M. K., Melo, F. V., Bruckner, C. H., Góes, H. T. P. D., Chagas, A. E., & Uchôa, S. C. P. (2011). Efeito da aplicação de pó de basalto no desenvolvimento inicial de mudas de camu-camu (*Myrciaria dubia* H.B.K. McVaugh). *Revista Brasileira de Fruticultura, 33*(3), 922-931. http://dx.doi.org/10.1590/S0100-29452011000300028

Yuyama, K. A. (2011). Cultura de camu-camu no Brasil. *Revista Brasileira de Fruticultura, 33*(2), 335-690. http://dx.doi.org/10.1590/S0100-29452011000200001

Yuyama, K., Mendes, N. B., & Valente, J. P. (2011). Longevidade de sementes de camu-camu submetidas a diferentes ambientes e formas de conservação. *Revista Brasileira de Fruticultura, 33*(2), 601-607. http://dx.doi.org/10.1590/S0100-29452011005000067

Yuyama, K., & Valente, J. P. (2011). *Camu-camu (Myrciaria dubia (Kunth) Mac Vaugh)* (p. 216). Curitiba: CRV.

Zanatta, C., & Mercadante, A. (2007). Carotenoid composition from the Brazilian tropical fruit camu-camu (*Myrciaria dubia*). *Food Chemistry, 101*(4), 1526-1532. http://dx.doi.org/10.1016/j.foodchem.2006.04.004

Genetic Diversity of Remaining Populations of Mangaba (*Hancornia speciosa* Gomes) in Restingas of Brazil

Ana Veruska Cruz da Silva[1], Julie Anne Espíndola Amorim[2], Marília Freitas de Vasconcelos Melo[3], Ana da Silva Ledo[1] & Allivia Rouse Carregosa Rabbani[4]

[1] Embrapa Coastal Tablelands, Aracaju, Sergipe, Brazil

[2] State University of São Paulo 'Júlio de Mesquita Filho', Jaboticabal, São Paulo, Brazil

[3] State University of São Paulo 'Júlio de Mesquita Filho', Botucatu, São Paulo, Brazil

[4] Federal Institute of Bahia, Porto Seguro, Bahia, Brazil

Correspondence: Ana Veruska Cruz da Silva, Embrapa Coastal Tablelands, Av. Beira mar, 3250, 49025040, Aracaju, SE, Brazil. E-mail: ana.veruska@embrapa.br

Abstract

Mangaba (*Hancornia speciosa* Gomes) is a fruit species that is native to Brazil, and has social, economic and cultural importance. Knowledge of the genetic relationships between the remaining populations is essential in order to promote conservation strategies for these genetic resources. In the present study, it was evaluated the genetic diversity of 35 individuals from three remaining restingas areas in the states of Ceará (Iguape and Cascavel) and Pernambuco (Tamandaré), located in the Brazilian Northeast. Nine ISSR primers were used to determine the genetic variability. Sixty-one fully polymorphic fragments (100%) were generated. The largest (10) and smallest (5) number of fragments were obtained with the primers HB14 and HB12, respectively. The Shannon index (I = 0.40), the genetic diversity (H = 0.30), and the percentage of polymorphic loci (%P = 73.77%) were also estimated. Both the methods of UPGMA and the Principal Coordinates Analysis (PCoA) clustered individuals according to their place of origin. Genetic divergence was greater within population (64%) than between them (36%). This may indicate a strong genetic structure, *i.e.*, the gene flow rate between populations is low, favoring inbreeding. ISSR markers were efficient for the analysis of genetic diversity, for the identification of clusters, and for the estimation of the genetic distance between and within populations.

Keywords: *Hancornia speciosa* Gomes, Apocynaceae, genetic variability, conservation

1. Introduction

Hancornia speciosa Gomes (Apocynaceae family) is a fruit species that is native to Brazil, popularly known as mangaba. It naturally occurs in coastal tablelands in in the Brazilian Northeast, and in the Brazilian Central-Western and Northern Cerrado. The fruit is typically tropical and highly appreciated by consumers due to its organoleptic characteristics and nutritional value. It is currently found in the list of endangered species, and has aroused interest for studies which could result in conservation strategies of this genetic resource.

Several factors, such as forest fragmentation, tourism, and increase in areas cultivated with coconut, pasture, and sugarcane have contributed to the significant reduction of naturally occurring growing areas. Therefore, studies examining the genetic diversity and structure of remnant mangaba populations are fundamental for establishing conservation strategies for germplasm and species preservation (Amorim et al., 2015).

Information on the development and genetic variation of native species are crucial, since domestication and incorporation of these species in regional production systems, and the development of efficient conservation strategies are closely related to the knowledge of the magnitude and of the distribution of genetic variability in natural populations. Molecular characterization is a form of diversity evaluation which allows, from genetic markers, regardless of environmental interference, inferring the diversity degree among individuals and populations (Costa et al., 2011).

Due to the economic potential of the species, several researches have been carried out in all areas of its occurrence. Mangaba Active Germplasm Bank is found in the state of Sergipe, with 271 individuals from 19 natural populations representative of eight states. Costa et al. (2011) evaluated the diversity of this BAG in its initial phase by means of RAPD markers (Random Amplified Polymorphic DNA), when there were only 55 genotypes, and observed low genetic similarity.

ISSR Molecular markers (Inter Simple Sequence Repeat) are widely used in genetic diversity studies. It is a simple, effective technique, which has high reproducibility and produces high polymorphism rates (Reddy et al. 2002). Most part of the published studies present results regarding genetic diversity in native populations, such as that of Moura et al. (2005) with populations of the Cerrado Biome, using RAPD; Costa et al. (2015) with populations from Rio Grande Norte, using ISSR markers; and Amorim et al. (2015) with populations from Ceará, Pernambuco and Sergipe, using microsatellites. Jimenez et al. (2015) characterized 38 individuals using ISSR markers from the coast of Pernambuco.

This study was carried out in order to know the genetic diversity of remaining restinga's areas of mangaba in the Northeast of Brazil using ISSR markers, aiming the genetic conservation of the species, seed collection for recovery of degraded areas and future breeding program.

2. Method

It was used 35 mangaba individuals collected in three remaining populations obtained in the cities of Iguape (Tapera, Ceará), Jacarecoara (Cascavel, Ceará), and Tamandaré (Pernambuco) (Table 1).

Table 1. Identification, geographical location and sample size of three populations of *H. speciosa* Gomes

Population	Geographical coordinates	N. individuals
Cascavel, Jacarecoara (CE)	4°07'10" S e 38°10'34" O	14
Iguape, Tapera (CE)	3°56'20" S e 38°20'18" O	06
Tamandaré (PE)	8°43'50" S e 35°6'10" O	15
Total		35

DNA extraction was based on the standard CTAB protocol (Costa et al., 2011). Quantification was conducted using a NanoDrop 2000c (Thermo Scientific) spectrophotometer at the ranges of 260-280 nm; while purity was calculated by the OD260/OD280 ratio. The evaluation of DNA quality was performed by electrophoresis on 1% agarose gel, and visualized in the Gel Doc L-pix HE (Loccus Biotechnology, Brazil) photo documentation device. After the measurement, samples were diluted in TE solution for the concentration of 25 ng/μL, and stored at -20 °C for subsequent use in ISSR reactions.

After preliminary tests using 28 ISSR primers, nine of them were selected (Table 2). PCR (polymerase chain reaction) consisted of 2 μL genomic DNA (25 ng/μL), 1 μL of each primer (5 mM), 14.4 μL sterile water MilQ, 2 μL 10X reaction buffer with $MgCl_2$ (Promega, Madison, South Dakota, EUA), 0.4 μL dNTP (10 nM), and 0.2 Taq polymerase (5 U/μL), with final volume of 20 μL reaction.

Amplification was carried out in a thermocycler (Axygen Maxygene®, Union City, EUA). Samples were subjected to denaturation at 95 °C for five min, followed by 45 amplification cycles. At each cycle, samples underwent denaturation at 94 °C for 1 min, annealing at different temperatures for 45 s, and extension at 72 °C for two min, and a final extension at 72 °C for 10 min, followed by cooling at 10 °C.

Fragments were visualized on 2% agarose (1x TBE 89 mM, Tris 89 mM, boric acid 2.5 mM, EDTA pH 8.3) in a horizontal electrophoresis system performed at 100 V for 90 min. The gel was stained with ethidium bromide solution (5 mg·mL^{-1}) for 30 min. Amplification products were visualized under UV light using the Gel doc (Loccus Biotecnologia, Cotia, SP) photo documentation device.

In the evaluation of the gels, the presence (1) and the absence (0) of bands were used for the construction of a binary matrix and subsequent statistical analyses. The polymorphic information content (PIC) was calculated according to Ghislain et al. (1999), and the marker index (MI) was determined as described by Zhao et al. (2007). The Shannon index - I (Brown & Weir, 1983), the genetic diversity - H (Lynch & Milligan, 1994; Maguire et al., 2002), and the analysis of molecular variance - AMOVA (Michalakis & Excoffier, 1996) were estimated using the Genalex v.6.3.

The estimate of the genetic similarity between each pair of individuals was calculated by the Jaccard coefficient, using the Free Tree software (Pavlicek et al., 1999), and its simplified representation was obtained by the dendrograms constructed by Unweighted Pair-Group Method with Arithmetic Mean (UPGMA) based on the

Treeview software (Page, 1996). In order to analyze the robustness of each clustering, analysis was performed by bootstrap resampling using the Treeview software (Page, 1996), at 10,000 x. The principal coordinates analysis (PCA) was performed using the Genalex v.6.3 software, based on the Jaccard coefficient (Peakall & Smouse, 2006).

3. Results

Nine primers (Table 2) generated 61 fully polymorphic fragments. The largest (10) and lowest (5) number of fragments were obtained with the primers HB14 and HB12 primers, respectively. The value of PIC was 0.65 (0.43 to 0.78), and MI had mean of 4.42 (2.56 to 7.33).

Table 2. Number of polymorphic fragments (NPF), polymorphic information content (PIC), and marker index (MI) of 37 individuals of *Hancornia speciosa* obtained from remaining population from the states of Ceará and Pernambuco

Locus	Sequence	NPF	PIC	MI
844A	CTCTCTCTCTCTC	6	0,43	2,56
844B	CTCTCTCTCTCTCTGC	7	0,59	4,12
17899B	CACACACACACAGG	7	0,78	5,48
HB10	GAGAGAGAGAGACC	8	0,67	5,35
HB12	CACCACCACGC	5	0,66	3,31
HB14	CTCCTCCTCGC	10	0,73	7,33
HB15	GTGGTGGTGGC	4	0,72	2,87
810	GAGAGAGAGAGAGAGAT	7	0,74	5,15
841	GAGAGAGAGAGAGAGATC	7	0,52	3,62
Total		61	0,65	4,42

The Shannon index ranged from 0 to 1, and the closer to 1, the greater is the genetic diversity (Estopa et al., 2006). In the populations studied, this value ranged from 0.37 to 0.44 (Table 3), which allowed to infer that the analyzed populations have genetic diversity reserves. The mean genetic diversity for the different areas was 0.28. The percentage of polymorphic loci was higher in the population CE1 (Cascavel, CE), with 85.25%, followed by PE (Tamandaré, PE), and CE2 (Iguape, CE), with 77.05 and 59.02%, respectively.

Table 3. Shannon Index (I), Genetic Diversity (H) and percentage of polymorphic loci (P%) in natural populations of *Hancornia speciosa* Gomes found in remaining populations from the states of Ceará and Pernambuco, using ISSR markers

Pop	I	H	P%
CE1	0.44 (±0.03)	0.30 (±0.02)	85.25%
CE2	0.37 (±0.04)	0.26 (±0.03)	59.02%
PE	0.40 (±0.04)	0.27 (±0.03)	77.05%
Total	0.40 (±0.02)	0.28 (±0.01)	73.77%

From the AMOVA (Table 4) it was possible to observe the pattern of genetic variability distribution (PhiPT 0.36 *** p < 0.001). Results showed that genetic diversity within populations was greater (64%) than between them (36%).

Table 4. Analysis of molecular variance (AMOVA) between and within different locations of accessions of *Hancornia speciosa* Gomes

	df	Variance	Variantion (%)	PhiPT
Among	2	4.6812	36	0.364***
Within	34	8.395	64	
Total	36	13.207	100%	

Note. df: degrees of freedom; *** p < 0.001.

According to the dendrogram, it was observed the formation of three main clusters: one consisting of the individual CE1-12 (CI); one consisting of the individuals CE1-11 and CE1-4 (CIII); and another consisting of the other individuals (CII), which can be subdivided into four subclusters. It was evident that the place of origin directly influenced the clusters. All individuals originated from Pernambuco were close, as well as CE2 and CE1. The dendrogram analysis also suggests that individuals from Tamandaré (PE) are genetically closer to those from Iguape (CE2) (Figure 1).

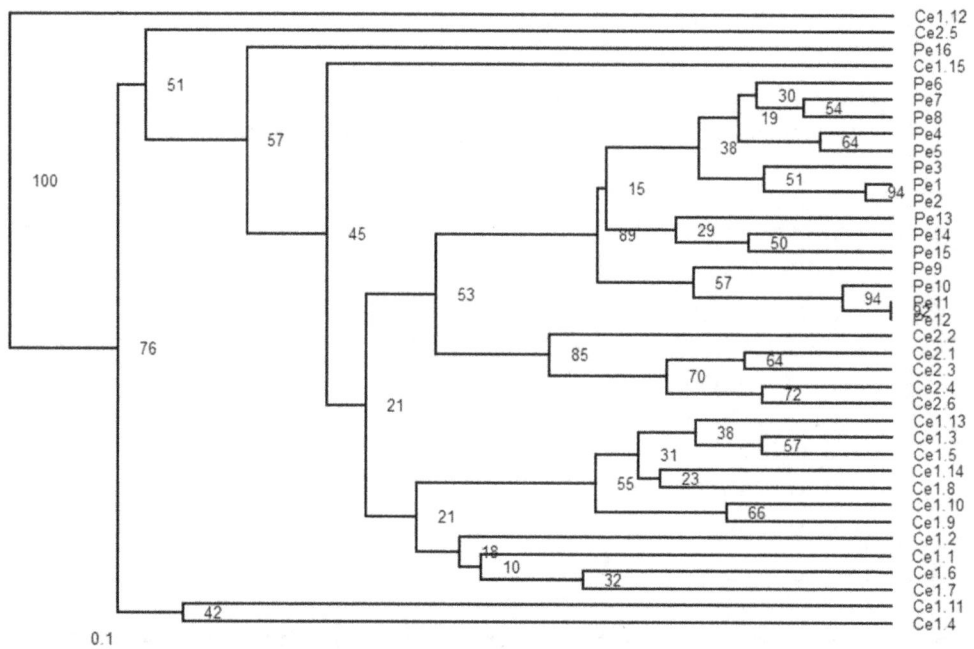

Figure 1. Similarity dendrogram by the Jaccard Coefficient, UPGMA (Unweighted Pair-Group Method with Arithmetic Mean) and bootstraps at 10,000x for mangaba remaining populations (*Hancornia speciosa* Gomes) evaluated by ISSR markers (CE1 - Cascavel; CE2 - Iguape; PE - Tamandaré)

Analysis of the main coordinates - PCoA (Figure 2) indicated that individuals formed four clusters, three of them separated by the origin. ISSRs efficiency was confirmed since the percentage of the variation accumulated in the first two axes was 66.76%, indicating genetic diversity between the individuals studied. Similar to the dendrogram analysis, individuals were clustered according to their origin. PE1 and CE2 are close to each other, as well as the individuals CE1-11 and CE1-14; CE1-12 and CE2-5, as shown in Figure 1.

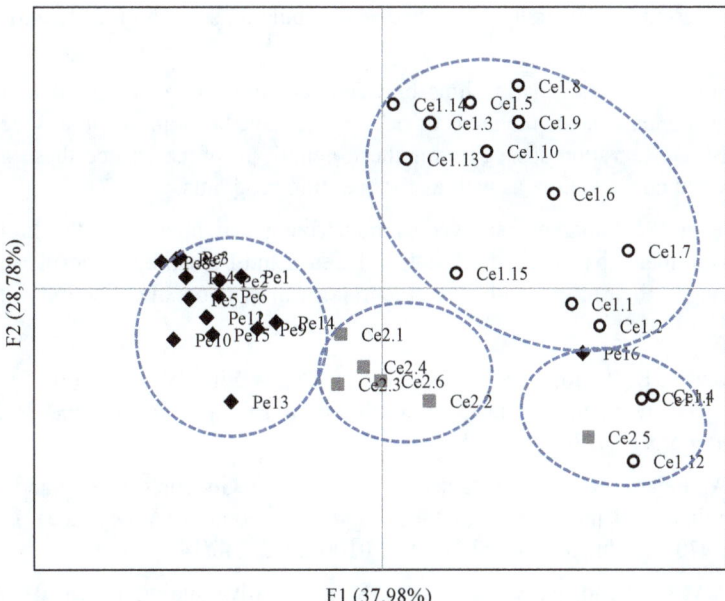

F1 (37,98%)

Figure 2. Principal Coordinates Analysis (PCoA) for remaining species of mangaba (*Hancornia speciosa* Gomes) evaluated by ISSR markers (CE1 - Cascavel; CE2 - Iguape; PE - Tamandaré)

4. Discussion

Selection of the characters of interest together with the genetic diversity data are important for the management of a germplasm bank, and they also assist in breeding programs. Knowledge of the genetic variability available for selection of superior plants is fundamental. The genetic relationships between remaining populations of mangaba will be important for the selection of individuals to be included in the germplasm bank, as well as for determining conservation strategies of these scarce genetic resources.

Regarding the number of polymorphic loci, we can check the work realized by Jimenez et al. (2015), where through six primers ISSR in mangaba, obtained a larger number of fragments (93), ranging from 9 to 21 fragments per primer. However, polymorphism percentage was lower (89.25%) than the 100% value observed in this study. The ten RAPD primers reported in mangaba by Silva et al. (2012) resulted in 85% polymorphism in the 60 amplified fragments. Fragments generated in this study can be considered very informative, unlike the results reported for mangaba germplasm (Costa et al., 2011) and for natural populations found in the state of Sergipe (Silva et al., 2012).

This index (I) is a useful tool for population analysis when using dominant markers, such as ISSRs. The Shannon index makes the effects of distortion caused by the inability of heterozygous loci detection be relatively neutral. The values obtained in this study are consistent with those reported by Silva et al. (2012) in natural population of mangaba from the state of Sergipe. In other species, the values of 'I' was lower, such as in cupuaçu – *Theobroma gradiflorum* (0.15-0.17) (Silva et al., 2016) and genipap – *Genipa americana* L. (mean value of 0.21) (Silva et al., 2014).

The value found for the genetic diversity is close to those found by Tambarussi et al. (2008) in *Piptadenia gonoacantha* Mart. (H = 0.29), and by Gois et al. (2014) in natural populations of *Ziziphus joazeiro* Mart. (H = 0.36). Research with *Theobroma speciosum* showed values well below - 0.076 (Gustina et al., 2014). However, in mangaba Silva et al. (2012) found mean value of 0.35. Under natural conditions, the value of 'H' is always different from zero, since individuals are susceptible to incorporation of new alleles by cross, even in small populations or fragments, in addition to losses due to genetic drift (Silva et al., 2014).

The lowest level of percentage of polimorfic locus in CE2 indicates that this population is isolated. Heterozygosity loss is usually associated with random events of colonization and genetic drift (Pinto & Carvalho, 2004). According to the results, the species presents genetic diversity in the evaluated remaining populations, which was proven by the value of total heterozygosity observed (0.28).

The greater genetic diversity within than between populations may indicate a strong genetic structure, i.e., the gene flow rate between populations is low, favoring inbreeding (Silva et al., 2014). Allogamous species with outcrossing generally present low diversity levels between populations and high differentiation within them (Sun et al., 2006).

The characterization of the genetic relationships between remaining populations of mangaba originated from these two states is valuable data, since the areas of occurrence have been almost totally devastated. Therefore, it is necessary to stablish conservation strategies for these genetic resources, since these genotypes can be used both for in situ and ex situ conservation, as well as for breeding programs.

In this work, the use of ISSR markers allowed characterizing and measuring the high genetic diversity of remaining populations of mangaba found in the states of Ceará and Pernambuco, northeastern Brazil. Genotypes were clustered according to their origin, and have potential for increments of collections aimed at conservation.

References

Amorim, J. A. E., Mata, L. R., Ledo, A. S., Azevedo, V. C. N., & Silva, A. V. C. (2015). Diversity and genetic structure of mangaba remnants in states of northeastern Brazil. *Genetics and Molecular Research, 14*, 823-833. https://doi.org/10.4238/2015.February.2.7

Costa, D., Vieira, F. A., Fajardo, C. G., & Chagas, K. P. T. (2015). Genetic diversity and issr initiators selection in a natural population of mangaba (*Hancornia speciosa* Gomes) (Apocynaceae). *Revista Brasileira de Fruticultura, 37*, 970-976. https://doi.org/10.1590/0100-2945-246/14

Costa, T. S., Silva, A.V. C., Ledo, A. S., Santos, A. R. F., & Silva Júnior, J. F. (2011). Genetic diversity of accessions of the mangaba germplasm bank in Sergipe, Brazil. *Pesquisa Agropecuária Brasileira, 46*, 499-508. https://doi.org/10.1590/S0100-204X2011000500007

Estopa, R. A., Souza, A. M., Moura, M. C. O., Botrel, M. C. G., Mendonça, E. G., Carvalho, D., et al. (2006). Genetic diversity in natural populations of candeia(*Eremanthus erythropappus* (DC.) MacLeish). *Scientia Forestalis, 70*, 97-106. Retrieved from http://www.ipef.br/publicacoes/scientia/nr70/cap10.pdf

Ghislain, M., Zhang, D., Fajardo, D., Huamán, Z., & Hijmans, R. J. (1999). Marker assisted sampling of the cultivated Andean potato *Solanum phureja* collections using RAPD markers. *Genetic Resources and Crop Evolution, 46*, 547. https://doi.org/10.1023/A:1008724007888

Gois, I. B., Ferreira, R. A., Mann, R. S., Pantaleão, S. M., Gois, C. B., & Oliveira, R. S. C. (2014). Genetic variability in natural populations of *Ziziphus joazeiro* Mart., by RAPD molecular markers. *Revista Árvore, 38*, 621-630. https://doi.org/10.1590/S0100-67622014000400005

Gustina, L. D., Luz, L. N., Vieira, F. S., Rossi, F. S., Soares-Lopes, C. R. A., Pereira, T. N. S., & Rossi, A. A. B. (2014). Population structure and genetic diversity in natural populations of *Theobroma speciosum* Willd. ex Spreng (Malvaceae). *Genetics and Molecular Research, 13*, 47-53. https://doi.org/10.4238/2014.february.14.5

Jimenez, H. J., Martins, L. S. S., Montarroyos, A. V. V., Silva Junior, J. F., Alzate-Marin, A. L., & Moraes Filho, R. M. (2015). Genetic diversity of the Neotropical tree *Hancornia speciosa* Gomes in natural populations in Northeastern Brazil. *Genetics and Molecular Research, 14*, 17749-17757. https://doi.org/10.4238/2015.December.21.48

Lynch, M., & Milligan, B. G. (1994). Analysis of population genetic structure with RAPD markers. *Molecular Ecology, 3*, 91-99. https://doi.org/10.1111/j.1365-294X.1994.tb00109.x

Maguire, T. L., Peakall, R., & Saeger, P. (2002). Comparative analysis of genetic diversity in the mangrove species *Avicennia marina* (Forsk.) Vierh (Avicenniaceae) detected by AFLPs and SSRs. *Theoretical and Applied Genetics, 104*, 388-398. https://doi.org/10.1007/s001220100724

Michalakis, Y., & Excoffier, L. (1996). A generic estimation of population subdivision using distances between alleles with special reference for microsatellite loci. *Genetics, 142*, 1061-1064. Retrieved from http://www.genetics.org/content/142/3/1061.

Moura, N. F., Chaves L. J., Vencovsky, R., Zucchi, M. I., Pinheiro, J. B., Morais, L. K., & Moura, M. F. (2005). Selection of RAPD markers to study genetic structure of *Hancornia speciosa* Gomes. *Bioscience Journal, 21*, 119-125. Retrieved from http://www.seer.ufu.br/index.php/biosciencejournal/article/viewFile/6615/4348

Page, R. D. M. (1996). TreeView: An application to display phylogenetic trees on personal computers. *Computer Applications in the Bioscience, 12*, 357-358.

Pavlicek, A., Hrda, S., & Flegr, J. (1999). Free-Tree: Freeware program for construction of phylogenetic trees on the basis of distance data and bootstrap/jackknife analysis of the tree robustness. Application in the RAPD analysis of genus Frenkelia. *Folia Biologica, 45*, 97-99. Retrieved from https://www.researchgate.net/publication/12585108

Peakall, R., & Smouse, P. E. (2006). Genalex6: Genetic analysis in Excel. Population genetic software for teaching and research. *Molecular Ecology Notes, 6*, 288‐295. https://doi.org/10.1111/j.1471-8286.2005.01155.x

Pinto, S. I. C., & Carvalho, D. (2004). Genetic structure in populations of pindaíba (*Xylopia brasiliensis* Sprengel) by isozymes. *Revista Brasileira de Botânica, 27*(3), 597-605. https://doi.org/10.1590/S0100-8404 2004000300019

Reddy P. M., Sarla, N., & Siddiq, E. A. (2002). Inter simple sequence repeat (ISSR) polymorphism and its application in plant breeding. *Euphytica, 12*8, 9-17. https://doi.org/10.1023/A:1020691618797

Silva, A. V. C., Freire, K. C. S., Ledo, A. S., & Rabbani, A. R. C. (2014). Diversity and genetic structure of jenipapo (*Genipa Americana* L.) brazilian. *Scientia Agricola, 71*, 345-355. https://doi.org/10.1590/0103-9016-2014-0038

Silva, A. V. C., Santos, A. R. F., Wickert, E., Silva Júnior, J. F., & Costa, T. S. (2012). Divergência genética entre acessos de mangabeira (*Hancornia speciosa* Gomes). *Revista Brasileira de Ciências Agrárias, 6*, 572-578. https://doi.org/10.5039/agraria.v6i4a943

Silva, B. M., Rossi, A. A. B., Dardengo, J. F. E., Araújo, V. A. A. C., Rossi, F. S., Oliveira, L. O., & Clarindo, W. R. (2016). Genetic diversity estimated using inter-simple sequence repeat markers in commercial crops of cupuassu tree. *Ciência rural, 46*, 108-113. https://doi.org/10.1590/0103-8478cr20141634

Sun, K., Chen, W., Ma, R., Chen, X., Li, A., & Ge, S. (2006). Genetic variation in Hippophaerhamnoidesssp. Sinensis (Elaeagnaceae) revealed by RAPD markers. *Biochemical Genetics, 44*, 186-197. https://doi.org/10.1007/s10528-006-9025-2

Tambarussi, E. V., Mori, E. S., Zimback, L., Mori, N. T., Pinto, C. S., & Fernandes, K. H. P. (2008). Genetic structure of *Piptadenia gonoacantha* (Mart.) Macbr. populations through molecular markers RAPD. *Revista Científica Eletrônica de Engenharia Florestal, 7*, 1-15. Retrieved from http://faef.revista.inf.br/imagens_arquivos/arquivos_destaque/llb0mWFbe71kwD5_2013-4-29-9-36-14.pdf

Zhao, K., Zhou, M. Q., & Chen, L. Q. (2007). Genetic diversity and discrimination of *Chimonanthus praecox* (L.) link germplasm using ISSR and RAPD markers. *HortScience, 42*, 1144-1148. Retrieved from http://hortsci.ashspublications.org/content/42/5/1144.full.pdf+html

3

Isolation, Characterization and Selection of Bacteria that Promote Plant Growth in Grapevines (*Vitis* sp.)

Gislaine Aparecida Denardi Biasolo[1], Daniel Antonio Kucmanski[1], Sabrina Pinto Salamoni[1], João Peterson Pereira Gardin[1,2], Elisandra Minotto[1] & Cesar Milton Baratto[1]

[1] Nucleus of Biotechnology, University of the West of Santa Catarina, UNOESC, Videira, SC, Brazil

[2] Santa Catarina Agency for Agricultural Research and Rural Extension (Epagri), Videira, Brazil

Correspondence: Sabrina Salamoni, Nucleus of Biotechnology, University of the West of Santa Catarina, Santa Catarina, Brazil. E-mail: sabrina.salamoni@unoesc.edu.br

Abstract

Certain bacteria can promote and stimulate plant growth, increasing the production of biomass and reducing damage caused by phytopathogens. With that in mind, this research effort set out to select these plant growth-promoting bacteria in order to evaluate their effects on the growth of grapevines (*Vitis* sp.). The bacteria were isolated from several vineyard soil samples, and evaluated based on their production of IAA (Indole-3-Acetic Acid), siderophores and cellulase, as well as their phosphate solubilization and nitrogen fixation capabilities. *In vivo* testing included six separate treatments with the following bacterial isolates: C12, O7, B3, I3, a control group and a blended group. The tests were performed in a greenhouse with bacterial suspension inoculation placed around the roots of Paulsen 1103 rootstock cuttings. The data collected included the following: number of leaves per plant, branch lengths, chlorophyll content, fresh and dry mass, and Carbon, Hydrogen, Nitrogen and Sulfur concentrations. Forty-six separate bacteria were isolated, of which 100% produced IAA, 65.21% produced siderophores, 63.04% solubilized phosphate, 34.78% produced cellulase, and 30.43% showed nitrogen fixation. The *in vivo* testing also revealed significant increases in the length of the branch and in percentages of Carbon and Nitrogen. The C12 isolate exhibited the highest increase in branch length (76.704 cm), whereas the O7 and C12 were identified as *Bacillus amyloliquefaciens* and *Bacillus thuringiensis*, respectively.

Keywords: bacteria, bioprospection, plant growth-promotion, viticulture

1. Introduction

Viticulture is an activity of great importance for Brazil's economy, especially for its leading producers, the states of Rio Grande do Sul, Pernambuco, São Paulo, Paraná, Bahia and Santa Catarina. According to IBGE (Portuguese initials for Brazilian Institute for Geography and Statistics), in 2015, Brazil produced 1,025,536 tons of grape, representing an increase of 6.74% over the previous year. The southern region states, specifically Rio Grande do Sul, Paraná, and Santa Catarina, were responsible for 90% of that production, the bulk of which was used to produce wine and grape juice. The state of Santa Catarina has lead the cultivation of grapes for the production of wines and sparkling wines. In 2015, this state's production levels reached 69,250 tons (IBGE, 2015), representing an increase of 2.86% over 2014. Currently, production levels of 54,262 tons make the mid-west region of Santa Catarina, known as Vale do Rio do Peixe, the leading producer of grapes, with Isabella, Niagara and Bordeaux being the main varieties cultivated (Desplobins & Silva, 2005). Within its cultivation cycle, several factors, whether abiotic (Tecchio, Teixeira, Terra, Moura, & Paioli-Pires, 2011; Brunetto et al., 2008) or biotic (Garrido, Sonêgo, & Gomes, 2004), can impact or jeopardize the quality and output of the crop. Bacteria that establish symbiotic relationships with the plants play a critical role in maintaining and/or increasing plant growth rates, and can be used to promote plant growth, significantly improving crop output. As the name implies, these plant growth-promoting bacteria (PGPB) can stimulate plant growth, increasing stem and root development, as well as the production of biomass, while, at the same time, reducing damages caused by phytopathogens (Gupta, Parihar, Ahirwar, Snehi, & Singh, 2015; Ahemad & Kibret, 2014; Lugtenberg & Kamilova, 2009; Van Loon & Bakker, 2005). Direct growth-promoting mechanisms are those that affect the plant's natural balance of growth regulators, improving its nutritional proficiencies and stimulating the processes that fight systemic diseases (for example, biological nitrogen fixation, phytohormones production, synthetization of enzymes, inorganic phosphate solubilization, and phosphate mineralization). Indirect growth regulator mechanisms, on the other hand, are the ones that reduce or inhibit the activities of pathogenic microorganisms through biocontrol, which includes the production of antibiotics and iron chelating agents (siderophores), and the synthetization of exoenzymes, such as cellulases and chitinases (Carvalhais et al., 2013; F. Ahmad, I. Ahmad, & Khan, 2008; Zahir, Asghar, Akhtar, & Arshad, 2005; Asghar, Zahir, Arshad, & Khaliq, 2002). Although several studies have reported the potential of different microorganisms to promote growth in plants such as wheat, soybeans, and potatoes, these are in short supply for the cultivation of grapevines, both in terms of the isolation of the microorganisms and when it comes to *in vitro* and *in vivo* testing (Dawwam, Elbeltagia, Emara, Abbas, & Hassan, 2013; Karagöz, Ates, Karagöz, Kotan, & Cakmakci, 2012; Smyth et al., 2011; Khalid, Arshad, & Zahir, 2004). In the last couple of decades, research efforts related to the development of biological consumables, such as inoculating agents, have mustered a lot of attention from researchers. Considering the importance of microorganisms and the attention focused on the search for alternatives that promote plant growth, the purpose of this research paper is to bioprospect plant growth-promoting bacteria by evaluating their physiological and enzymatic activities, and focusing on their application to grapevine cultivation.

2. Method

The isolation and characterization of the bacteria being studied were performed at UNOESC's (Universidade do Oeste de Santa Catarina, City of Videira Campus) Microbiology Laboratory, and the greenhouse experiment was carried out at EPAGRI (Portuguese acronym for State of Santa Catarina Agricultural, Livestock, and Rural Extension Research Company), also in the City of Videira.

2.1 Isolation of the Bacteria

The bacteria were isolated from several soil samples collected from grapevine growing properties located in the mid-west region of the state of Santa Catarina. The samples were homogenized and sieved to remove any coarser materials, and then re-suspended in 90 mL sterile peptone water. The suspended samples were then incubated for 30 minutes in a shaker unit, at room temperature. Decimal serial dilutions were performed following the homogenization. A 100 µL aliquot portion of the 10^{-3}, 10^{-4}, and 10^{-5} dilutions were seeded, using a Drigalsky agar nutrient medium. The plates were incubated for 24-72 hours, at a temperature of 30 °C. After this incubation period, the bacteria colonies were purified, preserved, and then used in the experiment.

2.2 Evaluation of Plant Growth-Promothing Agents

2.2.1 Production of Indole-3-Acetic Acid (IAA)

The isolates were cultivated in a King B medium, supplemented with L-tryptophan (5 mM·mL^{-1}), and incubated for 48 hours, at a temperature of 30 °C, as per methodology stipulated by Bric, Bostock, and Silverstone (1999), and adapted by Cattelan (1999). A 2 mL aliquot portion of the culture was centrifuged at 2,000 rpm, for 10 minutes. Subsequently, 1 mL of supernatant was transferred to a new test tube containing 1 mL of Salkowski solution (1.5 mL of 0.5 M of $FeCl_3 \cdot 6H_2O$ in 80 mL of 60% H_2SO_4), and left at room temperature, protected from light. After 30 minutes, a spectrophotometer reading was taken at 540 nm. The IAA concentration was determined based on an IAA standard curve (0, 1, 5, 10, 20, 40, 80 and 160 mg·mL^{-1}).

2.2.2 Production of Siderophores

An evaluation of the bacteria's siderophore production was performed in a chrome azurol S (CAS) reagent enriched King B medium (King et al., 1954). The isolates were cultured for 24 hours in that medium, under constant agitation and at a temperature of 30 °C. A 100 µL aliquot portion of that culture was transferred to plates of the same medium, which were then incubated for five to seven days, at a temperature of 30 °C. The formation of orange colored halos around the colony was proof positive of siderophore production. The rate of siderophore production was calculated based on the relationship between the total halo diameter (THD) and the colony halo diameter (CHD) (or THD/CHD, in millimeters).

2.2.3 Solubilization of Phosphate

The isolates' ability for phosphate solubilization was qualitatively evaluated according to Nautiyal et al. (1999). Using the pin prick method, the bacteria were inoculated in a culture medium containing tricalcium phosphate (10 g of glucose; 5 g of $Ca_5(OH)(PO_4)_3$; 5 g of $MgCl_2 6H_2O$; 0.25 g of $MgSO_4 7H_2O$; 0.2 g of KCl; 0.1 g of $(NH_4)_2SO_4$; 1.5% agar and pH 7.0). The plates were then incubated for seven days, at a temperature of 30 °C. Only the isolates showing clear halos around the colonies were considered to be proof positive for the solubilization of phosphate.

2.2.4 Asymbiotic Nitrogen Fixation

For this test, the bacteria were cultured in test tubes containing a nitrogen-free medium (NFM), which were kept in an incubator for 10 days, at a temperature of 30 °C. At the end of that period, the isolates were transferred to different test tubes, containing the same type of medium, and were incubated under the exact same conditions, for the same period of time. The same procedure was repeated yet a third time. Therefore, after 30 days in a NFM, an aliquot portion of 20 μL of each culture was transferred to a nutrient agar medium so as to confirm their viability. The bacteria's ability to asymbiotically fix nitrogen was assessed in accordance with a methodology proposed by Rennie (1981), and adapted by Cattelan (1999). Isolates that actually showed growth were considered to be proof positive of nitrogen fixation.

2.2.5 Production of Cellulase

For this phase, the bacteria were cultured in a mineral medium, to which a 5% carboxymethyl cellulose solution was added. After the inoculation, using the pin prick method, the plates were kept in an incubator at a temperature of 30 °C, for five days, at which time the plates were stained with Lugol's iodine solution. The formation of a colorless halo around the colony was proof positive of cellulase production.

2.3 Greenhouse Trial Design

Rooted and sprouted Paulsen 1103 grapevine rootstock cuttings were selected as the standard. The cuttings were washed in running water and transplanted to plastic bags containing an inert coco coir and vermiculite substrate (in a 3-to-1-proportion), and autoclaved for 20 minutes, at a temperature of 121 °C. These were kept in a climate controlled greenhouse for a period of 30 days at a maximum temperature of 25 °C, relative humidity of 60%, and a 12-hour photoperiod. The trial design consisted of repetition of six randomized blocks with five plants each. Treatments were also random within the blocks, as follows: T1 = control with no added bacteria; T2 = isolate C12; T3 = isolate B3; T4 = isolate I3; T5 = isolate O7, and T6 = isolates C12, B3, I3 and O7, together. All treatments consisted of inoculations using a 1.8×10^8 CFU/mL bacterial suspension. All selected bacteria were cultured in a brain-heart infusion (BHI) medium for 12-16 hours, under constant agitation of 125 r.p.m. and a temperature of 30 °C. A 20 mL bacterial suspension was added to each treatment. Then, for 60 days, all plants were kept in a controlled incubator with relative humidity of 80%, a maximum temperature of 25 °C, a 16-hour photoperiod, and a weekly wet down with Hoagland's nutrient solution (Hoagland & Arnon, 1950).

2.3.1 Characteristics Evaluations

The promotion of growth in grapevines (*Vitis* sp.) was evaluated using the following parameters: foliar chlorophyll content index, length of branches and number of leaves. To measure the chlorophyll content, a portable chlorophyll measuring device with diodes that emit light at 650 nm (red) and at 940 nm (infrared) was used; this because light at 650 nm is very close to the two main wavelengths associated with chlorophyll activity (645 nm and 663 nm). The length of the branches and number of leaves were measured and counted at regular intervals, (on the 1st day, 15th day, 30th day, 45th day and 60th day). On the 60th day, the plants were collected and separated into their three component parts (aerial, stem and roots). The roots were carefully manipulated in order to remove the coco coir/vermiculite substrate. After weighing them, the parts were packaged in previously dried paper bags, and stored in a plant incubator/dryer, at a temperature of 60 °C, where they were monitored on a daily basis until their weight stabilized. All samples were then macerated and sent to EPAGRI, in the City of Caçador, SC, for analysis, in order to determine their percent composition in terms of Carbon, Hydrogen, Nitrogen and Sulfur.

2.4 Identification of the Bacteria

Two of the bacteria were identified at the molecular level: the isolate, which yielded positive results in all in vitro tests; and the isolate, which showed the highest potential for promoting plant growth in the in vivo tests.

2.4.1 Molecular DNA Extraction

The bacteria were cultured in a BHI medium for 12-14 hours, at a temperature of 30 °C; their DNA was then extracted using the PureLink Quick Gel Extraction Kit, and quantified in keeping with Sambrook and Russel (2001).

2.4.2 Polymerase Chain Reaction (PCR)

In order to partially amplify the *rpoB* gene, an oligonucleotide primer pair was used, specifically AAR YTI GGM CCT GAA GAA AT and TGI ART TTR TCA TCA ACC ATG TG (Drancourt, Roux, Fournier, & Raoult, 2004). Amplification conditions matched those used by the forenamed authors. After amplification, the resulting material was purified and sent for sequencing at the Microbiology Department of the Universidade Federal do

Rio Grande do Sul (Federal University of the State of Rio Grande do Sul). The sequences were aligned, using ChromasPro 1.5, and then compared to reference species nucleotide sequences found in EMBL/GenBank's database, using NCBI's BLAST.

2.5 Statistical Analysis

Following the greenhouse trials, the collected data was submitted for analysis of variance (ANOVA), and their group means compared using the Scott Knott test ($p < 0.05$).

3. Results

From the seven soil samples that were processed, 46 bacteria were isolated, of which 28 (60.87%) were Gram-positive and 18 (39.13%) were Gram-negative. Table 1 below shows the profile for the 46 bacterial isolates based on the five enzymatic and physiological tests related to growth promotion of plants.

Table 1. Profiles of bacterial isolates based on the enzymatic and physiological *in vitro* tests related to plant growth-promotion

Isolate	Gram	IAA	Sid.	Phos.	Nit.	Cel.	Isolate	Gram	IAA	Sid.	Phos.	Nit.	Cel.
A10	-	+	-	-	+	-	I2	-	+	+	+	-	+
A11	+	+	-	-	+	+	I3*	-	+	+	+	-	-
A12	+	+	+	-	-	-	I4	+	+	+	-	-	-
A14	+	+	+	+	-	-	I5	+	+	-	-	+	-
A15	+	+	+	+	-	+	J1	+	+	+	+	+	+
A2	+	+	+	+	-	+	J2	+	+	-	-	-	-
A3	+	+	+	+	-	+	J3	+	+	+	+	-	+
A4	+	+	-	-	-	+	J4	-	+	+	+	-	-
A5	-	+	+	-	-	+	J6	+	+	+	+	+	-
A6	+	+	+	+	+	+	J7	+	+	-	+	-	-
A7	+	+	+	+	-	+	N3	-	+	+	+	-	-
A9	-	+	-	+	-	-	N4	-	+	-	+	-	-
B1	+	+	+	-	+	+	N7	-	+	-	+	+	-
B2	-	+	-	-	-	+	O11	+	+	-	+	-	-
B3*	+	+	+	+	+	-	O14	-	+	+	+	-	-
B7	-	+	+	-	+	-	O15	+	+	+	+	-	-
C11	+	+	+	-	-	+	O16	+	+	+	-	+	-
C12*	+	+	+	+	+	-	O2	+	+	-	-	-	-
C2	-	+	+	+	-	+	O3	+	+	-	+	-	-
C32	-	+	+	-	-	-	O4	+	+	-	-	-	-
C33	-	+	+	+	-	-	O7*	+	+	+	+	+	+
C35	-	+	+	+	-	-	O8	-	+	+	-	-	-
CX	+	+	-	+	+	-	O9	-	+	-	+	-	-

Note. IAA (mg/mL) = production of Indole-3-Acetic Acid; Nit = asymbiotic fixation of atmospheric nitrogen; Phos. = solubilization of phosphate; Sid. = production of siderophores; Cel. = production of cellulase; (-) NO – did not produce the enzyme/metabolite; (+) YES – did produce the enzyme/metabolite; * Isolates selected for the greenhouse *in vivo* tests.

As can be seen from this table, many of the isolates yielded positive results for more than one of the administered tests. Specifically, three of them tested positive for only one test, 14 (26.09%) were positive for two tests, 15 (32.60%) were positive for three tests, 11 (23.91%) were positive for four tests, and three of the isolates, A6, J1 and O7, were positive for all five tests. In a more overarching view, of the 46 bacterial isolates, all of them (100%) produced Indole-3-Acetic Acid, 29 (63.04%) showed solubilization of phosphate, 30 (65.21%) produced siderophores, 16 (34.78%) produced cellulase, and 14 (30.43%) fostered nitrogen fixation.

This study also revealed that levels of IAA production varied between 0.36 mg·mL^{-1} and 14.7 mg·mL^{-1}, while 30 (65.21%) of the microorganisms were capable of producing siderophores, with these production rates ranging between 2.33 mm and 22 mm (See Table 2).

Table 2. Indole-3-Acetic Acid (in mg/mL) and siderophores (in mm) produced by post-cultured bacterial isolates

Bacteria	IAA (mg·mL^{-1})	*Siderophores	Bacteria	IAA (mg·mL^{-1})	*Siderophores
A10	0.91	-	I2	1.84	15
A11	2.82	-	I3	14.77	10.33
A12	0.42	22.00	I4	1.51	3.67
A14	2.76	10.00	I5	0.91	-
A15	5.49	4.33	J1	0.45	6.0
A2	4.62	9.67	J2	10.62	-
A3	0.85	9.67	J3	5.49	20.0
A4	4.07	-	J4	3.64	4.33
A5	6.69	2.33	J6	0.69	7.67
A6	0.53	8.00	J7	9.09	-
A7	0.69	3.33	N3	1.89	13.33
A9	1.56	-	N4	1.18	-
B1	1.51	19.67	N7	0.69	-
B2	0.47	-	O11	0.69	-
B3	3.47	6.33	O14	4.51	10.33
B7	1.18	9.67	O15	0.63	16.33
C11	3.15	10.00	O16	0.47	3.33
C12	5.22	10.33	O2	0.74	-
C2	0.42	20.67	O3	3.47	-
C32	0.36	6.33	O4	0.91	-
C33	0.63	20.0	O7	2.44	8.0
C35	1.02	12.33	O8	0.53	20.33
CX	3.25	-	O9	0.63	-

Note. * Relationship between total halo diameter and the colony halo diameter.

Of the 46 isolates, 20 (43.48%) produced IAA levels below 1 mg·mL^{-1}, 19 (41.30%) produced levels between 1 mg·mL^{-1} and 5 mg·mL^{-1}, five (10.87%) produced between 5 mg·mL^{-1} and 10 mg·mL^{-1}, and isolates J2 and I3 produced 10.62 mg·mL^{-1} and 14.77 mg·mL^{-1} of IAA, respectively. Results also show that, among the bacteria isolates, A2, C2, O8, C33 and J3 produced siderophores, yielding rates of 22.00 mm, 20.67 mm, 20.33 mm, 20.00 mm and 20.00 mm, respectively.

Table 3 shows the average values for branch lengths, foliar chlorophyll content index, and number of leaves, measured and counted after each treatment.

Table 3. Average branch length (in cm), foliar chlorophyll content index, and number of leaves after treatment of the *Vitis* sp. soil with the various bacterial isolates

	Length (cm)	Chlorophyll	# of Leaves
Treat. 1	66.880[b]	24.9736[a]	28.840[a]
Treat. 2	76.704[a]	22.9924[b]	30.824[a]
Treat. 3	69.200[b]	24.1208[b]	29.624[a]
Treat. 4	58.288[c]	24.7944[a]	28.736[a]
Treat. 5	63.688[c]	24.3028[b]	27.112[a]
Treat. 6	68.328[b]	24.0056[b]	28.232[a]
CV	42.24%	14.9%	33.61%

Note. CV = Coefficient of Variation. Within each column, the averages with the same superscripted letter do not differ statistically from the Scott Knott test (p < 0.05). Treatment 1 = control group; Treatment 2 = bacteria C12; Treatment 3 = bacteria O7; Treatment 4 = bacteria B3; Treatment 5 = bacteria I3; and Treatment 6 = blend of bacteria C12, O7, B3 and I3.

In Treatment 2, the soil was inoculated with bacteria C12; results for this treatment tested positive for nitrogen fixation and phosphate solubilization, had a siderophore total halo/colony halo relationship of 10.33 mm, and produced 5.22 mg·mL^{-1} of IAA. The C12 treatment also showed an increase of 14.7% in branch length when compared to a control group, which shows good potential for plant growth-promotion and corroborates results obtained during the *in vitro* tests. It did not, however, show significant increase in the number of leaves and foliar chlorophyll content.

In Figure 1 below, one can see the variation in branch lengths between the different treatments, over time. The chart depicts a quadratic curve behavior, which shows that the biggest difference in branch lengths occurred in the first month.

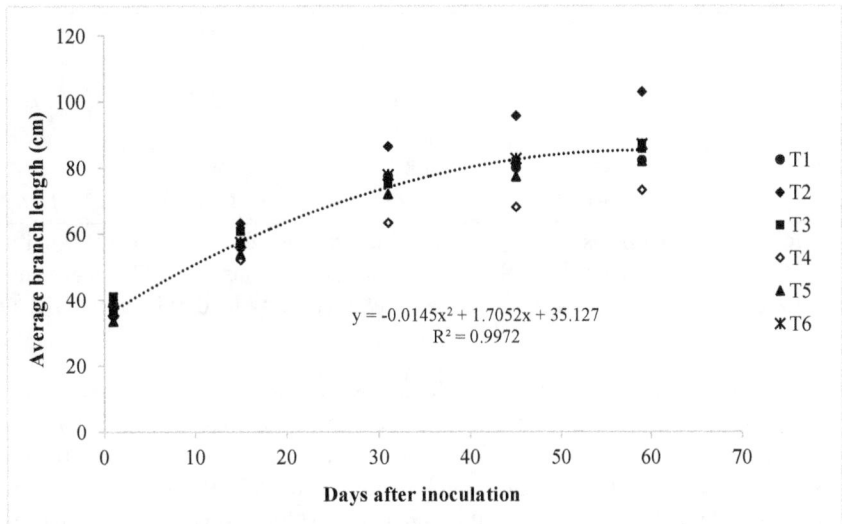

Figure 1. Average branch lengths of the *Vitis* sp. after the different treatments, during 60 days of incubation

Note that, basically, all treatments exhibited the same behavior, with Treatment 2 (T2) being the one that, statistically, most differed from the rest, showing the highest branch length increase, over time.

At the end of the experiment, fresh and dry masses (branches leaves and roots) were also determined, and those results are organized in Table 4.

Table 4. Average dry mass (in mg) and average fresh mass (in mg) for the roots, leaves and branches corresponding to each of the treatments, after 60 days

	DMR	DML	DMB	FMR	FML	FMB	TFM	TDM
Treat. 1	5.482	4.532	3.383	19.386	15.652	7.999	43.037	13.398
Treat. 2	5.312	5.495	3.987	19.195	18.400	8.888	46.484	14.795
Treat. 3	6.524	5.335	3.931	22.477	16.614	8.434	47.527	15.791
Treat. 4	5.093	4.708	2.858	18.159	14.852	6.521	39.532	12.660
Treat. 5	5.258	5.112	3.632	19.887	16.744	8.088	44.720	14.004
Treat. 6	6.285	4.874	3.500	22.348	15.158	7.766	45.273	14.660

Note. DMR = dry mass – roots; DML = dry mass – leaves; DMB = dry mass – branches; FMR = fresh mass – roots; FML = fresh mass – leaves; FMB = fresh mass – branches; TFM = total fresh mass; TDM = total dry mass; Treatment 1 = control group; Treatment 2 = bacteria C12; Treatment 3 = bacteria O7; Treatment 4 = bacteria B3; Treatment 5 = bacteria I3; and Treatment 6 = blend of bacteria C12, O7, B3 and I3.

The average total fresh mass (TFM) ranged from 39.532 mg to 47.527 mg, and the average total dry mass (TDM) from 12.660 mg to 15.791 mg. Both the TFM and the TDM were highest for Treatment 2 (46.484 mg and 14.795 mg) and Treatment 3 (47.527 mg and 15.791 mg). When comparing these against the control group, one will note an increase of 9.45% for the TFM and 15.15% for the TDM. Conversely, with results for TFM and TDM of

39.532 mg and 12.660 mg, respectively, Treatment 4 showed the lowest biomass production, yielding even lower levels than the control group. Though dry and fresh root biomass levels showed no significant difference between the control group and the other treatments, it is interesting to note that Treatments 3 and 6 yielded increases in DMR of 19% and 14.64%, respectively, with both showing increases in FMR above 15%.

On the 60th day, all samples were tested to determine content levels of Carbon, Hydrogen, Nitrogen and Sulfur, and results for each treatment are organized in Table 5.

Table 5. Average percentages of Carbon, Hydrogen, Nitrogen and Sulfur found in the biomass for all treatments, on the 60th day of the experiment

Treatment	C (%)	H (%)	N (%)	S (%)
Treat. 1	35.16^b	3.56^a	0.58^c	0.42^a
Treat. 2	38.88^a	4.38^a	0.94^b	0.70^a
Treat. 3	40.08^a	4.50^a	1.24^a	0.70^a
Treat. 4	39.20^a	4.38^a	1.28^a	0.48^a
Treat. 5	38.94^a	4.82^a	1.40^a	0.82^a
Treat. 6	30.34^c	4.16^a	1.28^a	0.60^a

Note. Within each column, the averages with the same superscripted letter do not differ statistically from the Scott Knott test ($p < 0.05$). Treatment 1 = control group; Treatment 2 = bacteria C12; Treatment 3 = bacteria O7; Treatment 4 = bacteria B3; Treatment 5 = bacteria I3; and Treatment 6 = blend of bacteria C12, O7, B3 and I3.

For the Carbon and Nitrogen contents, a significant difference was observed between treatments. When assessing the percentage content of Carbon, specifically, one can see that the control (Treat. 1) and blended (Treat. 6) groups have the lowest levels of Carbon (35.16% and 30.34%, respectively) when compared to Treatments 2 thru 5, for which the samples were inoculated with the individual bacteria C12, O7, B3 and I3, respectively, yielding an increase of Carbon content of 10%, 14%, 11.5% and 10.8%, also respectively. For Nitrogen, on the other hand, only the control group had a lower content percentage than the rest of the treatments, with the percentage increase in Nitrogen content ranging from 61.7% (Treat. 2) to 140% (Treat. 5).

Of the 46 bacteria, isolates O7 and C12 were identified based on their *rpoB* gene partial nucleotide sequence. Isolate O7 stood out for having yielded positive results in all tests, and isolate C12 for obtaining the best *in vivo* results, yielding longer branch lengths, and higher Carbon and Nitrogen contents, when compared to non-inoculated plants. The *rpoB* gene sequence amplification for isolate O7 produced 627 base-pairs, which was a 99% match to the *Bacillus amyloliquefaciens*. By the same token, isolate C12 produced a sequence amplification of 669 base-pairs, which was a 98% match to the *Bacillus thuringiensis*.

4. Discussion

This research showed that the majority of the bacteria isolated from vineyard soil samples have multiple capabilities. These results are similar to the ones achieved by Marasco et al. (2013) in an experiment that isolated 769 bacteria from soil and root samples of grapevines originally from Italy, Tunisia and Egypt. In that experiment, 95% of the isolates yielded positive results for more then one of the tests. On that occasion, the authors found that 82% of the isolates produced IAA, 61% solubilized phosphate, and 47% produced siderophores. Another group of authors, specifically Ahamad et al. (2008), reported isolating 72 bacteria, for which 80% of the *Azotobacter*, *Pseudomonas* and *Mesorhizobium* isolates produced IAA, while 56.68% showed solubilization of phosphate. In our study, all isolates tested positive for the production of IAA, results that are similar to those obtained by Kuss, Kuss, Lovato, and Flores (2007). According to Dobbelaere, Vanderleyden, and Okon (2003) the ability to synthesize phytohormones is widespread among bacteria associated with plants, and these hormones stimulate plant growth and promote an increase in root area, allowing for better nutrient absorption from the soil. In yet another study, Dawwam et al. (2013) verified that all seven bacteria isolated from the *Ipomoea batatas L.* rhizosphere produced IAA with concentrations ranging from 0.6 $\mu g \cdot mL^{-1}$ to 10.73 $\mu g \cdot mL^{-1}$; in this study, for which IAA concentrations ranged from 0.36 mg$\cdot mL^{-1}$ a 14.77 mg$\cdot mL^{-1}$.

Results for this study showed that 63.04% of the bacteria solubilized phosphate and 65.21% produced siderophores, which is comparable to results attained in other research efforts (Marasco et al., 2013). One of the main contributors to the solubilization of phosphate in the soil is its reduced pH value caused by the bacteria's production of organic acids (Karagöz et al., 2012). In terms of siderophores, according to Benite, Machado, and

Machado (2002), in the last three decades, over one hundred naturally occurring siderophores have been isolated and characterized, including those of bacterial (*Streptomyces*, *E. coli*, *Pseudomonas*, *Bacillus*, *Micobacterium*) and fungal origin (*Aspergillus*, *Penicillium*). Production of these iron chelating agents can be very diverse and have many benefits, the most important ones being that they not only act as biocontrol, biosensor and bioremediation agents, but also promote plant growth (Ahmed & Holmström, 2014). Of all isolated bacteria, 34.8% produced cellulase and, according to Asghar et al. (2002) and Glick (2012), plant growth can be indirectly promoted by reducing or inhibiting the activities of pathogenic microorganisms through the production of enzymes (such as cellulase and chitinases), antibiotics and siderophores by growth-promoting bacteria.

Fourteen of the 46 isolates tested positive for nitrogen fixation. According to Beneduzi et al. (2010), microorganisms present in the rhizosphere show great capacity for assimilating nitrogen. Chagas, Oliveira, and Oliveira (2009) go a step further and claim that they are also capable of producing phytohormones. An analysis of our *in vitro* test results revealed that the isolates had the potential to promote plant growth, especially C12, O7, B3 and I3, which were, therefore, selected for the *in vivo* tests.

The *in vivo* tests, in turn, showed a significant difference in branch lengths between the different treatments, with the C12 isolate promoting the most growth (706.704 cm). In general, all growth was more accentuated in the first 30 days. As claimed by Dias et al. (2009), the initial growth can be attributed to the metabolic substances produced after the bacterial inoculation. Under incubator conditions, the microorganisms *Bacillus* spp and *Sphingopyxis* sp. showed the potential to enhance root development, and increase branch length, dry weight, number of leaves, petiole length and aerial dry weight.

Although differences were observed between fresh mass and dry mass averages, they weren't statistically significant. Such results were comparable to those obtained by Passos et al. (2014) who also did not achieve any statistically significant results in terms of the root's dry mass when they inoculated apple seedlings with five bacterial isolates from the rhizosphere. Dawwam et al. (2013) observed an increase in nitrogen (50.5%) and phosphorus content (48.3%) in the *Ipomoea batatas* L.'s dry mass when compared to the control group; increases that were also observed in this study with respect to carbon (10% to 14%) and nitrogen (61.7% to 140%) contents. Passos et al. (2014) analyzed apple seedlings in terms of absorption of nitrogen, phosphorus and potassium, and determined that only plants inoculated with *Burkholderia sp.* showed high levels of phosphorus absorption. The element that showed significant differences in this study was nitrogen, with all treatments yielding a higher content percentage of it than the control group.

The O7 and C12 isolates, which produced positive results in all *in vitro* tests, were identified as *Bacillus amyloliquefaciens* and *Bacillus thuringiensis*, respectively. Studies about plant growth-promoting bacteria have demonstrated that rhizobacteria of the *Bacillus* genus are frequently found in soil; among these the *B. amyloliquefaciens* has stood out for its ability to promote plant growth and control phytopathogens (Fan et al., 2015; Cavalhais et al., 2013). With strawberry plants, Dias et al. (2009) established that bacteria of the *Bacillus spp.* genus have the potential to enhance root development, and increase branch length, dry weight, number of leaves, petiole length and aerial dry weight. Bobrowski, Fiuza, Pasqualis, and Bodanese-Zanettini (2003), on the other hand, assert that the *Bacillus thuringiensis* is a biotechnological alternative to ward off crop diseases and pesty insects.

5. Conclusions

The *in vitro* tests performed during this study determined that the majority of the bacterial isolates have multiple capabilities, many of which reveal their potential as promoters of plant growth. According to the results herein exposed, which include parameters such as branch length, chlorophyll content index, and carbon and nitrogen contents, one can establish that inoculation of *Vitis* sp. plants with specific bacteria showed an auspicious potential for plant development. Under the specific experimental conditions in which they were evaluated, the bacterial isolates C12, O7, B3 and I3 stood out for significantly increasing Carbon and Nitrogen contents in the plants. Additionally, the C12 isolate, for which 98% of its sequence matched the *Bacillus thuringiensis*, produced the highest branch length growth (76.704 cm). New studies are needed to evaluate the effects of *in vivo* inoculations and to optimize soil and tissue colonization, which, consequently, would promote higher plant growth activities in grapevines.

Acknowledgements

We would like to thank to University of the West of Santa Catarina (UNOESC), to Santa Catarina Agency for Agricultural Research and Rural Extension (EPAGRI-Videira), to National Council for Scientific and Technological Development (CNPq), and National Council for the Improvement of Higher Education (CAPES) for the scholarship and financial support for this work realization.

Reference

Asghar, H. N., Zahir, Z. A., Arshad, M., & Khaliq, A. (2002). Relationship between in vitro production of auxins by rhizobacteria and their growth-promoting activities in *Brassica juncea* L. *Biology and Fertility of Soils, 35*(4), 231-237. http://dx.doi.org/10.1007/s00374-002-0462-8

Ahemad, M., & Kibretb, M. (2014). Mechanisms and applications of plant growth promoting rhizobacteria: Current perspective. *Journal of King Saud University-Science, 26*(1), 1-20. http://dx.doi.org/10.1016/j.jksus.2013.05.001

Ahmad, F., Ahmad, I., & Khan, M. S. (2008). Screening of free-Living rhizospheric bacteria for their multiple plant growth promoting activities. *Microbiological Research, 163*(2), 173-181.

Ahmed, E., & Holmström, S. J. M. (2014). Siderophores in environmental research: Roles and applications. *Microbial Biotechnology, 7*, 196-208. http://dx.doi.org/10.1111/1751-7915.12117

Beneduzi, A., Costa, P. B., Parma, M., Melo, I. S., Bondanese-Zanettini, M. H., & Passaglia, L. M. (20101). *Paenibacillus riograndensis* sp. nov., a nitrogen-fixing species isolated from the rhizosphere of *Triticum aestivum*. *International Journal of Systematic and Evolutionary Microbiology, 60*(1), 128-133. http://dx.doi.org/10.1099/ijs.0.011973

Benite, A. M. C., Machado, S., & Machado, B. (2002). Sideroforos: "uma Resposta dos Microorganismos". *Química Nova, 25*(6), 1155-1164. http://dx.doi.org/10.1590/S0100-40422002000700016

Bric, J. M., Bostock, R. M., & Silverstone, S. E. (1991). Rapid in situ assay for indoleacetic acid production by bacteria immobilized on a nitrocellulose membrane. *Applied and Environmental Microbiology, 57*(20), 535-538.

Brunetto, G., Bongiorno, C. L., Mattiais, J. L., Deon, M., Melo, G. W., Kamisnki, J., & Ceretta, C. A. (2008). Produção, composição da uva e teores de nitrogênio na folha e no pecíolo em videiras submetidas à adubação nitrogenada. *Ciência Rural, 38*(9), 2622-2625.

Cattelan, A. J. (1999). Métodos quantitativos para determinação de características bioquímicas e fisiológicas associadas com bactérias promotoras de crescimento vegetal. *Documentos 139* (p. 36). Embrapa Soja, Londrina.

Carvalhais, L. C., Dennis, P. G., Fan, B., Fedoseyenko, D., Kierul, K., Becker, A., ... Borris, R. (2013). Linking Plant Status to Plant-Microbe Interactions. *Plos One, 8*(7), 1-19.

Chagas, Jr. A. F., Oliveira, L. A., & Oliveira, A. N. (2009). Produção de ácido-indol-acético por rizóbios isolados de caupi. *Revista Ceres, 56*(6), 812-817.

Dawwam, G. E., Elbeltagia, A., Emara, H. M., Abbas, I. H., & Hassan, M. M. (2013). Beneficial effect of plant growth promoting bacteria isolated from the roots of potato plant. *Annals of Agricultural Sciences, 58*(2), 195-201. http://dx.doi.org/10.1016/j.aoas.2013.07.007

Desplobins, G., & Silva, A. L. (2005). Construção de qualidade e de reconhecimento na vitivinicultura tradicional do vale do rio do peixe, em Santa Catarina. *Cadernos de Ciência & Tecnologia, 22*(2), 399-411.

Dias, A. C. F., Costa, F. E. C., Andreote, F. D., Lacava, P. T., Teixeira, M. A., Assumpção, L. C., Araújo, W. L., Azevedo, J. L., & Melo, I. S. (2009). Isolation of micropropagated strawberry endophytic bacteria and assessment of their potential for plant growth promotion. *World Journal of Microbiology and Biotechnology, 25*(2), 189-195. http://dx.doi.org/10.1007/s11274-008-9878-0

Dobbelaere, S., Vanderleyden, J., & Okon, Y. (2003). Plant growth-promoting effects of diazotrophs in the rhizosphere. *Critical Reviews in Plant Sciences, 22*(2), 107-149. http://dx.doi.org/10.1080/713610853

Bobrowski, V. L., Fiuza, L. M., Pasqualis, G., & Bodanese-Zanettini, M. H. (2008). Genes de *Bacillus thuringiensis*: uma estratégia para conferir resistência a insetos em plantas. *Ciência Rural, 34*(1), 843-850. http://dx.doi.org/10.1590/S0103-84782003000500008

Drancourt, M., Roux, V., Fournier, P. E., & Raoult, D. (2004). *rpo*B gene sequence-based identification of aerobic Gram-positive cocci of the genera *Streptococcus*, *Enterococcus*, *Gemella*, *Abiotrophia*, and *Granulicatella*. *Journal of Clinical Microbiology, 42*(2), 497-504. http://dx.doi.org/10.1128/JCM.42.2.497-504.2004

Fan, B., Li, L., Chao, Y., Forstoner, K., Vogel, J., Borross, R., & Wu, X. Q. (2015). dRNA-Seq Reveals Genomewide TSSs and Noncoding RNAs of Plant Beneficial Rhizobacterium *Bacillus amyloliquefaciens* FZB42. *Plos One, 10*(11), 0142002. http://dx.doi.org/10.1371/journal.pone.0142002

Garrido, L. R., Sônego, O. R., & Gomes, V. N. (2004). Fungos associados com o declínio e morte de videiras no Estado do Rio Grande do Sul. *Fitopatologia Brasileira, 29*(3), 322-324. http://dx.doi.org/10.1590/S0100-4 1582004000300016

Glick, B. R. (1995). The enhancement of plant growth by free-living bacteria. *Canadian Journal of Microbiology, 41*(2), 109-117. http://dx.doi.org/10.1139/m95-015

Glick, B. R. (2012). Plant growth-promoting bacteria: Mechanisms and applications. *Hindawi Corporation Scientifica,* 1-15. http://dx.doi.org/10.6064/2012/963401

Gupta, G., Parihar, S. S., Ahirwar, N. K., Snehi, S. K., & Singh, V. (2015). Plant growth promoting rhizobacteria (PGPR): Current and future prospects for development of sustainable agriculture. *Journal of Microbial and Biochemical Tecnolog, 7*(2), 96-102. http://dx.doi.org/10.1016/j.plaphy.2013.01.020

Hoagland, D. R., & Arnon, D. I. (1959). *The water culture method for growing plants without soil* (p. 347). Calif. Agr. Exp. STA. Cir.

Karagöz, K., Ates, F., Karagöz, H., Kotan, R., & Cakmakci, R. (2012). Characterization of plant growth-promoting traits of bacteria isolated from the rhizosphere of grapevine grown in alkaline and acidic soils. *European Journal of Soil Biology, 50,* 144-150. http://dx.doi.org/10.1016/j.ejsobi.2012.01.007

Khalid, A., Arshad, M., & Zahir, Z. A. (2004). Screening plant growth-promoting rhizobacteria for improving growth and yield of wheat. *Journal of Applied Microbiology, 96,* 473-480. http://dx.doi.org/10.1046/j.1365-2672.2003.02161.x

King, E. O., Ward, M. K., & Raney, D. E. (1954). Two simple media for the demonstration of pyocyanin and fluorescein. *The Journal of Laboratory and Clinical Medicine, 44*(2), 301-307.

Kuss, A. V., Kuss, V. V., Lovato, T., & Flôres, M. L. (2007). Fixação de nitrogênio e produção de ácido indolacético in vitro por bactérias diazotróficas endofíticas. *Pesquisa Agropecuária Brasileira, 42*(10), 1459-1465. http://dx.doi.org/10.1590/S0100-204X2007001000013

Lugtenberg, B., & Kamilova, F. (2009). Plant-Growth-Promoting Rhizobacteria. *Annual Review of Microbiology, 63,* 541-556. http://dx.doi.org/10.1146/annurev.micro.62.081307.162918

Marasco, R., Rolli, E., Fusi, M., Cherif, A., Abou-Hadidi, A., El-Bahairy, U., ... Daffonhio, D. (2013). Plant growth promotion potential is equally represented in diverse grapevine root-associated bacterial communities from different biopedoclimatic environments. *BioMed Research International,* 1-17. http://dx.doi.org/10.1155/2013/491091

Nautiyal, C. S. (1999). An efficient microbiological growth medium for screening phosphate solubilizing microorganisms. *FEMS Microbiology Letters, 170*(1), 265-270. http://dx.doi.org/10.1111/j.1574-6968.1999.tb13383.x

Nautiyal, C. S., Srivastava, S., Chauhan, P. S., Seem, K., Mishra, A., & Sopory, S. K. (2013). Plant growth-promoting bacteria *Bacillus amyloliquefaciens* NBRISN13 modulates gene expression profile of leaf and rhizosphere community in rice during salt stress. *Plant Physiology and Biochemistry, 66,* 1-9. http://dx.doi.org/10.1016/j.plaphy.2013.01.020

Passos, J. F., da Costa, P. B., Costa, M. D., Zaffari, G. R., Nava, G., Boneti, J. I., ... Passaglia, L. M. (2014). Cultivable bacteria isolated from apple trees cultivated under different crop systems: Diversity and antagonistic activity against *Colletotrichum gloeosporioides. Genetics and Molecular Biology, 37*(93), 560-572.

Smyth, E. M., MCcarthy, J., Nevin, R., Khan, M. R., Dow, J. M., O'Gara, F., & Doohan1, F. M. (2011). In vitro analyses are not reliable predictors of the plant growth promotion capability of bacteria; a *Pseudomonas fluorescens* strain that promotes the growth and yield of wheat. *Journal of Applied Microbiology, 111,* 683-692. http://dx.doi.org/10.1111/j.1365-2672.2011.05079.x

Tecchio, M. A., Teixeira, L. A. J., Terra, M. M., Moura, M. F., & Paioli-Pires, E. J. (2011). Extração de nutrientes pela videira niagara rosada enxertada em diferentes porta-enxertos. *Revista Brasileira de Fruticultura, 33,* 736-742. http://dx.doi.org/10.1590/S0100-29452011000500103

Van Loon, L. C., & Bakker, P. (2005). Induced systemic resistance as a mechanism of disease suppression by rhizobacteria. *PGPR: Biocontrol and Biofertilization*, 39-66.

Zahir, Z. A., Asghar, H. N., Akhtar, M. J., & Arshad, M. (2005). Precursor (L-tryptophan) inoculum (*Azotobacter*) interaction for improving yields and nitrogen uptake of maize. *J. Plant Nut, 28*, 805-817. http://dx.doi.org/10.1081/PLN-200055543

4

Tissue-Specific Expression Profiling of Seedling Stage in Early-Maturity Mutant Induced by Carbon Ion Beam in Sweet Sorghum

Xicun Dong[1,†], Xia Yan[2,3,†], Wenjian Li[1], Ruiyuan Liu[1] & Wenting Gu[1]

[1] Department of Radiobiology, Institute of Modern Physics, Chinese Academy of Sciences, Lanzhou, China

[2] Key Laboratory of Inland River Ecohydrology, Cold and Arid Regions Environmental and Engineering Research Institute, Chinese Academy of Sciences, Lanzhou, China

[3] Key Laboratory of Stress Physiology and Ecology in Cold and Arid Regions of Gansu Province, Cold and Arid Regions Environmental and Engineering Research Institute, Chinese Academy of Sciences, Lanzhou,China

Correspondence: Xicun Dong, Radiobiology Department, Institute of Modern Physics, Chinese Academy of Sciences, Lanzhou, China. E-mail: dongxicun@impcas.ac.cn

† These authors contributed equally to this work.

The research is financed by the STS project (KFJ-EW-STS-086) and Western Light Co-scholar (29Y406020) Program of the Chinese Academy of Sciences.

Abstract

An early-maturity mutant KFJT-1 has been screened out after carbon ion irradiation in sweet sorghum (*Sorghum bicolor* (L.) Moench). In this study, tissue specific digital gene expression analysis was performed between the KFJT-1 mutant and the wild type KFJT-CK at seedling stage. The results showed that a total of 717, 2160 and 2,331 tags-mapped genes were differently expressed in roots, stems and leaves of young seedling, respectively. In KFJT-1, 557 (77.7%) genes were up-regulated and 160 (22.3%) genes were down-regulated in young root; 1,232 (57.0%) genes were up-regulated and 928 (43.0%) were down-regulated in young stem; and 1,577 (67.7%) genes were up-regulated and 754 (32.3%) genes were down-regulated in young leaf. Functional annotation revealed that most induced genes functioned as "binding", "synthase activity", "transferase" and "transporter activity" which involved in the biological processes of metabolic and response to stimulus. Surprisingly, the up-regulated genes in KFJT-1 were classified into four KEGG pathways: "alpha-Linolenic acid metabolism", "flavonoid biosynthesis", "inositol phosphate metabolism" and "fatty acid biosynthesis", which related to the stress resistance and supported the outstanding agronomic traits of KFJT-1 in the process of plant growth and development. Among the DEGs, a critical photoreceptor from photoperiod pathway *PHYA* gene was significantly up-regulated in leaf and root of KFJT-1, suggesting the mutation could occur on the genomic upstream of *PHYA*. This work may provide helpful insights to further understand the mutation mechanism in sweet sorghum.

Keywords: tissue-specific expression profiling, young seedling, different expressed genes, sweet sorghum

1. Introduction

Sweet sorghum (*Sorghum bicolor* (L.) Moench) is a useful energy crop because of high photosynthetic efficiency, high biomass- and sugar- yielding (Billa et al., 1997). However, being a short-day plant, the grains cannot mature under long day condition. We previously isolated an early-maturity mutant KFJT-1 from wild type plants KFJT-CK by heavy ion beam irradiation. Resistance experiment showed that proline content was increased by 11.05% with drought stress, which showed that the tolerance of KFJT-1 to the stress is advantage to KFJT-CK (Dong & Li, 2012).

The biological effects of heavy-ion radiation encompass a wide range of alterations, including developmental abnormalities (Kranz, 1994), chromosomal aberrations (Kawat et al., 2001; Kikuchi et al., 2009; Wei et al., 2006) and genomic structural variation (Mei et al., 2011; Xu et al., 2006). Many studies has shown that the carbon ions beam induce more effective structural alterations in DNA than other radiation (Shikazono et al., 2005), sequence analyses of radiation-induced mutations have been widely carried out in plants (Bruggenmann et al., 1996; Shikazono et al., 2000; Shikazono et al., 2003). These genetic variations could directly induce expression differentiation of the plenty of genes which involved in the biological processes. Nowadays, Digital gene expression (DGE) tag profiling has been widely utilized to monitor the differences in transcriptional level to elucidate the genome-wide expression profiling among different tissues and organs. It directly quantify the transcript abundance of the uniquely tagged corresponding genes with ultra-high-throughput sequencing of cDNA fragments (Hong et al., 2011), which could be conveniently detected for the organisms without prior annotations of genomic information, such as cotton (Wei et al., 2013), spruce (Albouyeh et al., 2010), *Brassica napus* (Jiang et al., 2013), and moss (Nishiyama et al., 2012).

This study aimed to gain comprehensive understanding of DGEs in KFJT-1 compared to KFJT-CK at seedling stage and improve our current understanding of the molecular mechanism of KFJT-1 induced by carbon ions beam.

2. Materials and Methods

2.1 Plant Materials and Growth Conditions

The dry seeds of equal size from KFJT-1 and KFJT-CK, which showed no moldy and lesion, were selected and placed in a 90 mm Petri dish containing double-layer wet filter paper, respectively. The seeds were germinated at 25 ± 2 °C in a growth chamber under a 16 h light photoperiod provided by fluorescent light tubes (50 μmol m^{-2}s^{-1}). Each genotype was replicated three times and 100 seeds were employed for each replication. After 30 days, the samples of roots, stems and leaves were harvested, respectively and quickly frozen in liquid nitrogen for RNA isolation.

2.2 RNA Isolation and Library Preparation for DGE

RNA extraction was performed according to the manufacturer's instructions of TRIzol reagent (Invitrogen, USA), followed by RNase-free DNase treatment (TaKaRa, Dalian, China). The total RNA was checked for quality and quantity using a Biophotometer Plus (Multiskan Spectrum, German), and a minimum of 6 ug of total RNA was used for Illumina sequencing. The total RNA samples isolated from the three parallels tissues were pooled for libraries preparation, in which RC and RF, SC and SF, LC and LF represented the transcripts of roots, stems and leaves from control KFJT-CK ("C" characterized) and mutant KFJT-1 ("F" characterized), respectively.

Over 6 μg from each total RNA samples were constructed as the DGE libraries using Illumina gene expression kit (IllumingaInc; San Diego, CA, USA) according to the manufacturer's protocol (version 2.1B), mRNA was purified using biotin-Oligo (dT) magnetic bead adsorption. The first- and second-strand cDNA synthesis was performed after the RNA was bound to the beads. The double stranded cDNA were digested with *NlaIII* to produce cohesive end. After purification with Dynabeads, and the digestion was ligated to GEX adapter 1 which contains *MmeI* restriction CATG site, and downstream 17 bp then cut with the *NlaIII*. The 21 bp tags containing adapter I were ligated to GEX adapter 2 to generate a tag library. These tag fragments were amplified by liner PCR for 15 cycles using PCR primers anneral to the adapter ends. The 85 bp amplicons were seperated on 6% TBE PAGE gel, purified and denatured to produce single strand molecules. These molecules were anchored to Solexa sequencing array and sequenced on Illuminga GA II at BGI- Shenzhen, Shenzhen, China. Raw sequence data were generated by Illuminga pipeline.

2.3 Sequence Annotation and DGEs Pathways Identification

Raw sequences were transformed into clean tags by filtering off adapter-only tags and low-quality tags as described (Li et al., 2013). All the clean tags were mapped to the reference sequences of *Sorghum bicolor* and only 1 bp of mismatch was considered. The remaining clean tags were designed as unambiguous clean tags. In order to compare the expression abundance among the samples, the number of unambiguous clean tags for each gene was calculated and then normalized to TPM (number of transcripts per million clean tags). The final assembled transcripts ($\geq$ 100 bp) were submitted for homology and annotation searches using Blast2GO software v2.4.4 (Wei et al., 2013). For BLASTX against the NR database, the threshold was set to E-value lower than 10^{-5}. However, most of the gene information of sorghum was hypothetical or putative. Therefore, all the putative sorghum genes were BlastX with was performed against *Sorghum* genes. GO classification was achieved using WEGO software together with David Bioinformatics Resources 6.7 (http://david.abcc.ncifcrf.gov/home.jsp). Enzyme codes were extracted and Kyoto Encyclopedia of Genes and Genomes (KEGG) pathways were retrieved from KEGG web server (http://www.genome.jp/kegg/). We used a rigorous algorithm to identify differentially expressed genes between the KFJT-1 and KFJT-CK in this study. FDR ranking, FDR (False Discovery Rate) was used applied to adjust the p-value in multiple tests and analyses (Qin et al., 2011). The transcripts with at least two-fold differences (absolute values of log2 (Ratio) $\geq$ 1 with FDR $<$ 0.001) were regarded as significantly different expressed genes.

2.4 Real-Time Quantitative RT-PCR (q RT-PCR) Analysis

Real-time quantitative RT-PCR (qRT-PCR) analysis was used to verify the DGE results. The RNA samples used for the qRT-PCR assays were the same as in the DGE experiments. Gene specific primers were designed according to the reference unigene sequences using Primer Premier 5.0. Seven genes were selected from the DEGs for quantitative qRT-PCR assays. QRT-PCR was performed according to the manufacturer's specifications (). The following SYBR Green PCR cycling conditions were used: denaturation at 95 C for 10 s, followed by 40 cycles of 94 C for 5 s and 60 C for 20 s. The PCR experiments were performed using an iQ 5 Multicolor real-time PCR detection system (BioRAD, USA). Sorghum actin gene (forward: GCCGAGCGAGAAATTGTAAG and reverse: ATCATGGATGGCTGGAAGAG) was used as a normalizer. The relative gene expression levels were calculated using $2^{-\Delta\Delta CT}$.

3. Result

3.1 Construction of Digital Gene Expression (DGE) Library for KFJT-1 and KFJT-CK at Seedling Stage

To obtain a global view of the tissue specific characteristics at the transcriptional level between KFJT-1 and KFJT-CK at seedling stage, total six DGE libraries from roots, stems and leaves were sequenced with Solexa/Illumina DGE analysis, respectively. Among the libraries, we got the total numbers of tags ranging from 5.8 to 6.7 million, which composing distinct tags with 125840 and 131973,140208 and 156400, 147227 and 137977 in young roots, stems and leaves libraries for KFJT-CK and KFJT-1, respectively (Table 1). The number of the tags and unambiguous tags mapping to genes was almost the same for about 70%, for example in RC, 62208 distinct tags (72.07% of clean tag), 62089 unambiguous tags (71.99% of clean tag) was mapping to gene, which means that the tags was well matched to the specific genes. Those distinct tags matched to the genes occupied about 50% of the clean tags. The distribution of total clean tags and distinct clean tags over different tag-abundance categories were shown in Figures 1A and 1B. In terms of the total clean tags, the percentage of 2-5 copies ranged from 3.31-3.89%, 6-10 copies from 2.3-2.71%, 11-20 copies from 3.20-3.81%, 21-50 copies from 6.44-7.47%, 51-100 copies from 7.7-8.83%. The largest percentage was constituted by the copies over 100 which ranged from 73.58-76.90%. However, compared to the total clean reads, the distinct clean tags displayed different distribution among all the six DGE libraries. The largest proportion was constituted by the 2-5 copies for about 54%. The second largest part was the 6-10 copies which occupied about 14%, approximately 10% had copy numbers higher than 100. The smallest proportion was constituted by 51-100 copies for about 7%. The numbers of the tag-mapped genes or unambiguous tag-mapped gene were decreased sharply compared to the distinct tags. For example, the number of matched gene in RC was 15768, however, the distinct tags was 62208. Finally, 15786 (RC), 16007 (RF), 16179 (SC), 16780 (SF), 15378 (LC) and 15200 (LF) tag-mapped genes were generated against sorghum reference genome between KFJT-1 and KFJT-CK. The saturation analysis showed that the number of the genes was not increased proportionally with the number of sequences (total tag number) when the sequencing counts reached 4M (Figure S1). Thus, these tag-mapped genes were completely satisfying the further analysis.

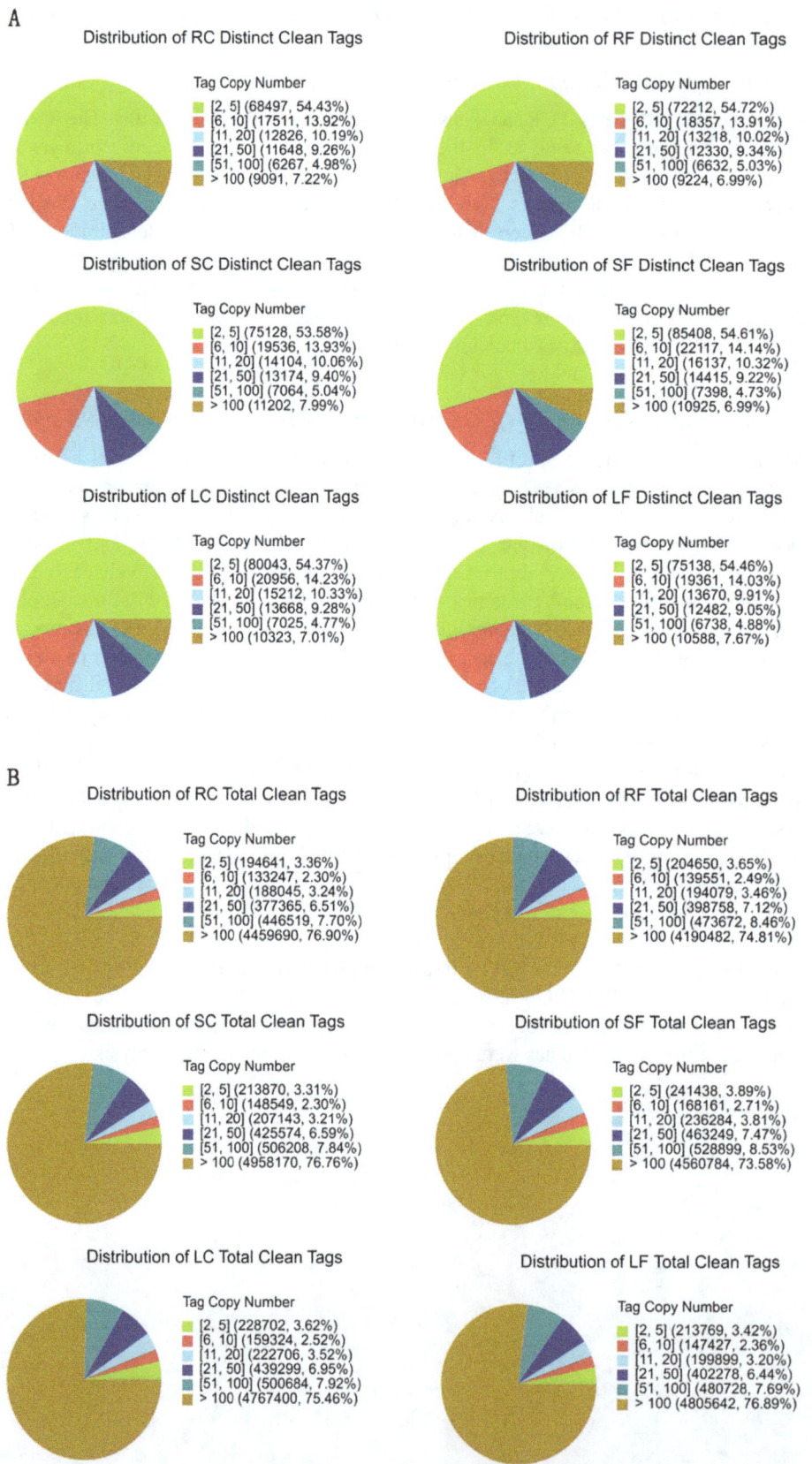

Figure 1. Distribution of distinct clean tags and total clean tags. (A) Distribution of distinct clean tags. (B) Distribution of total clean tags. RC, SC, LC, and RF, SF, LF represented roots, stems and leaves of wild type and mutant, respectively. Numbers within square brackets indicate the range of copy number for a specific category of tags. Numbers within parentheses indicate the types of tag in A or copy number of tags in B

Table 1. Statistics of DGE sequencing

Summary		Roots		Stems		Leaves	
		RC	RF	SC	SF	LC	LF
Raw data	Total	6091059	5890651	6746137	6493952	6694215	6603642
	Distinct tag	298254	305052	333748	343180	341594	328619
Clean tag	Total number	5799507	5601192	6459514	6198815	6318115	6249743
	Distinct tag number	125840	131973	140208	156400	147227	137977
All tags mapping to gene	Total % of clean tag	72.07%	71.15%	71.61%	71.10%	66.34%	68.92%
	Distinct tag number	62208	62402	66229	75306	68097	65630
	Distinct tag % of clean tag	49.43%	47.28%	47.24%	48.15%	46.25%	47.57%
Unambiguous tags mapping to gene	Total % of clean tag	71.99%	71.08%	71.54%	71.00%	66.29%	68.87%
	Distinct tag number	62089	62276	66098	75159	67968	65523
	Distinct tag % of clean tag	49.34%	47.19%	47.14%	48.06%	46.17%	47.49%
All tag-mapped genes	Number	15786	16007	16179	16780	15378	15200
	% of ref genes	53.61%	54.36%	54.94%	56.98%	52.22%	51.62%
Unambiguous tag-mapped genes	Number	15747	15963	16132	16730	15327	15166
	% of ref genes	53.47%	54.21%	54.78%	56.81%	52.05%	51.50%

3.2 Tissue-Specific Gene Expression in the Development of the Seedling between KFJT-1 and KFJT-CK

To compare differential expression patterns between KFJT-1 and KFJT-CK, we normalized tag distribution for gene expression level in each library to make an effective library size and extracted significance of differentially expressed transcripts (DETs) with FDR $\leq$ 0.05 and log2 fold-change $\geq$ 1 by edgeR (Empirical analysis of Digital Gene Expression in R). The regulated genes were shown in Figure 2. The red dots and green dots represent transcripts higher or lower in abundance for more than two fold, and the blue dots represented the transcripts that differed less than two fold between the KFJT-1 and the wild type. In root, a total of 557 genes were up-regulated (77.7%, red dot in Figure 2A) and 160 genes were down-regulated (22.3%, green dot in Figure 2A). In stem, a total of 1,232 genes were up-regulated (57.0%, red dot in Figure 2B) and 928 genes were down-regulated (43.0%, green dot in Figure 2B). In leaf, total 1,577 genes were up-regulated (67.7%, red dot in Figure 2C) and 754 genes were down-regulated (32.3%, green dot in Figure 2C). An increasing trend in the number of differently expressed genes was observed in young stem and leaf compared to young root. The total numbers of the tags-mapped genes were 717, 2160 and 2331 in root, stem and leaf, respectively (Figure 3A). Venn analysis revealed that 66 genes were differently expressed in the all young root, stem and leaf. In addition, 219 genes were differently expressed both in leaf and root, 489 genes in leaf and stem, and 148 genes in root and stem. Numbers of 284, 1457 and 1557 genes were differentially expressed specific to the root, stem and leaf, respectively (Figure 3B). To be mentioned, about 99% unique tags were expressed within five-fold difference (red bar in Figure 4) between KFJT-1 and KFJT-CK, covered 99.85% in young root, 99.39% in young stem and 98.77% in young leaf, respectively. Only 0.1-1.01% of the DEGs over five folds was up-regulated (green bar in Figure 4), while 0.05-0.23% was down-regulated.

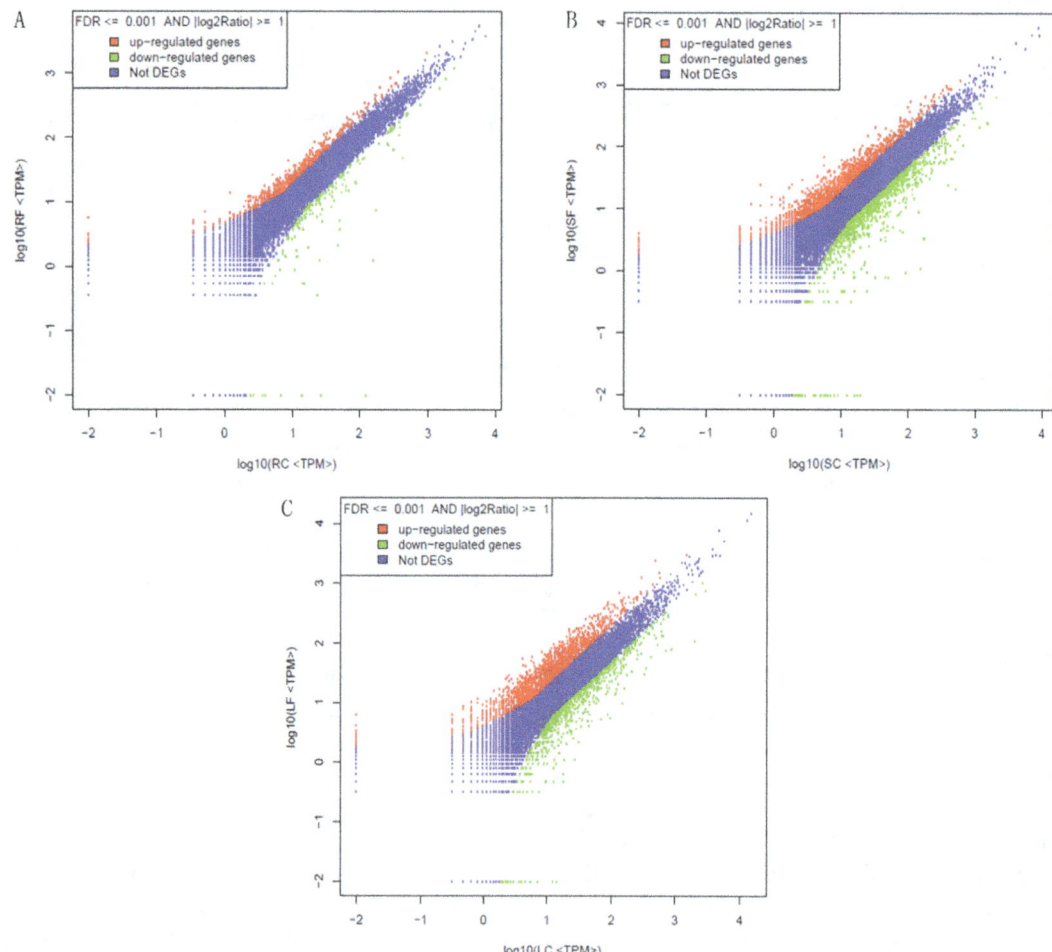

Figure 2. Comparison of gene expression between different libraries. Blue dots represent the transcripts with no significant expression. Red dots and green dots represent transcripts more abundant between early-maturity mutant and wild type in roots, stems and leaves, respectively. "FDR < 0.001" and "absolute value of log2 Ratio > 1" were used as the thresholds to judge the significance of gene expression difference

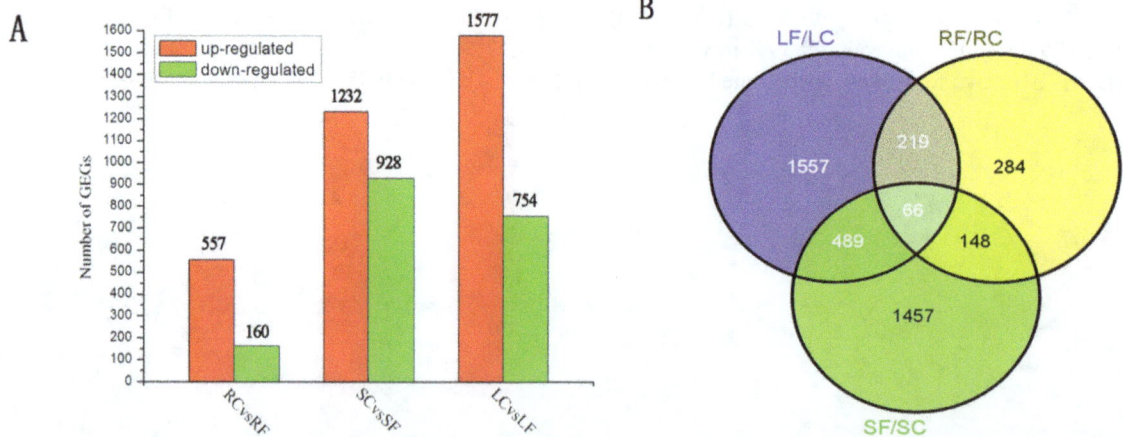

Figure 3. Distribution of differentially expressed genes. (A) Changes in gene expression profile of roots, stems and leaves. (B) Venn diagram to illustrate the number of genes regulated in the roots, stems and leaves between EM mutant and the wild type in sweet sorghum

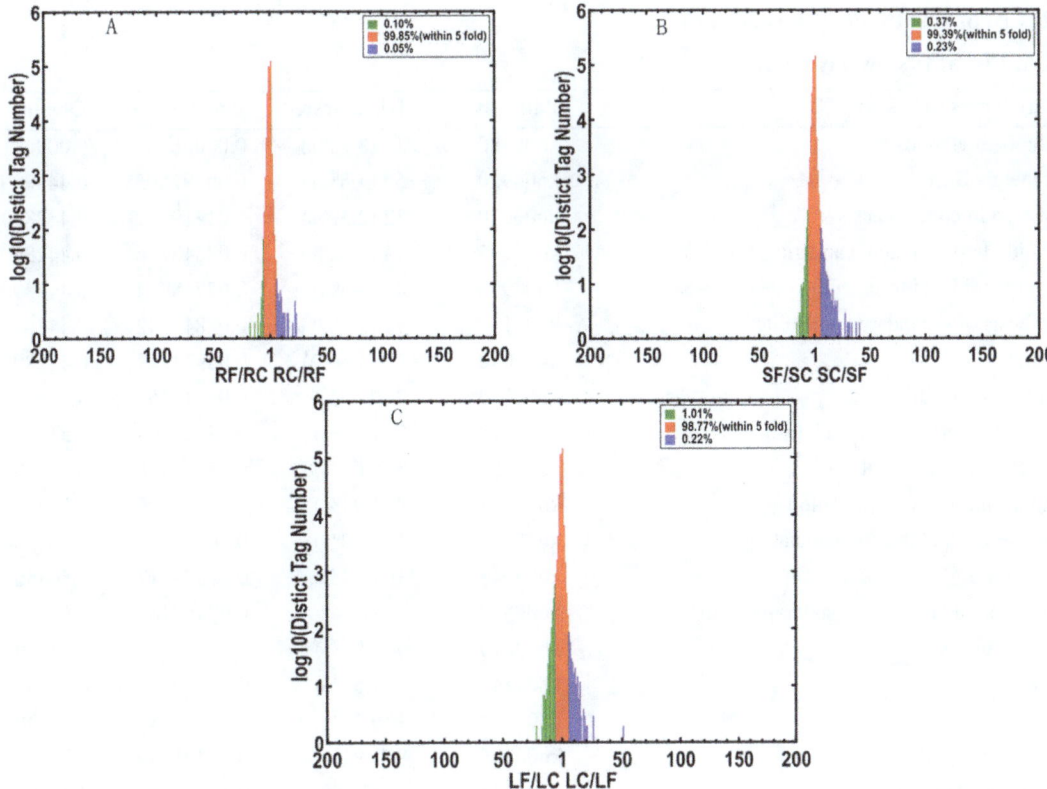

Figure 4. Tags with different expression in early-maturity compared to wild type sample. Red region represents the differentially expressed tags with differentia expressionless than 5 folds. Blue and green region represent the up- and down-regulated tags for more than 5 folds, respectively

3.3 Functional Analysis of the Differentially Expressed Genes at Seedling Stage

Gene ontology (GO) assignments were used to classify the functions of differentially expressed genes with at least two-fold differences (Figure 5). According to biological process, 193 genes were related to "small molecule metabolic process" with p-value lower than 0.05. Accordingly, 55 genes related to "structural molecule activity" for molecular function, and total 1131 genes cluster into main 9 category were over-represented with p-value lower than 0.05 (Table 2). Regarding to regulated genes was about 99% within five fold, we used genes with expression level over five fold as more strict criteria to assign GO annotation on DAVID Bioinformatics Resource 6.7. Among the DEGs over five folds, 331 genes were related to "metabolic process" and 148 related to "response to stimulus", respectively. The genes response to stimulus including "cold", "water deprivation", "salt" "wounding" and "osmotic" stress. Furthermore, the term "fatty acid biosynthetic process" which involve in stress resistance was also enriched, which suggested that KFJT-1 mutants were more adaptable to stress environment than KFJT-CK since seedling stage. "Lignin biosynthetic process" also enriched to support high compact stem structure in KFJT-1. Genes involved flowering time were also enriched such as "response to far red light", "pollen development" including "auxin transport", which consist with the short growth cycle of KFJT-1 compared to KFJT-CK. Specially, in young leaf, "polysaccharide metabolic process" "carboxylic acid biosynthetic process" and "developmental maturation" founded which suggested that KFJT-1 fixed higher carbon than KFJT-CK. In young root, genes involved in "nitrate metabolic process" were high enriched, which suggested that KFJT-1 harbor more potential than KFJT-CK in the nitrogen fixation to facilitate development in barren soil. In stem, besides of genes involved in stress response, "sugar mediated signaling" and "response to carbohydrate stimulus" were enriched which might be correlated with the high sugar content in KFJT-1. According to the molecular function, those genes were related to the terms "binding", "synthase activity", "transferase" and "transporter activity", which facilitate to the biological process above.

Table 2. List of first twenty pathways for DEGs

Table 2A. List of first twenty pathways in root

Pathway term (RC vs RF)	Pathway ID	DEGs tested	P value	Q value
Glutathione metabolism	ko00480	15 (3.23%)	0.0000858	0.009091637
Sulfur metabolism	ko00920	5 (1.08%)	0.009928955	0.442890563
Glycolysis/Gluconeogenesis	ko00010	12 (2.59%)	0.01419812	0.442890563
Stilbenoid, diarylheptanoid and gingerol biosynthesis	ko00945	14 (3.02%)	0.02643276	0.442890563
Sesquiterpenoid and triterpenoid biosynthesis	ko00909	4 (0.86%)	0.02738994	0.442890563
Biosynthesis of secondary metabolites	ko01110	82 (17.67%)	0.02848042	0.442890563
Ribosome	ko03010	17 (3.66%)	0.02924749	0.442890563
Vitamin B6 metabolism	ko00750	3 (0.65%)	0.03542163	0.469336598
Cysteine and methionine metabolism	ko00270	7 (1.51%)	0.04333564	0.510397538
Selenocompound metabolism	ko00450	3 (0.65%)	0.05098186	0.540407716
Limonene and pinene degradation	ko00903	9 (1.94%)	0.06588606	0.564560846
Glycine, serine and threonine metabolism	ko00260	6 (1.29%)	0.06739679	0.564560846
Spliceosome	ko03040	16 (3.45%)	0.07073317	0.564560846
Amino sugar and nucleotide sugar metabolism	ko00520	10 (2.16%)	0.07456464	0.564560846
ABC transporters	ko02010	9 (1.94%)	0.1039357	0.671309224
Peroxisome	ko04146	7 (1.51%)	0.1053809	0.671309224
Caffeine metabolism	ko00232	1 (0.22%)	0.1076628	0.671309224
Carotenoid biosynthesis	ko00906	9 (1.94%)	0.1189165	0.700286056
Flavonoid biosynthesis	ko00941	12 (2.59%)	0.1282596	0.715553558
Proteasome	ko03050	4 (0.86%)	0.1386467	0.73482751

Table 2B. List of first twenty pathways in stem

Pathway term (SC/SF)	Pathway ID	DEGs tested	P value	Q value
Fatty acid metabolism	ko00071	16 (1.19%)	0.000612223	0.07346681
Ribosome	ko03010	44 (3.28%)	0.005398256	0.27306652
Taurine and hypotaurine metabolism	ko00430	6 (0.45%)	0.006826663	0.27306652
Other types of O-glycan biosynthesis	ko00514	5 (0.37%)	0.01560009	0.3743676
Regulation of autophagy	ko04140	14 (1.04%)	0.01770056	0.3743676
Amino sugar and nucleotide sugar metabolism	ko00520	26 (1.94%)	0.02162359	0.3743676
Carotenoid biosynthesis	ko00906	25 (1.86%)	0.02409221	0.3743676
Phenylalanine, tyrosine and tryptophan biosynthesis	ko00400	11 (0.82%)	0.02495784	0.3743676
Glycerophospholipid metabolism	ko00564	23 (1.71%)	0.02836514	0.37820187
Porphyrin and chlorophyll metabolism	ko00860	11 (0.82%)	0.03293996	0.39527952
Peroxisome	ko04146	18 (1.34%)	0.03923865	0.41932133
Ubiquinone and other terpenoid-quinone biosynthesis	ko00130	10 (0.74%)	0.05126163	0.41932133
Butanoate metabolism	ko00650	8 (0.6%)	0.05197998	0.41932133
Pyruvate metabolism	ko00620	17 (1.27%)	0.0522923	0.41932133
alpha-Linolenic acid metabolism	ko00592	15 (1.12%)	0.05507743	0.41932133
C5-Branched dibasic acid metabolism	ko00660	2 (0.15%)	0.05590951	0.41932133
Lysine degradation	ko00310	9 (0.67%)	0.06190139	0.43469967
Tyrosine metabolism	ko00350	13 (0.97%)	0.06520495	0.43469967
beta-Alanine metabolism	ko00410	7 (0.52%)	0.07620636	0.47301936
Biosynthesis of secondary metabolites	ko01110	212 (15.79%)	0.07883656	0.47301936

Table 2C. List of first twenty pathways in stem

Pathway term (LC/LF)	Pathway ID	DEGs tested	P value	Q value
Ribosome	ko03010	52 (3.52%)	0.000502993	0.06035914
Amino sugar and nucleotide sugar metabolism	ko00520	30 (2.03%)	0.007302478	0.31563592
Regulation of autophagy	ko04140	16 (1.08%)	0.007890898	0.31563592
Glyoxylate and dicarboxylate metabolism	ko00630	15 (1.02%)	0.01449332	0.4347996
Nicotinate and nicotinamide metabolism	ko00760	4 (0.27%)	0.02350194	0.47046309
Riboflavin metabolism	ko00740	9 (0.61%)	0.02450513	0.47046309
Glutathione metabolism	ko00480	23 (1.56%)	0.02744368	0.47046309
Selenocompound metabolism	ko00450	6 (0.41%)	0.04639775	0.65290427
beta-Alanine metabolism	ko00410	8 (0.54%)	0.04896782	0.65290427
Protein processing in endoplasmic reticulum	ko04141	45 (3.05%)	0.07631821	0.74382171
Alanine, aspartate and glutamate metabolism	ko00250	14 (0.95%)	0.07877682	0.74382171
Pentose phosphate pathway	ko00030	11 (0.75%)	0.07967838	0.74382171
Peroxisome	ko04146	18 (1.22%)	0.08199375	0.74382171
Thiamine metabolism	ko00730	4 (0.27%)	0.09704446	0.74382171
alpha-Linolenic acid metabolism	ko00592	15 (1.02%)	0.1034048	0.74382171
Endocytosis	ko04144	21 (1.42%)	0.104978	0.74382171
Carbon fixation in photosynthetic organisms	ko00710	15 (1.02%)	0.1090852	0.74382171
Spliceosome	ko03040	41 (2.78%)	0.1177536	0.74382171
Glycine, serine and threonine metabolism	ko00260	13 (0.88%)	0.1213056	0.74382171
Cutin, suberine and wax biosynthesis	ko00073	15 (1.02%)	0.127228	0.74382171

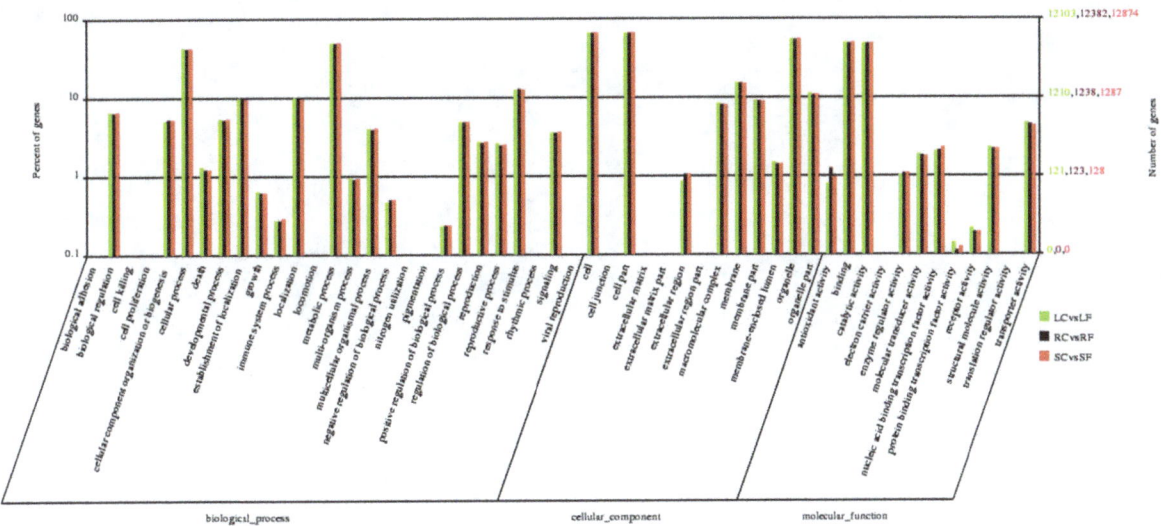

Figure 5. Gene ontology classification. The results are the percentage of each GO category of genes in terms of biological process, cellular component and molecular function. The y-axis indicates the percentage of a specific category of genes and number of genes in that main category

3.4 Metabolism Pathway Involved in the Development of Seedling from the DEGs

To shed more lights into the functional roles of DEGs between KFJT-1 and KFJT-CK, biological metabolic pathways were investigated by the enrichment analysis of DEGs among the different tissues samples at seedling stage. In young roots, it is revealed that 106 metabolic pathways were affected by DEGs (Table S1), in young stems were 120 (Table S2), in young leaves were 120 (Table S3). With hypergeometric test, there is only 9 pathway was significantly enriched in young root with the p-value lower than 0.05. The most enriched pathway was glutathione metabolism (ko00480) with p-value adjustment (q-value lower than 0.05, Table 2A). Similarly, there were 11 pathways in young stems with p value lower than 0.05. The most enriched pathway in stems was the fatty acid metabolism (ko00071) (q-value less than 0.05, Table 2B). However, it was hard to judge which pathway was the most significantly enriched with q-value control in young leaves (q-value less than 0.05, Table 2C). With genes over five fold change, four KEGG term involved in the stress tolerance were enriched: "alpha-Linolenic acid metabolism", "flavonoid biosynthesis", "inositol phosphate metabolism" and "fatty acid biosynthesis", which also suggested that KFJT-1 should have advantage of stress resistance in the process of plant growth and development.

In this study, the circadian rhythm (ko04712) was over-represented at seedling stage. In young root (Figure 6A), the genes involved in the circadian rhythm were mostly up-regulated including (PHYA, PHYB, TOC1, APR3, GI, LHY, CCA1 and WNK1). Whereas, in young stem (Figure 6B), most of the genes were down-regulated including (PHYA, PHYB, ARP7, CHS, TOC1, CCA1). In young leaf, six genes were involved, three of which were up-regulated (PHYA, TOC1 and APR3) which marked with red borders, the other three genes were down-regulated (WNK1 and CK2α, CK2β) which marked with green borders in the Figure 6C. The gene GI which is a typical disruption of the PHYB signal transduction pathway was also slightly up-regulated in young root.

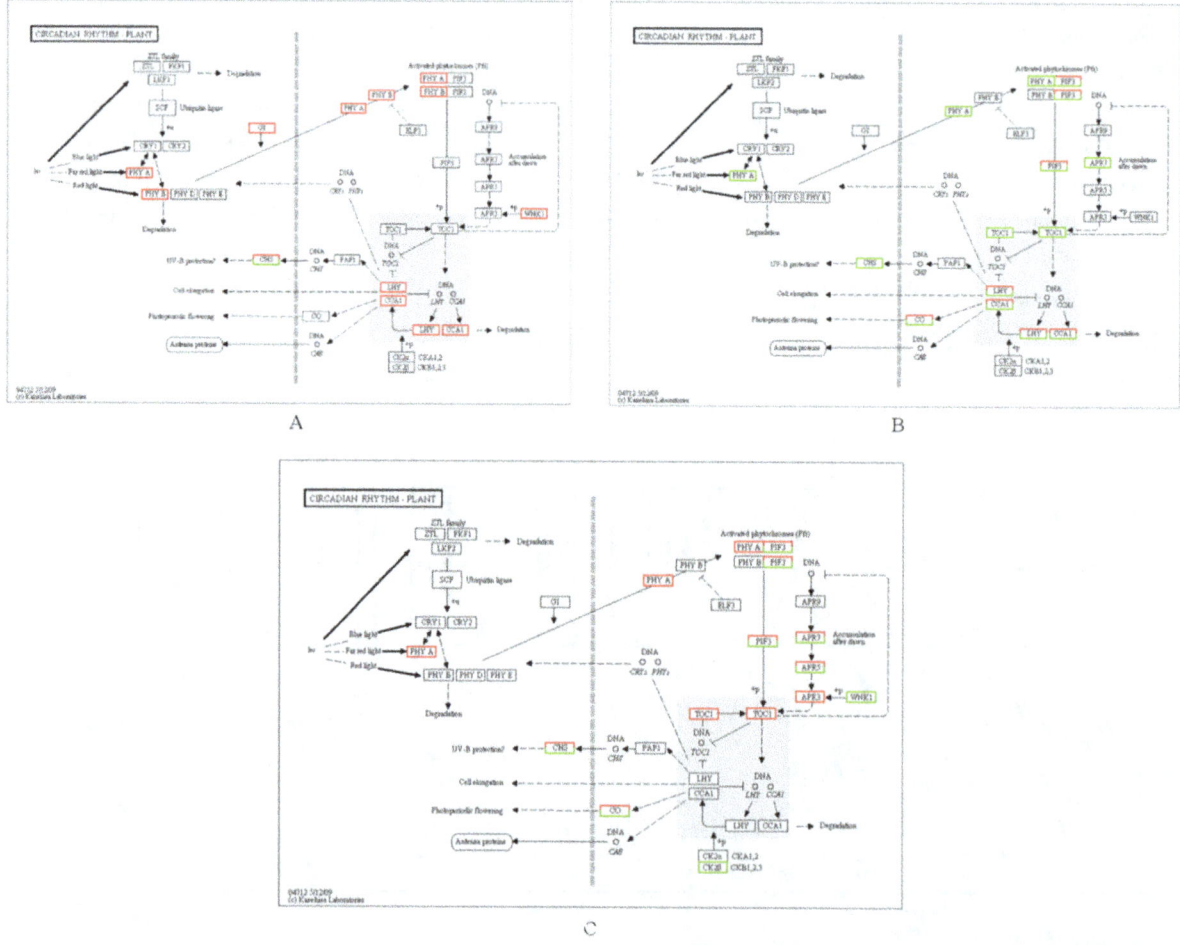

Figure 6. Change in gene expression of the circadian rhythm pathway in roots, stems and leaves at seedlings stage between early maturity mutant and wild type in sweet sorghum. Genes that up-regulated are marked with red borders while genes that down-regulated are marked with green borders. Genes that did not change are marked with black borders. A: roots, B: stems, C: leaves

Regarding the high sugar content in stems at maturation stage in the future, we checked the starch and sucrose metabolism pathway (ko00010) in different tissues at seedling stage. In young root, 8 genes were up-regulated which involved in the synthesis of β-D-Fructose, α-D-Glucose, pectin and trehalose; three genes were down-regulated which involved in the α- and β-D-Glucose (Figure 7A). In young stem, 9 genes were up-regulated (Figure 7B). 7 genes involved in the synthesis of sugar such as pectin, sucrose, α- and β-D-Glucose, D-Xylose, whereas, the up-regulated gene SS4 was for the decrease of the ADP-glucose which result in the less starch synthesis. The other gene was for the pectate degradation. 9 genes were down-regulated in this pathway, too. In young leaf, 11 genes were up-regulated, 7 genes were down-regulated. These genes involved in accumulation of glucose and sucrose, or limit their consumption (Figure 7C). Those genes were functioned to increase sugar content or reduce its consumption. For example in young stem, the down-regulated gene 3.2.1.26 decreased ADP-glucose, whereas, the up-regulated gene 2.4.1.21 accelerated the ADP-glucose consumption, which limited synthesis of the starch. For the sucrose, the up-regulated gene 2.4.1.13 increase the sucrose, whereas, the down-regulated gene 3.2.1.26 limited its consumption. The process was similar sugars, which resulted in the high sugar content.

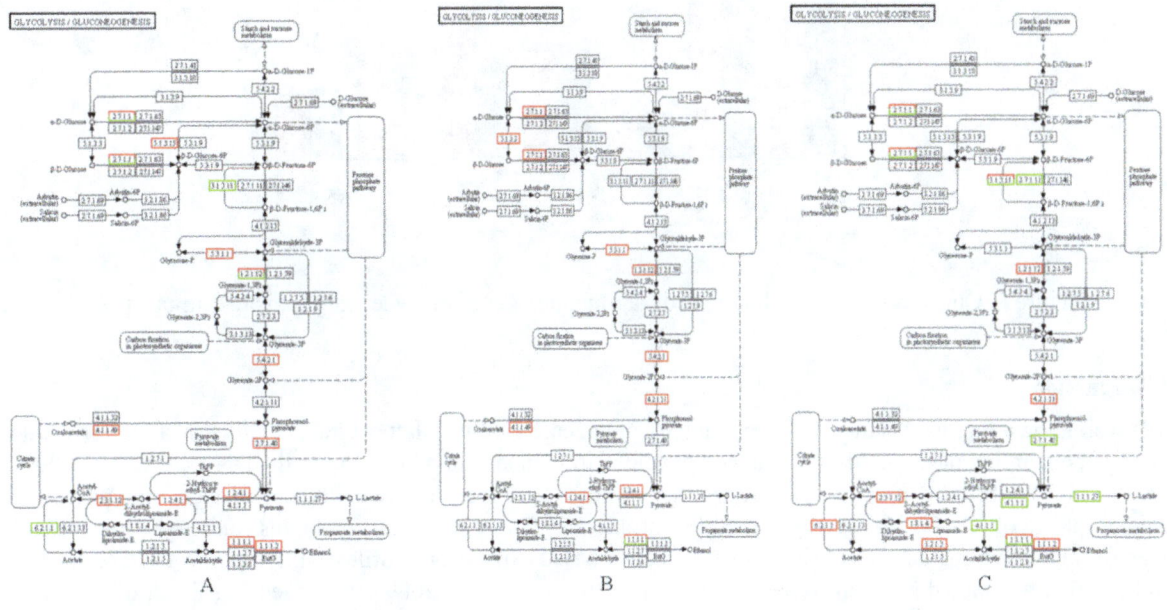

Figure 7. Change in gene expression of the Glycolysis pathway in roots, stems and leaves at seedlings stage between early maturity mutant and wild type in sweet sorghum. Genes that up-regulated are marked with red borders while genes that down-regulated are marked with green borders. Genes that did not change are marked with black borders. A: roots, B: stems, C: leaves

3.5 Confirmation of Differentially Expressed Genes by qRT-PCR

Seven genes were selected for qRT-PCR analysis to validate the DGE data. Expression of six genes detected by qRT-PCR matched the DGE data. Moreover, the fold-changes obtained by DGE were generally greater than those obtained by qRT-PCR (Figure 8). The corresponding primers are listed in Table S4.

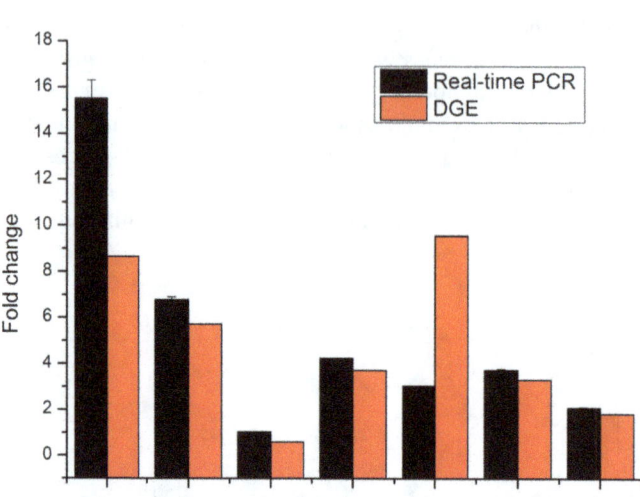

Figure 8. Quantitative PCR validations of tag-mapped genes of sweet sorghum at seedling stage

4. Discussion

It is well known that ion beams can have dramatic mutagenic effects (Matuo et al., 2006; Kawata et al., 2004) through penetrating the seed coat and the cell envelope to induce strand breaking in the DNA (Abe et al., 2002; Liu et al., 2008). We previously analyze the agronomic character, physiological change and its genetic polymorphism of KFJT-1 induced by carbon ions irradiation (Dong & Li, 2012). In this study, we developed and tested a framework for analysis of short-read, sequence-based expression profiles using Illumina DGE technology and the first assembled sorghum reference genome (Paterson et al., 2009). A sequencing depth of 5.8 to 6.7 million tags per library was reached (Table 1). Our results demonstrated that 20- to 21-nucleotide DGE tags can be used to successfully resolve genome-wide expression profiles in sweet sorghum and detect differences in transcript abundance over a broad dynamic range with sufficient sequencing depth to cover the transcriptome (Figure S1). About 70% of clean tags mapped to 55% of reference genes in the six libraries from the three different tissues due to two main reasons: Firstly, reference gene annotations in sorghum were still not completely finished and may contain some mis-annotations. Second, *NlaIII* site that is required for detection by DGE technology is contained by only 88% of reference genes (Hegedus et al., 2009), which means that some clean tags was not identified. However, this study provided much valuable information of the different expressed genes induced by carbon ion beam irradiation between KFJT-1 and KFJT-CK. Those unmapped tags could represent information for the novel genes which could be developed in the future.

In addition, we identified 717, 2160 and 2331 tags-mapped genes were differently expressed in roots, stems and leaves between KFJT-1 and KFJT-CK at seedling stage, respectively. More regulated genes enriched in young leaf and stem (the number of young stems and leaves were over three times the number of young roots) and most of them were up-regulated (*i.g.* 1577 up-regulated and 754 down regulated in young leaf), which suggested that the genes were activated by the carbon ion beam irradiation during the development of the seedling for KFJT-1. 66 regulated genes were shared in young roots, stems and leaves, which we suspected that the regulator elements in the promoter regions could be mutated by carbon ion beam irradiation. However, due to the genome sequence was not performed in KFJT-1 yet, the relationship between the mutation sites and the different expressed genes were not easy to be concluded. Further experiment or analysis on the genome sequences should be conducted to clarify the effect of carbon ion beam irradiation.

Regarding to the characteristic of KFJT-1, the expression of the circadian rhythm genes was analyzed at seedling stage. The genes PHYA, PHYB, TOC1, APR3, GI, LHY, CCA1 and WNK1, CK2α and CK2β genes was differently expressed in three tissue in KFJT-1 compare to KFJT-CK. In *Arabidopsis*, when the CK2 was over-expressed, the period of rhythmic expression of the CCA1/LHY genes was shorted and caused early flowering in both long-and short-day conditions (Sugano et al., 1999). However, in early flower mutant KFJT-1, the CK2α and CK2β were all down-regulated in young leaf. This suggested that there were no strong connection between the CK2 and CCA1/LHY in sorghum. The gene TOC1 has dual role in the control of circadian and photomorphogenic responses in *Arabidopsis* (Mas et al., 2003), mostly it was negatively regulated by CCA1/LHY, and positively regulated by the WNK1 gene (Wang & Tobin, 1998). Interestingly, the genes all above were assigned to late flower, which was contradicted against the early flower phenotypic of KFJT-1. This might be contributed to the complex regulatory module for photoperiodic flowering signaling. We proposed that the earlier flower characteristic of KFTJ-1 was related to the up-regulation of PHYA in young root and leaf. Previous study indicated that PHYA are the major day-length sensors in Arabidopsis (Mockler et al., 2003) and thought to promote flowering. The PHYA mutant flowers significantly later than the wild type in response to day-length extensions with a far-red-enriched white light (Johnson et al., 1994), differed between long-days and short-days (Weller et al., 1997). Interestingly, the PHYA is tissue-specific expressed in KFJT-1. Previous study on Arabidopsis suggested that sucrose could depress expression of PHYA (Dijkwel et al., 1997). Thus, we conferred that the down-regulated PHYA in young stem in KFJT-1 was due to high content of sucrose which mutual adjusted by the gene 3.2.1.26 and 2.4.1.13.

5. Conclusion

To our best knowledge, this study gained comprehensive understanding of DGEs between KFJT-1 and KFJT-CK, which was the first genome-wide effort to investigate the transcription dynamics of sweet sorghum induced by carbon ion beam at seedling stage. Furthermore, this work provides some useful information to develop function genes for the industry process of energy crop by carbon ion beam.

References

Abe, T., Matsuyama, T., Sekido, S., Yamaguchi, I., Yoshida, S., & Kameya, T. (2002). Chlorophyll-deficient mutants of rice demonstrated the deletion of a DNA fragment by heavy-ion irradiation. *Journal of Radiation Research, 43*, 157-161. http://dx.doi.org/10.1269/jrr.43.S157

Albouyeh, R., Farzaneh, N., Bohlmann, J., & Ritland, K. (2010). Multivariate analysis of digital gene expression profiles identifies a xylem signature of the vascular tissue of white spruce (*Piceaglauca*). *Tree Genet. Genomes, 6*, 601-611. http://dx.doi. 10.1007/s11295-010-0275-0

Billa, E., Koullas, D. P., & Monties, B. (1997). Structure and composition of sweet sorghum stalk components. *Ind Crops Prod., 6*, 297-302. http://dx.doi.org/10.1016/S0926-6690(97)00031-9

Bruggenmann, E., Handwerger, K., Essex, C., & Storz, G. (1996). Analysis of fast neutron-generated mutants at the *Arabidopsis thaliana* HY4 locus. *Plant J., 10*, 755-760. http://dx.doi.org/10.1046/j.1365-313X.1996.100 40755.x

Dijkwel, P. P., Huijster, C., Weisbeek, P. J., Chua, N. H., & Smeekens, S. C. (1997). Sucrose control of phytohcrome A signaling in *Arabidopsis*. *Plant Cell, 9*, 583-595. http://dx.doi.org/10.1105/tpc.9.4.583

Dong, X. C., & Li, W. J. (2012). Biological features of an early-maturity mutant of sweet sorghum induced by carbon ions irradiation and its genetic polymorphism. *Advances in Space Research, 50*, 496-501. http://dx.doi.org/10.1016/j.asr.2012.04.028

Hegedus, Z. A., Agoston, V. C., Ordas, A., Racz, P., & Mink, M. (2009). Deep sequencing of the zebrafish transcriptome response to mycobacterium infection. *Mol. Immunol., 46*, 2918-2930. http://dx.doi.org/ 10.1016/j.molimm.2009.07.002

Hong, L. Z., Li, J., Schmidt, K. A., Warren, W. C., & Barsh, G. S. (2011). Digital gene expression for non-model organisms. *Genome Res., 21*, 1905-1915. http://dx.doi.org/10.1101/gr.122135.111

Jiang, J. J., Shao, Y. L., Du, K., Ran, L. P., Fang, X. P., & Wang, Y. P. (2013). Use of digital gene expression to discriminate gene expression differences in early generations of resynthesized *Brassica napus* and its diploid progenitors. *BMC Genomics, 14*, 72. http://dx.doi.org/10.1186/1471-2164-14-72

Johnson, E., Bradley, M., Harberd, N. P., & Whitelam, G. C. (1994). Photoresponses of Light-Grown phyA Mutants of *Arabidopsis* (Phytochrome A Is Required for the Perception of Daylength Extensions). *Plant Physiol., 105*, 141-149. http://dx.doi.org/10.1104/pp.105.1.141

Kawata, T., Ito, H., George, K., Durante, M., Furusawa, Y., Wu, H., & Cucinotta, F. A. (2001). G2-chromosome aberrations induced by high-LET radiations. *Advances in Space Research, 27*, 383-391. http://dx.doi.org/ 10.1016/S0273-1177(01)00006-0

Kawata, T., Ito, H., Uno, T., Saito, M., Yamamoto, S., Furusawa, Y., ... Cucinotta, F. A. (2004). G2 chromatid damage and repair kinetics in normal human fibroblast cells exposed to low- or high-LET radiation. *Cytogenetic and Genome Research, 104*, 211-215. http://dx.doi.org/10.1159/000077491

Kikuchi, S., Saito, Y., Ryuto, H., & Fukunishi, N. (2009). Effects of heavy-ion beams on chromosomes of common wheat *Triticumaestivum*. *Mutation Research, 669*, 63-66. http://dx.doi.org/10.1016/j.mrfmmm. 2009.05.001

Kranz, A. R. (1994). Heavy ion and cosmic radiation effects in different targets of the *Arabidopsis* seed. *Acta Astronaut, 33*, 201-210. http://dx.doi.org/10.1016/0094-5765(94)90126-0

Li, J. Q., Wang, L. H., Zhan, Q. W., Liu Y. L., Fu, B. S., & Wang, C. M. (2013). Sorghum *bmr6* mutant analysis demonstrates that a shared MYB1 transcription factor binding site in the promoter links the expression of genes in related pathways. *Funct. Integr. Genomics, 13*, 445-453. http://dx.doi.org/10.1007/s10142-013 -0335-2

Liu, B. M., Wu, Y. J., Xu, X., Song, M., Zhao, M., & Fu, X. D. (2008). Plant height revertants of dominant semidwarf mutant rice created by low-energy ion irradiation. *Nuclear Instruments and Methods in Physics Research B, 266*, 1099-1104. http://dx.doi.org/10.1016/j.nimb.2008.02.045

Más, P., Alabadí, D., Yanovsky, M. J., Oyama, T., & Kay, S. A. (2003). Dual role of TOC1 in the control of circadian and photomorphogenic responses in *Arabidopsis*. *Plant Cell, 15*, 223-236. http://dx.doi.org/ 10.1105/tpc.006734

Matuo, Y., Nishijima, S., Hase, Y., Sakamotob, A., Tanakab, A., & Shimizu, K. (2006). Specificity of mutations induced by carbon ions in budding yeast *Saccharomyces cerevisiae*. *Mutation Research, 602*, 7-13. http://dx.doi.org/10.1016/j.mrfmmm.2006.07.001

Mei, M., Deng, H., Lu, Y., Zhuang, C., Liu, Z., Qiu, Q., ... Yang, T. C. (2011). Mutagenic effects of heavy ion radiation in plants. *Advanced Space Research, 14*, 363-372. http://dx.doi.org/10.1016/0273-1177(94) 90489-8

Mockler, T. C., Yang, Y., Yu, H., Parikh, D., Cheng, Y. X., Dolan, S., & Lin, C. T. (2003). Regulation of photoperiodic flowering by *Arabidopsis* photoreceptors. *Proc. Natl. Acad. Sci. USA, 100*, 2140-2145. http://dx.doi.org/10.1073/pnas.0437826100

Nishiyama, T., Miyawaki, K., Ohshima, M., Thompson, K., & Nagashima, A. (2012). Digital gene expression profiling by 5-end Sequencing of cDNAs during reprogramming in the moss physcomitrella patens. *PLoS One, 7*, 36471. http://dx.doi.org/10.1371/journal.pone.0036471

Paterson, A. H., Bowers, J. E., Bruggmann, R., Doreen, W., Peter, W., Klaus, M., ... Daniel, R. (2009). The sorghum bicolor genome and the diversification of grasses. *Nature, 457*, 551-556. http://dx.doi.org/10.1038/ nature07723

Qin, Y. F., Fang, H. M., Tian, Q. N., Liang, F., Zhang, Y. Z., & Zhang, S. T. (2011). Transcriptome profiling and digital gene expression by deep-sequencing in normal/regenerative tissues of planarian *Dugesia japonica*. *Genomics, 97*, 364-371. http://dx.doi.org/10.1016/j.ygeno.2011.02.002

Shikazono, N., Suzuki, C., Kitamura, S., Watanabe, H., Tano, S., & Tanaka, A. (2005). Analysis of mutations induced by carbon ions in *Arabidopsis thaliana*. *Journal of Experimental Botany, 56*, 587-596. http://dx.doi.org/10.1093/jxb/eri047

Shikazono, N., Tanaka, A., Watanabe, H., & Tano, S. (2000). Rearrangements of the DNA in carbon ion-induced mutants of *Arabidopsis thaliana*. *Genetics, 157*, 379-387.

Shikazono, N., Yokota, Y., Kitayama, S., Chihiro, S., & Hiroshi, W. (2003). Mutation rate and novel *tt* mutants of *Arabidopsis thaliana* induced by carbon ions. *Genetics, 163*, 1449-1455.

Sugano, S., Andronis, C., Ong, M. S., Green, R. M., & Tobin, E. M. (1999). The protein kinase CK2 is involved in regulation of circadian rhythms in *Arabidopsis*. *Proc. Natl. Acad. Sci. USA, 96*, 12362-12366. http://dx.doi.org/10.1073/pnas.96.22.12362

Wang, Y., & Tobin, E. M. (1998). Constitutive expression of the *CIRCADIAN CLOCK ASSOCIATED 1(CCA1)* gene disrupts circadian rhythms and suppresses its own expression. *Cell, 93*, 1207-1217. http://dx.doi.org/ 10.1016/S0092-8674(00)81464-6

Wei, L. J., Yang, Q., Xia, H. M., Furusawa, Y., Guan, S. H., Xin, P., & Sun, Y. Q. (2006). Analysis of cytogenetic damage in rice seeds induced by energetic heavy ions on-ground and after spaceflight. *Journal of Radiation Research, 47*, 273-278. http://dx.doi.org/10.1269/jrr.0613

Wei, M. M., Song, M. Z., Fan, S. L., & Yu, S. X. (2013). Transcriptomic analysis of differentially expressed genes during anther development in genetic male sterile and wild type cotton by digital gene-expression profiling. *BMC Genomics, 14*, 97. http://dx.doi.org/10.1186/1471-2164-14-97

Weller, J. L., Murfet, I. C., & Reid, J. B. (1997). Pea mutants with reduced sensitivity to far-red light define an important role for phytochrome A in day-length detection. *Plant Physiol., 114*, 1225-1236. http://dx.doi.org/10.1104/pp.114.4.1225

Xu, J. L., Wang, J. M., Sun, Y. Q., Wei, L. J., Luo, R. T., Zhang, M. X., & Li, Z. K. (2006). Heavy genetic load associated with the subspecific differentiation of japonica rice (*Oryza sativa* ssp. *japonica* L.). *Journal of Experimental Botany, 57*, 2815-2824. http://dx.doi.org/10.1093/jxb/erl046

Notes

Please download the supplementary files (Tables S1-S4 and Figure S1) at http://ccsenet.org/journal/index.php/jas/article/view/63224/35097

5

Determining Optimal Dose of Chemical Fertilizer on Biofortified Bean in Sud-Kivu Highlands

Casinga Clérisse[1,2], Haminosi Ghislain[2] & Cirimwami Legrand[3]

[1] International Institute of Tropical Agriculture, Bukavu, Sud-Kivu, Democratic Republic of Congo

[2] Faculté des Sciences Agronomiques et Environnement, Université Evangélique en Afrique, Bukavu, Sud-Kivu, Democratic Republic of Congo

[3] Faculté des Sciences, Université de Kisangani, Kisangani, Province de la Tshopo, Democratic Republic of Congo

Correspondence: Casinga Clérisse, International Institute of Tropical Agriculture, Bukavu, Sud-Kivu, Democratic Republic of Congo. E-mail: c.casinga@cgiar.org

The research is financed by International Centre of Tropical Agriculture through its project Harvest Plus-Beans/DR Congo.

Abstract

Rational application of chemical fertilizer increases crop yield of biofortified bean. This study aimed at determining the optimal dose of chemical fertilizer to apply on two biofortified bean varieties used in the community in order to maximize their yield. Following a split-plot design, a field experiment was carried out on CODMLB001 and HM21-7 varieties, in Kashusha (Kabare territory) in Sud-Kivu Highlands, after a strategic application of increasing doses of chemical fertilizer NPK 17-17-17 (D_0: Control; D_1: 50 kg·ha^{-1}; D_2: 75 kg·ha^{-1}; D_3: 100 kg·ha^{-1}; D_4: 125 kg·ha^{-1} and D_5: 150 kg·ha^{-1}). The said doses were applied on the sowing day in a parallel gutter at 5cm from the sowing line. The germination rate, the number of days at both the flowering stage and the stage of physiological maturity, as well as the number of harvested crops and beans per plant, number of beans per pod, weight per 1000 grains and yield were observed. Positive and negative interaction between different increasing doses of chemical fertilizers regarding the two varieties were observed. This strategic application allows increased performance according to considered varieties and doses. For instance, the HM21-7 variety gave the best performance with the D_5 dose, while the CODMLB001 variety did better under D_2.

Keywords: biofortified bean, Sud-Kivu highlands, increasing doses, chemical fertilizers, yield

1. Introduction

Africa has a complexity of climate conditions and a range of soils (Goudie, 1996; Griffiths, 2005; ODINAFRICA, 2007; UNEP, 2008; Casinga et al., 2015b, 2016). However, the latter have generally low fertility because of their ageing due to the lack of volcanic substrata for the establishment of new structures and their rejuvenation (Bationo et al., 2006). Nearly 16% of Africa's soils are of the best quality, 13% are of average quality, 16% of the least quality and 55% of poor quality although supporting a diversity of crops (Eswaran et al., 1996). On the one hand, approximately 900 million hectares of soils of high and average quality bear 400 million of persons or 45% of the African population. On the other hand, nearly 30% of the population (about 250 millions) live or depend on soils with low production potentiality and so leads to a situation of food insecurity (Eswaran et al., 1996; Bationo et al., 2006, 2008).

Most of sub-Saharan Africa's soils as well as those of the Democratic Republic of Congo, and particularly those of Sud-Kivu highlands are less fertile and productive because of increased degradation due to overexploitation (Casinga et al., 2015c). Besides, inadequate conservation and improvement agricultural practices affecting soil fertility correlate there and lead to low yield of crops, including bean (Blondel, 1971; Westelaar & Ganry, 1982; Ganry et al., 1990; McCan, 2005; Casinga et al., 2015b). Furthermore, an exponential increase in population exerting pressure on this resource contributes to it (FAO, 2005, 2010; Casinga et al., 2015a, 2015c). In the southern countries, the annual mean loss of the arable top soils in macronutrients, including Nitrogen,

Phosphorus and Potassium are respectively about 22 kg·ha^{-1}, 2.5 kg·ha^{-1} and 15 kg·ha^{-1} (Steiner 1996). These bad farming practices have led some writers to characterize the agriculture of these countries as "mining agriculture" (Stoorvogel & Smaling, 1990) since outputs exceed inputs in these exploitation systems (Van Keulen & Breman, 1990; Van Reuler & Prins, 1993; Van Reuler, 1996). Due to rapid demographic explosion and dietary needs that follow, agricultural production must increase significantly in order to feed the population and eradicate malnutrition (Useni et al., 2012; Casinga et al., 2015a, 2015b, 2015c, 2016). Recourse to chemical fertilizers proves a key factor in the modernization of agriculture in developing countries due to the existence of a positive correlation between yield and the amount of chemical fertilizers used properly (Useni et al., 2012). The other advantage of fertilizers is that they not only improve efficiency but also crop residues serve as organic fertilizer from the previous crop (Batiano et al., 2006; Casinga et al., 2015b). In Sud-Kivu, the yields of the bean crop are low and do not exceed 500 kg·ha^{-1} in farming without fertilization (Casinga et al., 2015b). In large farms, with the use of improved varieties and the use of mineral fertilizers, yields varying from 800 to 1200 kg are achieved, HarvestPlus (2013). The present study aims to determine the optimal dose of chemical fertilizers NPK 17-17-17 able of inducing and enhancing increase in the yield of biofortified beans and the identification of the variety that best responds to mineral fertilization.

2. Method

2.1 Location

The experiment was carried out during the growing season 2011A-2012A, in the experimental site of the "Université Evangélique en Afrique (UEA)" at Kashusha, Kabare territory, Sud-Kivu in the Democratic Republic of Congo. The geographical coordinates of the station are 028°47′ East longitude and 02°19′ South latitude while the altitude is 1712 m. Kashusha has an AW$_3$ climate type, according to Köppen's classification. The average annual rainfall reaches 1450 mm while the average annual temperature is 19.5 °C. The soil of the experimental site belongs to the class of Ferralsols, according to FAO-UNESCO classification (Baert, 1995; Beernaert, 1999; Baert et al., 2012; Botula et al., 2012; Casinga et al., 2015b); whereas its texture silty-clay (Casinga et al., 2015a).

2.2 Materials

2.2.1 Biological Material

Two biofortified beans varieties were used as biological material: the CODMLB001 and the HM21-7 varieties.

2.2.2 Mineral Fertilizers

The ternary chemical NPK 17-17-17 was used as fertilizer.

2.3 Method

The experiment was conducted on a split-plot design. Varieties of biofortified beans randomized and sown in line in the subplots at the depth of 4 cm with two seeds per hole, with a distance of 0.40 × 0.20 m were the main factor, while six doses of chemical fertilizer (D$_0$: Control; D$_1$: 50 kg·ha^{-1}; D$_2$: 75 kg·ha^{-1}; D$_3$: 100 kg·ha^{-1}; D$_4$: 125 kg·ha^{-1} and D$_5$: 150 kg·ha^{-1}) represented the secondary factor. Fertilizers were spread at sowing and applied in a gutter parallel to the sowing line, at a distance of more or less 0.5 cm, and a depth of more or less 5 cm. The latter were immediately covered with a little soil after application. At the beginning of vegetation germination rate was determined with the ratio of number of plants raised per the number of grains sown multiplied by one hundred. During vegetation, the numbers of days to flowering were determined by the difference in number of days between the date of the appearance of inflorescences (at least 50% on a plot) and sowing date. The yield components (number of pods, number of grains per pod and weight of 1000 grains) and the yield obtained by the formula were determined.

$$\text{Yield (kg·ha}^{-1}) = \frac{\text{Poids des Graines SU} \times 10000}{\text{Surface Utile (SU)}} \tag{1}$$

The experimental results were assessed by the Analysis of Variance, correlation and multiple regression while the averages were separated by the test LSDα = 0.05. Excel 2013 software, R 3.3.0 and Assistat 9.5.1 were used as calculation tool.

3. Results

Statistical analysis revealed significant differences in the parameters of the number of days at the flowering stage, the number of harvested plants and yield for the different increasing doses of mineral fertilizers. However, they revealed the lack of significant differences in the parameters of the germination rate, the number of days at the stage of physiological maturity, number of pods per plant, number of grains per pod and weight of 1000 grains, depending on the different increasing doses of mineral fertilizers.

Table 1. Effect of increasing doses of chemical fertilizers on the growth of bean

	Treatment	Rate of Germination (%)	Days to the flowering stage (n°)	Days to physiological maturity stage (n°)
CODMLB001	D_0	$83 \pm 6{,}5$	36	78
	D_1	$75 \pm 11{,}9$	36	78
	D_2	$78 \pm 4{,}1$	36	78
	D_3	$82 \pm 8{,}6$	36	$82{,}7 \pm 6{,}5$
	D_4	$78 \pm 5{,}7$	36	78
	D_5	$80 \pm 13{,}4$ *	36 ***	78
HM21-7	D_0	$59 \pm 8{,}5$ *	$38 \pm 1{,}4$ **	80
	D_1	$53 \pm 10{,}8$ **	39 ***	80
	D_2	$57 \pm 4{,}7$ **	39 ***	80
	D_3	$59 \pm 0{,}8$ *	39 ***	80
	D_4	$47 \pm 3{,}2$ *	39 ***	80
	D_5	$59 \pm 8{,}4$ *	39 ***	80

Note. D_0: Control, D_1: 50 kg·ha⁻¹, D_2: 75 kg·ha⁻¹, D_3: 100 kg·ha⁻¹, D_4: 125 kg·ha⁻¹, D_5: 150 kg·ha⁻¹.

Table 2. Effect of doses of increasing chemical fertilizers on the yield and its components

	Treatment	Number of plants harvested (n°)	Number of pods per plant (n°)	Number of grains per pod (n°)	Weight of 1000 grains (gram)	Yield kg·ha⁻¹
CODMLB001	D_0	$45 \pm 3{,}3$	5 ± 1	4 ± 1	$396 \pm 75{,}8$ *	$584 \pm 104{,}3$
	D_1	$41 \pm 2{,}2$	6	4 ± 1	$420 \pm 24{,}5$	$814 \pm 127{,}9$
	D_2	$46 \pm 3{,}6$	5 ± 1	4 ± 1	$420 \pm 78{,}7$ *	$1196 \pm 222{,}7$
	D_3	$46 \pm 3{,}6$	6 ± 2	4	$403 \pm 52{,}5$	$1052 \pm 401{,}6$
	D_4	$39 \pm 4{,}5$	5 ± 2	4 ± 1	$366 \pm 26{,}5$	$725 \pm 142{,}9$
	D_5	$44 \pm 6{,}5$	6 ± 2	4 ± 1	$427 \pm 105{,}8$	$883 \pm 203{,}4$
HM21-7	D_0	$27 \pm 9{,}5$ *	5 ± 1	3 *	$436 \pm 68{,}5$	$687 \pm 146{,}8$ *
	D_1	$34 \pm 11{,}6$	6 ± 2	4 ± 1	$413 \pm 65{,}9$	$1018 \pm 226{,}2$
	D_2	$26 \pm 7{,}3$ **	7 ± 1	4 ± 1	$537 \pm 83{,}4$	$1016 \pm 277{,}3$ *
	D_3	$21 \pm 9{,}6$ **	$7 \pm \; 1$	4	$423 \pm 79{,}3$	$837 \pm 470{,}4$
	D_4	$27 \pm 1{,}2$ *	6 ± 2	4 ± 1	$460 \pm 29{,}4$	$991 \pm 68{,}4$
	D_5	$36 \pm 4{,}4$ *	7 ± 2	4 ± 1	$477 \pm 61{,}8$	$1250 \pm 277{,}1$ *

Note. D_0: Control, D_1: 50 kg·ha⁻¹, D_2: 75 kg·ha⁻¹, D_3: 100 kg·ha⁻¹, D_4: 125 kg·ha⁻¹, D_5: 150 kg·ha⁻¹.

The CODMLB001 variety presented a better rate of germination in comparison to the HM21-7 variety. Furthermore, it happened earlier than the latter since it reached flowering and physiological maturity earlier. At the harvest, the number of harvested plants for the CODMLB001 variety was greater than that of the HM21-7 variety. The number of pods per plant and the number of seeds per pod were similar for the two varieties, the increasing doses of chemical fertilizers and their interactions. The weight of 1000 seeds was positively correlated with the increasing doses of chemical fertilizers for the two varieties. The HM21-7 variety had the better yield in comparison to the CODMLB001 variety.

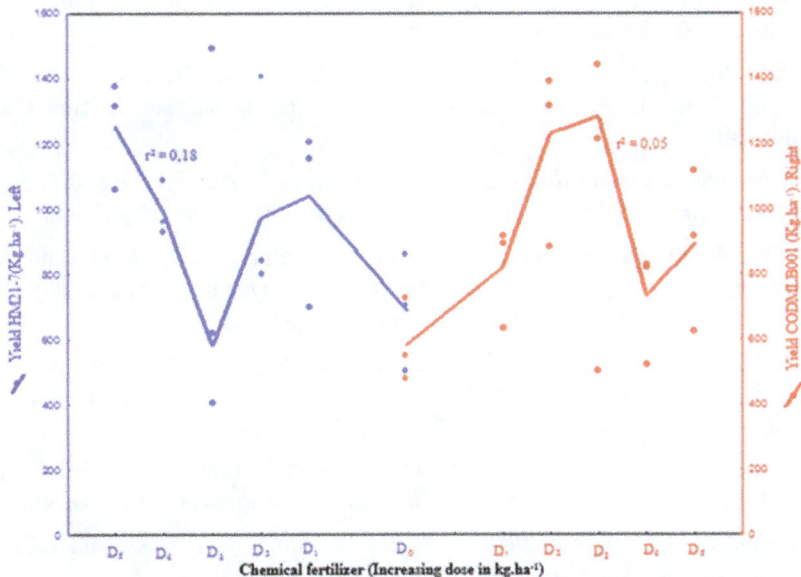

Figure 1. Correlation between HM21-7 and CODMLB001 yields varieties and their treatment

This figure highlights the interaction between the varieties and certain increasing doses of chemical fertilizers, where the HM21-7 variety got the better yield under D_5 treatment while for the CODMLB001 variety it was under D_2 treatment.

4. Discussion

According to FAO (2003) land degradation remains a major global concern because of its adverse impacts on agricultural production, food security and the environment. Responses to these mineral fertilization varieties depended on the dose applied and so confirmed de Nascente et al. (2015)'s results stigmatizing that the application the increasing doses of chemical fertilizer at sowing greatly increases the performance of beans. Indeed, the strategic application of increasing doses of chemical fertilizer induced an increase in the number of pods per plant for the different varieties in different treatments, thus confirming the results of Abdel-Mawgoud et al. (2005) and Mahmoud et al. (2010) who found similar results. Moreover, the number of seeds per pod did not vary according to treatment in both genotypes contradicting the work of Kamanu et al. (2012) who found that increasing fertilizer application results in an increase in the number of seeds per pod in some varieties. This is accounted for by the fact that it is a highly stable genetic characteristic of these genotypes not to react to this environmental component (Beebe et al., 2000). This increasing application of doses of chemical fertilizers induced an increase in yield, respectively 204.7% for CODMLB001 variety because of the dose D_2, comparatively to D_0 and 211% for the variety HM21-7 due to D_5 dose compared to D_0, confirming the work of Abdel-Mawgoud et al. (2005) and Tantawy et al. (2009) as well as Zucareli et al. (2011) and Kamanu et al. (2012) who found similar results compared to the control. Declining results observed for doses ranging from D_2 to D_5 in the CODMLB001 variety confirm Mitscherlich (1909)'s law of less proportional surpluses highlighting that increasing quantities of fertilizers, crop surpluses obtained are lower and lower, producing a depressing, let alone toxic effect.

5. Conclusion

The application of increasing doses of chemical fertilizer induced significant interactions on crop varieties under study. The variety HM21-7 gave the better performance under treatment D_5 while for CODMLB001 variety was under treatment D_2. We recommend that the Sud-Kivu population adopt these doses depending on the varietal preference. Extensive studies of these doses based on physiological critical stages must be considered in determining the value end cost.

References

Abdel-Mawgoud, A. M. R., El-Desuki, M., Salman, S. R., & Abou-Hussein, D. (2005). Performance of some snap bean varieties as affected by different levels of mineral fertilizers. *Journal of Agronomy, 4*(3), 242-247. http://dx.doi.org/10.3923/ja.2005.242.247

Baert, G. (1995). *Properties and chemical management aspects of soils on different parent rocks in the Lower Zaire* (PhD Thesis, p. 320). Ghent University, Belgium.

Baert, G., Van Ranst, E., Ngongo, M., & Verdoodt, A. (2012). *Soil Survey in DR Congo-from 1935 until today* (p. 17). Research Unit of Soil Degradation and Soil Conservation, Department of Soil Management & Soil Care, Ghent University.

Bationo, A. (2008). *Integrated soil fertility management option for agriculture intensification in the Sudano Sahelien zone in West Africa* (p. 356). Academy of Science Publishers, Nairobi, Kenya.

Bationo, A., Hartemink, A., Lungu, O., Naimi, M., Okoth, P., Smaling, E., & Thiombiano, L. (2006). *African Soils: their productivity and profitability of fertilizer use* (p. 52). Document de base présenté à l'occasion du Sommet africain sur les engrais, 9-13 Juin 2006, Abuja, Nigeria.

Beebe, S., Skroch, P. W., Tohme, J., Duque, M. C., Pedraza, F., & Nienhuis. (2000). Structure of genetic diversity among common bean landraces of Middle American origin based on correspondence analysis of RAPD, *Crop. Sci., 40*, 264-273. http://dx.doi.org/10.2135/cropsci2000.401264x

Beernaert, F. R. (1999). Feasibility Study of a Production Project of Lime and/or Ground Travertine for the Management of Acid Soils in Rwanda. *PRO-INTER Project Consultants* (p. 287). Belgium.

Blondel, D. (1971). Contribution à la connaissance de la Dynamique de l'Azote Minéral en sol Ferrugineux Tropical à Nioro du Rip (Sénégal). *Dot. Mult., 7.*

Botula, Y.-D., Cornelis, W. M., Baert, G., & Van Ranst, E. (2012). Evaluation of pedotransfer functions for predicting water retention of soils in Lower Congo (D.R. Congo). *Agricultural Water Management.* http://dx.doi.org/10.1016/j.agwat.2012.04 .006

Casinga, C. M. (2015a). Etude comparative des réponses de quatre variétés d'haricots bio fortifiés à trois régimes hydriques dans le Sud-Kivu montagneux: Cas de Hogola. *Mémoire de maitrise* (p. 82). Université Evangélique en Afrique.

Casinga, C. M., Cirimwami, L. T., Amzati, G. S., Katembera, J. I., Kanyenga, A. L., & Mushagalusa, G. N. (2015c). Effect of the environment on the adaptability of biofortified bean genotypes in the eastern Democratic Republic of Congo: Case of South-Kivu. *European Journal of Agriculture and Forestry Research, 3*(9), 38-47.

Casinga, C. M., Cirimwami, L. T., Bisimwa, E. B., & Mushagalusa, G. N. (2015b). The impact of leguminous culture system and sowing dates on the cereal yield in mountainous South-Kivu: Burhale Case. *International Journal of Innovation and Scientific Research, 18*(2), 297-303.

Casinga, C. M., Kanyenga, A. L., & Mambani, P. B. (2016). Les haricots biofortifiés sous stress hydrique au Sud-Kivu montagneux. *Editions Universitaires Européenes, 76.*

Eswaran, H., Kimble, J., Cook, T., & Beinroth, F. H. (1996). Soil diversity in the tropics: Implications for agricultural development. In R. Lai & P. A. Sanchez (Eds.), *Myths and science of soils of the tropics* (pp. 1-16). SSSA Spec.

FAO. (2003). *Gestion de la fertilité des sols pour la sécurité alimentaire en Afrique subsaharienne* (p. 66). Rome, Italie. Retrieved from http://www.fao.org

FAO. (2005). *Gestion de la fertilité des sols pour la alimentaire en Afrique subsaharienne* (p. 63). Rome, Italie. Retrieved from http://www.fao.org

FAO. (2010). *Enjeux et possibilités pour l'agriculture et la sécurité alimentaire en Afrique* (p. 26). Rome, Italie. Retrieved from http://www.fao.org

Ganry, F. (1990). Application de la méthode isotopique à l'étude des bilans azotés en zone tropicale sèche. *Université de Nancy, I, 351.*

Ganry, F., Guiraud, G., & Dommergues, Y. (1978). Effect of straw incorporation on the yield and nitrogen balance in the sandy soil-pearl millet cropping system of Senegal. *Plant Soil, 50*, 647-662. http://dx.doi.org/10.1007/BF02107216

Goudie, A. S. (1996). *Climate: Past and Present* (pp. 34-59). Oxford University Press, Oxford.

Griffiths, J. F. (2006). Climate of Africa. *Encyclopedia of World Climatology* (p. 24). Springer, Berlin.

HarvestPlus. (2013). *Briging the delta: Annual report 2012.* Retrieved from http://www.harvestPlus.org

Kamanu, J. K., Chemining'wa, G. N., Nderitu. J. H., & Ambuko, J. (2012). Growth, yield and quality response of snap bean (*Phaseolus vulgaris* L.) plants to different inorganic fertilizers applications in central Kenya. *Journal of Applied Biosciences, 55*, 3944-3952.

Mahmoud, A. R., El-Desuki, M., & Abdel-Mouty, M. M. (2010). Response of snap bean plants to bio-fertilizer and nitrogen level application. *International Journal of Academic Research, 2*(3), 2004-2014.

McCann, J. C. (2005). Maize and Grace. *Africa's Encounter with a new World Crop* (pp. 500-200). Harvard University Press, Cambridge. http://dx.doi.org/10.4159/9780674040748

Mitscherlich, E. A. (1909). Das Gesetz des Minimums und das Gesetz des abnehmenden Bodenertrages. *Landwirtschaftliche Jahrbücher T., 38*, 537-552.

Nascente, A. S., Lacerda, M. C., Carvalho, M. C. S., & Vas Mondo, V. H. (2015). Broadcast fertilizer rates impact common bean grain yield in a no-tillage system. *African Journal of Agricultural Research, 10*(14), 1773-1779. http://dx.doi.org/10.5897/AJAR2014.8525

ODINAFRICA. (2007). African Marine Atlas. *International Oceanographic Data and Information Exchange (IODE)*. Intergovernmental Oceanographic Commission's (IOC). Retrieved from http://www.africanmarin eatlas.net/index.htm

Steiner, K.G. (1996). *Causes de la dégradation des sols et approches pour la promotion d'une utilisation durable des sols* (p. 97). Suisse.

Stoorvogel, J. J., Smaling, E. M. A., & Janssen, B. H. (1993). Calculating soil nutrient balances in Africa at different scales: I. Supra-national scale. *Fert. Res., 35*, 227-235. http://dx.doi.org/10.1007/BF00750641

Tantawy, A. S., Abdel-Mawgoud, A. M. R., Hoda, A. M., Habib & Magda, M. H. (2009). Growth, Productivity and Pod Quality Responses of Green Bean Plants (*Phaseolus vulgaris*) to foliar application of Nutrients and Pollen Extracts. *Research Journal of Agricultural and Biological Sciences, 5*(6), 1032-1038.

UNEP. (2008). *Africa: Atlas of Our Changing Environment* (p. 374). Division of Early Warning and Assessment (DEWA). Retrieved from https://wedocs.unep.org/rest/bitstreams/11689/retrieve

Useni, S. Y., Nyembo, K. L., Mpundu, M. M., Bugeme, M. D., Kasongo, L. E., & Baboy, L. L. (2012). Effets des apports des doses variées de fertilisants inorganiques (NPKS et Urée) sur le rendement et la rentabilité économique de nouvelles variétés de Zea mays L. à Lubumbashi, Sud-Est de la RDCongo. *Journal of Applied Biosciences, 59*, 4286-4296.

Van Keulen, H., & Breman, H. (1990). Agricultural development in the West Africa Saharan region: A cure against land hunger. *Agri. Ecos. Envir., 32*, 177-197. http://dx.doi.org/10.1016/0167-8809(90)90159-B

Van Reuler, H. (1996). Nutrient management over extended cropping periods in the shifting cultivation system of South-West Cote d'Ivoire, *Wageningen Agric. Univ.*, 196.

Van Reuler, H., & Prins, W. H. (1993). The role of plant nutrients for sustainable food crop production in sub-Saharan Africa. *Vereniging van Kunstmest Producenten* (pp. 3-11).

Wetselaar, R., & Ganry, F. (1982). Nitrogen balance in tropical agrosystems, In Y. R. Dommergues & H. G. Diem (Eds.), *Microbiology of tropical soils and plant productivity* (pp. 1-35). Martinus Nijhoff/Dr 31. W. Junk, the Hague, the Netherlands.

Zucareli, C., Ramos Junior, E. U., Oliveira, M. A., Cavariana, C., & Nakagawa, J. (2011). Physiological and biometric indices in bean under different doses of phosphorus seminal, *Ciên. Agrár., 31*(1), 1313-1324.

6

Combining Ability Analysis of Blast Disease Resistance and Agronomic Traits in Finger Millet [*Eleusine coracana* (L.) Gaertn]

Lawrence Owere[1,2], Pangirayi Tongoona[1], John Derera[1] & Nelson Wanyera[3]

[1] African Centre for Crop Improvement, School of Agricultural, Earth and Environmental Sciences, College of Agriculture, Engineering and Science, University of KwaZulu Natal, Scottsville, Pietermaritzburg, Republic of South Africa

[2] Buginyanya Zonal Agricultural Research and Development Institute, Mbale, Uganda

[3] National Semi-Arid Resources Research Institute, Serere, Private Bag, Soroti, Uganda

Correspondence: Lawrence Owere, Buginyanya Zonal Agricultural Research and Development Institute, P.O. Box, 1356, Mbale, Uganda. E-mail: labowere@gmail.com

Abstract

Blast disease is the most important biotic constraint to finger millet production. Therefore disease resistant varieties are required. However, there is limited information on combining ability for resistance and indeed other agronomic traits of the germplasm in Uganda. This study was carried out to estimate the combining ability and gene effects controlling blast disease resistance and selected agronomic traits in finger millet. Thirty six crosses were generated from a 9×9 half diallel mating design. The seed from the 36 F_1 crosses were advanced by selfing and the F_2 families and their parents were evaluated in three replications. General combining ability (GCA) for head blast resistance and the other agronomic traits were all highly significant ($p \leq 0.01$), whereas specific combining ability (SCA) was highly significant for all traits except grain yield and grain mass head^{-1}. On partitioning the mean sum of squares, the GCA values ranged from 31.65% to 53.05% for head blast incidence and severity respectively, and 36.18% to 77.22% for the other agronomic traits measured. Additive gene effects were found to be predominant for head blast severity, days to 50% flowering, grain yield, number of productive tillers plant^{-1}, grain mass head^{-1}, plant height and panicle length. Non-additive gene action was predominant for number of fingers head^{-1}, finger width and panicle width. The parents which contributed towards high yield were *Seremi 2*, *Achaki*, *Otunduru*, *Bulo* and *Amumwari*. Generally, highly significant additive gene action implied that progress would be made through selection whereas non-additive gene action could slow selection progress and indicated selection in the later generations.

Keywords: combining ability, finger millet, grain yield, gene action, head blast disease

1. Introduction

Finger millet production is faced with many biotic challenges; the most important of them being blast disease caused by *Pyricularia grisea* (Cooke) Sacc. There have been attempts to address this challenge resulting in some ephemeral solutions. *Pyricularia grisea* can cause yield losses as high as 50% on finger millet (Lenne et al., 2007) and in favourable seasons the losses can be as high as 90% (Esele, 1993). In Uganda, finger millet blast is endemic to all growing areas although some cultivars are more susceptible than others (Takan et al., 2004) and more severe in some areas than others depending on weather conditions. Despite its wide prevalence very little is actually known about host plant resistance and its inheritance compared to rice for instance. Blast appears on all plant parts damaging leaves, stems, peduncle and heads, with head blast the most destructive as it directly reduces yield (Prabhu, Filippi, & Zimmermann, 1996). Although chemical control has been shown to be effective (Bua & Adipala, 1995; Seetharam & Ravikumar, 1993), its use on a field scale is not practical because of resource constraints of the farmers growing finger millet making exploitation of host plant resistance an extremely important option in preventing yield loss and enhancing yields.

The gene action conditioning resistance to finger millet blast disease is not fully understood and similarly no information exists on the combining abilities of finger millet lines adapted to tropical conditions in Uganda under finger millet blast pressure. There however, exists some information especially from India and extensive work on rice. Generation of such information would be useful in selecting parents in a breeding programme and choosing appropriate breeding procedures. Studies elsewhere have identified finger millet genotypes with resistance to *Pyricularia grisea* (Cooke) Sacc. (Shailaja, Thirumeni, Paramasivam, & Ramanadane, 2010; Krishnappa, Ramesh, Chandraprakash, Bharathi, & Doss, 2009; Takan et al., 2004) indicating that breeding for resistance is a realistic option. This can form the basis for initiating studies to determine the genetics of resistance to blast disease pathogen and later be able to incorporate this resistance in new cultivars with appropriate agronomic and farmer preferred attributes.

The main objectives of the current study were to assess the nature and magnitude of gene action controlling blast disease inheritance and other agronomic traits important to yield determination and to suggest breeding strategies for finger millet improvement. The specific objectives were to: (i) estimate the general combining ability (GCA) of selected parents and the specific combining ability (SCA) of a parent in a cross with another parent, and (ii) determine the genetic effects which control the inheritance of blast disease resistance and selected agronomic traits in finger millet.

2. Materials and Methods

2.1 Selection of Parental Materials

The experimental material consisted of nine finger millet varieties (Table 1) as parents. The varieties selected were adapted landraces, bred and released varieties and introductions from ICRISAT. The landraces and released cultivars used are highly popular among the farmers and are being used in various production systems. Owing to their already high adaptability, acceptability, resistance to blast disease (in some cases) and yielding ability, these were chosen for hybridization to exploit the existing variation for finger millet improvement in Uganda. Among the nine varieties, five had green pigmentation whereas four had purple pigmentation at the nodes and leaf margin (Table 1). These were deliberately selected so that the F_1s could easily be identified as the purple pigmentation is known to be dominant over the green pigmentation (Shailaja, Thirumeni, Paramasivam, & Ramanadane, 2010; Krishnappa, Ramesh, Chandraprakash, Bharathi, & Doss, 2009) which served as a useful marker in identifying true crosses at the seedling stage where the parents had different nodal and head pigmentation. Other added markers were plant height, head shapes and seedling vigour.

2.2 Crossing Procedures

Finger millet is predominantly a self-pollinated crop with bisexual flowers (florets) which are small in size making artificial hybridization a difficult process. Emasculation without injury to floral parts is extremely difficult hence two methods were adopted for this study to improve chances of success.

Table 1. Parental lines with entry numbers, reaction to head blast disease, nodal pigmentation, head shapes and germplasm source

Entry	Type of disease reaction†	Nodal pigmentation	Head shapes	Source
01 (E11)	S	Purple	Open	ICRISAT
02 (ACF 5)	S	Green	Incurved	Introduction – world collection
03 (Seremi 2)	R	Purple	Semi compact	Released cultivar
04 (ACF 19)	R	Green	Tips curved	Introduction – world collection
05 (Achaki)	R	Green	Compact	Landrace – Tororo
06 (Abao)	MR	Purple	Compact	Landrace – Lira
07 (Otunduru)	MR	Purple	Compact	Landrace – Kaberamaido
08 (Bulo)	MR	Green	Tip curved	Landrace – West
09 (Amumwari)	R	Green	Open	Landrace – Busia

Note. † S = resistant, R = resistant, MR = moderately resistant.

The two methods were: 1) the polythene bag method (in which emasculation was obtained using a 7.5 cm × 10 cm polythene bag lined with moist filter paper inverted over the flower and plugged with absorbent cotton wool. This creates high humidity inside the bag. Under such humidity, the florets open, the anthers emerge but shed no pollen. Pollen was collected from the designated male parents by tapping the bag before dehiscence of anthers. The pollen collected from the bag was dusted on the emasculated head and again covered with a pollination bag and labeled; 2) The contact method of crossing as described by Ravikumar (1988) and successfully used by

Ratnakar, Mallikarjuna, Naveen-Kumar, and Jayarame-Gowda (2009) were adopted to obtain F_1 seed. In this second method the heads of the male and female parents were brought together and finger to finger contacts were made by tying them together with a thread just before anthesis. Anthesis is known to take place from 1 am to 4 am and ends by 11 am (Ratnakar, Mallikarjuna, Naveen-Kumar, & Jayarame-Gowda, 2009). After pollination, ear heads were separated and seeds collected only from the female parent. This method is known to enhance the frequency of out-crossing by providing an opportunity for the pollen of male parents to come in close contact with the stigmatic surface of female parents.

2.3 Diallel Crosses and Evaluation of the Parents and Progenies

The nine selected parents were crossed in a green house at NaSARRI (Latitude 1°29′39N Longitude 33°27′19E 1085 m.a.s.l.) using the 9 × 9 half diallel mating design. The successful F_{1S} were identified in the field during the following season by comparing the crosses with the maternal parents. This was done by sowing the F_1 seed between rows of both parents and among the crosses, plants similar to female parents were identified and removed based on the morphological markers. The true F_1 plants were then advanced to obtain F_2 seed. The F_2 seed was sown under natural infestation in the field alongside the parents in an alpha-lattice design of 5 × 9 by adopting a spacing of 30 cm × 10 cm between rows and plants in a single row. Basal application of diammonium phosphate fertilizer and top-dressing with urea was used to boost the nitrogen levels to facilitate disease development (Prabhu, Filippi, & Zimmermann, 1996; Seetharam & Ravikumar, 1993; Russell, 1978). Fourty competitive plants were labelled per plot from which data were recorded.

2.4 Data Collection

Data was collected on the following traits: head blast incidence and head blast severity under natural infestation, days to 50% flowering, number of productive tillers per plant, finger number per head, grain mass per head, plant height, finger length, finger width, panicle length, panicle width and grain yield ha^{-1}. Data on these traits were collected using finger millet descriptors (IBPGR, 1985) as a guide.

Grain yield (tons ha^{-1}): measured as grain mass was taken from the fourty plants, post-harvest and converted to tons ha^{-1}. Using the formula:

$$\text{Grain yield (tons ha}^{-1}) = \frac{333{,}333 \times \text{Yeild of the 40 plants (Kg)}}{40 \times 1000} \qquad (1)$$

Head blast incidence and severity were recorded at the time of grain maturity. The disease incidence was calculated as the number of diseased plants divided by the total number of plants sampled per plot, whereas for severity, all heads from the fourty plants were used to determine blast severity at maturity. For each head, proportions of spikelets affected by the disease were estimated and a Standard Evaluation System (SES) (IRRI, 1996) was adopted based on the number of heads, and head blast severity. The plants were then categorised as: 0 = no disease or immune, less than 10% = highly resistant, 11-20% = resistant, 20-30% = moderately resistant, 30-50% = susceptible and more than 50% highly susceptible.

2.5 Analysis

Data were analysed as a randomized complete block design (RCBD) since preliminary Lattice analysis resulted in no gain in accuracy due to blocking over RCBD analysis. Genetic analysis for blast disease resistance and other agronomic traits were performed as fixed effects model for the 45 entries (36 crosses and nine parents) in three replications. Diallel SAS05 programme was used to perform Griffings method 2, model I diallel analysis (Zhang, Kang, & Lamkey, 2005). This model was most suitable for the present study where only parents and one set of F_{1S} (without reciprocals) were included and treated as fixed effects in the analysis. From the mean sums of squares, estimates of GCA effects (g_i) for each parent and SCA effect (s_{ij}) for each cross combination were also determined. The statistical model for the mean value of a cross (i × j) is as follows:

$$Y_{ij} = \mu + g_i + g_j + s_{ij} + 1/b \ \Sigma_k\Sigma_l e_{ijkl} \qquad (2)$$

Where,

Y_{ij} = Mean of (i × j)th cross over replications k (k = 1, 2, … b);

μ = The population (general) mean;

g_i and g_j = General combining ability (g.c.a.) effects of ith and jth parents, respectively;

s_{ij} = Specific combining ability (s.c.a) effect of ij^{th} cross such that $s_{ij} = s_{ji}$;

e_{ijkl} = Environmental effect associated with $ijkl^{th}$ observation in kth replication.

Restrictions are imposed on combining ability effects, such that $\Sigma_i g_i = 0$ and $\Sigma_i s_{ij} = 0$ (for each j) therefore, $1/b \cdot \Sigma_k \Sigma_l e_{ijkl} =$ Mean error effect.

The relative importance of general and specific combing ability in determining progeny performance was assessed by calculating the proportion of GCA: GCA + SCA sum of squares. The GCA: GCA + SCA sum of square ratio was proposed by Sprague and Tatum (1942) (cited and used by Simmonds & Smartt, 1999).

3. Results

The mean of the parental lines for blast disease incidence, severity and grain yield plant[-1] are presented in Table 2. The nine parental lines showed significant differences in the reaction to head blast disease indicated by both incidence and severity, and grain yield. There was a whole range of reaction from resistance based on classification used here to susceptible being exhibited by parental lines *E 11* and *ACF 5* both of which were introductions from ICRISAT and collections at University of KwaZulu Natal respectively.

Table 2. Means of parental lines for head blast incidence, severity and grain yield (tons ha[-1])

Entry	Head blast incidence (%)	Head blast severity (%)	Type of disease reaction‡	Grain yield (tons ha[-1])
01 (E11)	68.7	34.0	S	1.36
02 (ACF 5)	52.6	57.0	HS	1.41
03 (Seremi 2)	31.0	16.7	R	2.61
04 (ACF 19)	30.7	18.0	R	2.94
05 (Achaki)	25.3	12.7	R	4.17
06 (Abao)	38.7	26.0	MR	2.30
07 (Otunduru)	24.3	26.7	MR	2.89
08 (Bulo)	27.0	28.7	MR	3.46
09 (Amumwari)	30.0	19.7	R	3.40
Mean	36.5	26.9		2.73
Minimum	18.0	11.00		1.11
Maximum	93.0	50.90		4.49
LSD (0.05)	8.58	5.71		0.26
C.V. (%)	31.7	28.2		10.8

Note. ‡ type of disease reaction: S = Susceptible, HS = Highly susceptible, R = Resistant and MR = Moderately resistant.

3.1 Combining Ability Estimates

Results of mean sum of squares for blast disease incidence, severity and agronomic traits are presented in Table 3. The mean sum of squares for entry, GCA effects and SCA effects for head blast incidence and severity were highly significant ($p \leq 0.01$) and partitioning the cross sum of squares the GCA effects of head blast incidence and severity accounted for 31.65% and 53.05% respectively.

Mean sum of squares for the other agronomic traits were all highly significant ($p \leq 0.01$) for entry and GCA effects, whereas SCA effects were highly significant for all traits except panicle width which was just significant ($p \leq 0.05$). Specific combining ability effects were non-significant ($p \leq 0.05$) for grain mass head[-1] and grain yield ha[-1]. On partitioning the mean sums of squares, the GCA effects ranged from 36.18%-77.22%, whereas SCA effects contributed 22.78-63.82% of the total variance among the crosses. The contribution of GCA effects was highest for days to 50% flowering and lowest in panicle width, contrary to SCA effects. Considering all the agronomic traits; SCA effects were predominant for: number of fingers head[-1], finger width and panicle width, whereas GCA effects were predominant for grain yield ha[-1], days to 50% flowering, number of productive tillers plant[-1], grain mass head[-1], plant height and panicle length.

Table 3. Mean sum of squares for blast disease incidence, severity and other agronomic traits of finger millet in half diallel cross evaluated at NaSARRI

Source of variation	DF	FBI	FBS	Grain yield (tons ha^{-1})	Days to 50% flowering	Tillers $plant^{-1}$	Finger number $head^{-1}$	Grain mass $head^{-1}$	Plant height	FL	FW	PANW	PANL
Rep	2	112.13^{ns}	45.97^{ns}	0.83^{ns}	11.34^{*}	0.92^{*}	0.39^{ns}	0.30^{ns}	173.41^{**}	0.10^{ns}	0.02^{**}	1.32^{**}	0.26^{ns}
Entry	44	382.53^{***}	345.54^{***}	1.93^{***}	53.67^{**}	1.20^{***}	1.15^{**}	0.70^{***}	161.94^{**}	1.20^{**}	0.02^{**}	0.12^{**}	1.46^{**}
GCA	8	665.97^{**}	1008.08^{***}	6.85^{***}	227.94^{**}	3.57^{**}	2.74^{**}	2.50^{**}	653.37^{**}	3.23^{**}	0.03^{**}	0.25^{**}	4.41^{***}
SCA	36	319.54^{***}	198.31^{**}	0.82^{ns}	14.95^{**}	0.67^{**}	0.79^{**}	0.29^{ns}	52.73^{**}	0.75^{**}	0.01^{**}	0.10^{*}	0.80^{**}
Error	88	85.50	37.97	0.61	2.65	0.09	0.13	0.22	9.52	0.11	0.002	0.06	0.11
CV		27.47	22.83	11.25	2.22	11.49	5.41	24.33	4.07	5.57	5.22	11.99	5.25
R^2		0.69	0.82	0.73	0.91	0.87	0.82	0.62	0.90	0.84	0.82	0.61	0.87
Corrected total	134												
Contribution of GCA		31.65	53.05	65.53	77.22	54.14	43.47	65.40	73.36	49.03	36.46	36.18	54.98
Contribution of SCA		68.35	46.95	34.57	22.78	45.86	56.53	34.60	26.64	50.97	63.54	63.82	45.02

Note. *, **, and *** indicates the term is significant at $p \leq 0.05$, $p \leq 0.01$ and $p \leq 0.001$ respectively; ns – not significant ($p > 0.05$), FBI = finger blast incidence, FBS = finger blast severity, FL = finger length, FW = finger width, PANW = panicle width, PANL = panicle length.

3.2. General Combining Ability Effects of the Parental Materials

The GCA effects for the nine parental lines for head blast disease and other agronomic traits are presented in Table 4. For head blast disease the desirable GCA effect for the parents should be negative. The GCA effects for head blast disease incidence were significantly positive for *E 11* ($p \leq 0.01$), *ACF 5*, ($p \leq 0.001$) and *ACF 19* ($p \leq 0.05$), while negative, significant effects were shown for *Achaki* ($p \leq 0.001$) and *Otunduru* ($p \leq 0.01$), whereas, *Seremi 2*, *Abao*, *Bulo* and *Amumwari* were negative though non-significant ($p \leq 0.05$). For blast severity, positive significant effects were recorded for parents: *E 11* ($p \leq 0.001$), *ACF 5* ($p \leq 0.001$) and ($P \leq 0.05$). The parental materials produced similar effects (in terms of sign) for both incidence and severity. Negative significant effects were observed for *Seremi 2* ($p \leq 0.001$), *Achaki* ($p \leq 0.001$), and *Amumwari* ($p \leq 0.001$) while *Otunduru* and *Bulo* showed non-significant ($p \leq 0.05$), negative effect and *Abao* a positive, non-significant ($p \leq 0.05$) effect. The results therefore indicated that the desirable parents were *Seremi 2*, *Achaki*, *Amumwari* and to some extent *Otunduru* and *Bulo*.

For grain yield ha^{-1}, grain mass $head^{-1}$, tillers $plant^{-1}$, number of fingers $head^{-1}$, finger length, finger width, panicle length and panicle width the desirable GCA effect was positive. Whereas desirable GCA effects for days to 50% flowering and plant height is negative. Parents with significant, positive GCA effects for grain yield ha^{-1} were *Seremi 2*, *Achaki*, *Otunduru*, *Bulo* and *Amumwari*, whereas, for grain mass $head^{-1}$ were *Seremi 2*, *Achaki*, *Otunduru* and *Bulo*. Desirable combiners for productive tillers $plant^{-1}$ were *E 11*, *Achaki* and *Amumwari*; for number of fingers $head^{-1}$, *Seremi 2*, *ACF 19*, *Achaki*, *Abao* and *Bulo*; while finger length had *E 11*, *Seremi 2*, *Achaki* and *Bulo*. Parents that showed negative, significant GCA effects for days to 50% flowering and therefore desirable were *E 11* and *Seremi*, whilst negative, high significant GCA effects ($p \leq 0.01$) on plant height were recorded for *E 11*, *ACF 5*, *Seremi 2*, and *Abao*, while *Bulo* had no significant ($p \leq 0.05$) GCA effect.

Table 4. General combining ability effects for blast disease and other agronomic traits

Parent	FBI	FBS	Grain yield (tons ha⁻¹)	Days to 50% flowering	Tillers plant⁻¹	Finger number head⁻¹	Grain mass head⁻¹	Plant height	FL	LFW	PANW	PANL
1	4.21**	3,67***	-0.72***	-5.57***	0.37***	-0.60***	-0.43***	-4.01***	0.22***	-0.06***	0.13**	0.57***
2	7.91***	11.02***	-0.53***	0.67*	-0.42***	0.11	-0.32***	-4.58***	-0.20**	-0.03***	0.05	-0.35***
3	-2.99	-5.70***	0.29*	-3.05***	-0.32***	0.21**	0.17*	-6.44***	0.28***	0.01	-0.04	-0.07
4	3.15*	2.20*	-0.26*	1.19***	-0.19**	0.16*	-0.15	2.09***	-0.15**	0.01	-0.07	-0.04
5	-6.15***	-7.05***	0.54***	0.86**	0.48***	0.16**	0.33***	4.79***	0.47***	0.02*	0.16**	0.57***
6	0.91	1.80	-0.31	-0.02	-0.09	0.16**	-0.19*	-1.85**	-0.53***	-0.02**	-0.06	-0.37***
7	-4.45**	-1.23	0.35**	1.91***	-0.05	-0.33***	0.21**	5.19***	-0.12*	0.01	-0.03	-0.28**
8	-1.73	-0.54	0.42**	1.46***	-0.17**	0.23***	0.25**	0.11	0.20**	0.04***	0.07	0.19**
9	-0.87	-4.18***	0.21	2.55***	0.39***	-0.09	0.13	4.71***	-0.17**	0.03***	0.05	-0.21**
SE	9.24	6.15	0.28	1.63	0.30	0.36	0.47	3.08	0.34	0.04	0.24	0.33

Note. *, **, and *** indicates the term is significant at $p \leq 0.05$, $p \leq 0.01$ and $p \leq 0.001$ respectively; FBI = finger blast incidence, FBS = finger blast severity, FL = finger length, LFW = longest finger width, PANW = panicle width, PANL = panicle length. Parents 1, 2, 3, 4, 5, 6 and 7 are: E 11, ACF 5, Seremi 2, ACF 19, Achaki, Abao, Otunduru, Bulo and Amumwari respectively.

4. Discussion

The results indicated a high range for both blast disease incidence and severity which probably implied continuous variation exhibited by the genotypes in terms of head blast resistance. This may point to polygenic control, coupled with the fact that no cultivar showed or approached immunity. Some of the varieties showed high levels of resistance which may provide economically acceptable control of the disease and therefore could be used as sources of resistance in combination with other analysis results.

4.1 Combining Ability Effects and Gene Action

The significant GCA and SCA effects observed for both head blast severity and incidence showed that both additive and non-additive gene effects were important to head blast resistance. The GCA effects accounted for most of the head blast severity variance whereas the SCA effects contributed most of the head blast incidence variance based on cross sums of squares, an indication that selection of parents can contribute to progress for blast severity. Similar findings were reported by Seetharam and Ravikumar (1993) on severity, but completely in contrast to that of Selvaraj, Nagarajan, Thiyagarajan, Bharathi, and Rabinddran (2011) on rice panicle blast. The variance to the results of Seetharam and Ravikumar (1993) on incidence, and Selvaraj, Nagarajan, Thiyagarajan, Bharathi, and Rabinddran (2011) on both incidence and severity may point to the fact that the mechanisms of resistance depend on the germplasm used and environment where investigations are carried out as was also reported by Ravikumar (1988). The current results showed that additive gene action was more predominant for head blast severity while non-additive gene action was more predominant for head blast incidence, an indication of severity being fairly heritable whereas incidence is less heritable and making progress would be slow. The presence of greater additive genetic variance for severity would also suggest that disease reaction for progeny families is predictable based on the GCA estimates of its parents (Falconer & Mackay, 1996; Dhillon, 1975). In contrast, the presence of greater non-additive genetic variance as exhibited in incidence makes it less predictable and would slow progress to selection for incidence.

The results further showed preponderance of additive gene action for grain yield, days to 50% flowering, tillering ability, grain mass head⁻¹, plant height and panicle length except for finger number head⁻¹, finger width and panicle width; a suggestion that both additive and non-additive gene actions and/or variations are important. Similar results were obtained by Parashuram, Gowda, Satish, and Mallikarjun (2011) for number of fingers head⁻¹, finger width, and panicle width but contrary for the other agronomic traits in the current study, and completely contrary to report by Shailaja, Thirumeni, Paramasivam, and Ramanadane (2010) whose report indicated non-significance for these traits under salinity conditions further augmenting the importance of environmental conditions on expression of these traits in finger millet.

Based on the results of these investigations, additive gene effects were more important in transmission of blast resistance, number of productive tillers, days to 50% flowering, grain mass per head, plant height and panicle length. This implies that breeding progress can be achieved through selection for these traits. Selection for these traits therefore would involve breeding methods that entail selection in the early generations such as single seed descent, pedigree selection and modified pedigree as suggested in rice by Hammoud, Sedeek, El-Rewainy, and El-Namaky (2012). In finger millet specifically, Andrews (1993) suggested a method that involves bulking before evaluation as an appropriate method. In situations where non-additive gene effects are more important, selection should be delayed until later generations as the case was for finger and panicle width. For these traits repeated crossing in the segregating populations may be useful to pool all the desirable genes in one genotype as proposed by Selvaraj, Nagarajan, Thiyagarajan, Bharathi, and Rabinddran (2011).

4.2 General Combining Ability Effects of Parents for Blast Reaction and Yield Traits

The selection of parents based on *per se* performance may not always result in producing superior crosses (Simmonds & Smartt, 1999; Falconer & Mackay, 1996), and they pointed out that combining ability of parents gives useful information on the choice of parents in terms of expected performance of their progenies. This was clearly shown in cases where the magnitude and sign of the effect of each parent was not in agreement with individual performance. In the current investigations, most resistant parents to blast disease infection included *Achaki, Seremi 2, ACF 19*, and *Amumwari*. Of these parental materials, *Achaki, Seremi 2*, and *Amumwari* had negative, significant GCA effects which were desirable for blast resistance showing their capacity to transmit resistance for head blast disease. However, *ACF 19* in spite of being resistant, showed positive, significant (p ≤ 0.05) GCA effects for both head blast incidence and severity implying it would contribute towards susceptibility in most of the progeny families for which it is involved unlike the other three parental lines, therefore, it is not appropriate for incorporation in blast resistance breeding. Furthermore, in the current study, *Otunduru,* which exhibited moderate resistance had a negative, highly significant (p ≤ 0.01) GCA effect for head blast incidence and significant, negative SCA effect for blast disease in its progeny families with *E 11* a susceptible material, *Seremi 2* and *Amumwari* whilst positive, significant SCA effect in crosses with *ACF 5*, and *Abao*. It is suffice to suggest that *Otunduru*, unlike *ACF 19* is appropriate for incorporation in blast resistance breeding.

Parents that had positive significant GCA effects for grain yield contributed towards higher yields in most of the progenies in which they were part. For days to 50% flowering, negative, significant GCA effects indicated early maturity and these were observed for *E 11* and *Seremi 2*. Likewise desirable height was depicted by significant, negative GCA effects as was observed with *E 11, ACF 5, Seremi 2* and *Abao*. Positive, significant GCA effects for days to 50% flowering indicated late maturity; however, overall these results are indications that parents with good combining ability for grain yield per plant but were late maturing as depicted by positive, significant GCA effects for days to 50% flowering may be suited for high resource (potential) areas. It is also possible to select lines with positive GCA effects for yield and negative GCA effects for days to 50% flowering for limited resource (low potential) areas as they may also escape drought. Moreover they could also be used to generate early maturing cultivars suitable for increasing cropping intensity for the high potential areas. Meanwhile desirability of negative effects for height is to avoid lodging, which would even further be enhanced in high potential areas.

Knowledge of combining ability with mean performance of parents is therefore of great value in selecting suitable parents for hybridization programme (Selvaraj, Nagarajan, Thiyagarajan, Bharathi, & Rabinddran, 2011; Simmonds & Smartt, 1999). In the current study, high values for mean performance (in terms of grain yield) and GCA effects observed in some parental lines is clearly evident and this was also observed by Parashuram, Gowda, Satish, and Mallikarjun (2011). Parents *Seremi 2, Achaki, Otunduru* and *Amumwari* recorded high mean performance and GCA effects for yield contributing traits studied and blast disease resistance and, therefore, will be pertinent in the hybridization programme for selection of superior recombinants. Parashuram, Gowda, Satish, and Mallikarjun (2011); Tamilcovane and Jayaraman (1994); and Ravikumar, Shankare-Gowda & Seetharam (1986) also identified good general combiners in finger millet in India.

5. Conclusion

In conclusion parental materials that were resistant to head blast disease observed in the study included *Achaki, Seremi 2, ACF 19, Otunduru* and *Amumwari*, the parents *Achaki, Seremi 2*, and *Amumwari* had negative GCA effects and contributed negative SCA effects in most of the crosses involving them indicating that they are potential parents for head blast resistance breeding. Parental materials *Achaki, Seremi 2, Otunduru, Bulo* and *Amumwari* contributed towards high grain yield and with exception of *Seremi 2* were late in maturity. General combining ability contributed 31.65% and 53.05% of the crosses sums of squares for blast incidence and severity respectively while SCA effects contributed 68.35% and 46.95% respectively. The GCA effects for grain yield, days to 50% flowering and plant height accounted for 65.5%, 77.22% and 73.36% respectively of the crosses sums of squares. This indicated the predominance of genes with additive gene effects for grain yield ha^{-1}, days to 50% flowering and plant height in the parental lines and by extension high heritability for these traits in finger millet. Overall, highly significant additive effects implied that progress in high grain yield and blast disease resistance would be made through methods such as pedigree breeding and modified pedigree.

References

Andrews, D. J. (1993). Principles for breeding small millets. In K. W. Riley, S. C. Gupta, A. Seetharam, & J. Mushonga (Eds.), *Advances in small millets* (pp. 411-423). International Science Publisher, New York, USA.

Bua, B., & Adipala, E. (1995). Relationship between head blast severity and yield of finger millet. *International Journal of Pest Management, 41*, 55-59. http://dx.doi.org/10.1080/09670879509371922

Dhillon, B. S. (1975). The application of partial-diallel crosses in plant breeding: A review. *Crop Improvement, 2*, 1-8.

Esele, J. P. (1993). The current status of research on finger millet blast disease (*Pyricularia grisea*) at Serere research station. In K. W. Riley, S. C. Gupta, A. Seetharam, & J. Mushonga (Eds.), *Advances in small millets* (pp. 467-468). International Science Publisher, New York.

Falconer, D. S., & Mackay, T. F. C. (1996). *Introduction to quantitative genetics* (4th ed.). Longman Group Ltd, United Kingdom.

Griffings, B. (1956). Concept of general and specific combining ability in relation to diallel crossing systems. *Australian Journal of Biological Sciences, 9*, 463-493. http://dx.doi.org/10.1071/BI9560463

Hammoud, S. A. A., Sedeek, S. E. M., El-Rewainy, I. O. M., & El-Namaky, R. A. (2012). Genetic behaviour of some agronomic traits, blast disease and stem borer resistance in two rice crosses under two nitrogen levels. *Journal of Agricultural Research of Kafer El-Sheik University, 38*, 83-105.

IBPGR (International Board for Plant Genetic Resources). (1985). *Descriptors of finger millet (Eleusine coracana (L.) Gaertn)* (p. 20). IBPGR Secretariat, Rome, Italy.

IRRI. (1996). *Standard Evaluation System for Rice* (4th ed.). IRRI, INGER Genetic Resources Center, Manila, Philippines.

Krishnappa, M., Ramesh, S., Chandraprakash, J., Bharathi, J. G., & Doss, D. D. (2009). Genetic analysis of economic traits in finger millet. *SAT eJournal, 7*, 1-5.

Lenne, J. M., Takan, J. P., Mgonja, M. A., Manyasa, E. O., Kaloki, P., Wanyera, N., ... Sreenivasaprasad, S. (2007). Finger millet blast disease management. A key entry point for fighting malnutrition and poverty in East Africa. *Outlook on Agriculture, 36*, 101-108. http://dx.doi.org/10.5367/000000007781159994

Parashuram, D. P., Gowda, J., Satish, R. G., & Mallikarjun, N. M. (2011). Heterosis and combining ability studies for yield and yield attributing characters in finger millet (*Eleusine coracana* (L.) Gaertn.). *Electronic Journal of Plant Breeding, 2*, 494-500.

Prabhu, A. S., Filippi, M. C., & Zimmermann, F. P. (1996). Genetic control of blast in relation to nitrogen fertilization in upland rice. *Pesq. Agropec. Bras., 31*, 339-347.

Ratnakar, M. S., Mallikarjuna, N. M., Naveen-Kumar, K. S., & Jayarame-Gowda, M. V. (2009). Cytomorphological studies in inter-specific hybrids of finger millet. *Gregor Mendel Foundation Proceedings, 2009*, 69-74.

Ravikumar, R. L. (1988). *Genetic and biochemical basis of blast resistance in finger millet (Eleusine coracana Gaertn.)* (PhD Thesis). University of Agricultural Sciences, Bangalore, India.

Ravikumar, R. L., Shankare-Gowda, B. T., & Seetharam, A. (1986). Studies on Heterosis in finger millet. *Millets Newsletter, 5*, 26-27.

Russell, G. E. (1978). *Plant breeding for pest and disease resistance*. Butterworth and Company Publisher, Ltd. London, UK.

Seetharam, A., & Ravikumar, R. L. (1993). Blast resistance in finger millet – Its inheritance and biochemical nature. In K. W. Riley, S. C. Gupta, A. Seetharam, & J. Mushonga (Eds.), *Advances in small millets* (pp. 449-465). International Science Publisher, New York, USA.

Selvaraj, C. I., Nagarajan, P., Thiyagarajan, K., Bharathi, M., & Rabinddran, R. (2011). Studies on heterosis and combining ability of well known blast resistant rice genotypes with high yielding varieties of rice (*Oryza sativa* L.). *International Journal of Plant Breeding and Genetic, 5,* 111-129. http://dx.doi.org/10.3923/ijpbg.2011.111.129

Shailaja, H. B., Thirumeni, S., Paramasivam, K., & Ramanadane, T. (2010). Combining ability analysis in Finger millet *Eleusine coracana* (L.) Gaertn.) under salinity. *Electronic Journal of Plant Breeding, 1,* 129-139.

Simmonds, N. W., & Smartt, J. (1999). *Principles of Crop Improvement* (2nd ed.). Blackwell Science Ltd., Oxford, UK.

Takan, J. P., Akello, B., Esele, P., Manyasa, E. O., Obilana, A., Audi, P. O., … Sreenivasaprasad, S. (2004). Finger millet blast pathogen diversity and management in East Africa: a summary of project activities and outputs. *International Sorghum and Millets Newsletter, 45,* 66-69.

Tamilcovane, S., & Jayaraman, N. (1994). Association between yield components in ragi. *Journal of Phytological Research, 7,* 194-194.

Zhang, Y., Kang, M. S., & Lamkey, K. R. (2005). DAILLEL-SAS05. A Comprehensive programme for Griffings and Gardner-Eberhart analyses. *Agronomy Journal, 97,* 1097-1106. http://dx.doi.org/10.2134/agronj2004.0260

7

Maize Response to Nitrogen: Timing, Leaf Variables and Grain Yield

Adilson Nunes da Silva[1], Evandro Luiz Schoninger[2], Paulo Cesar Ocheuze Trivelin[2], Durval Dourado-Neto[1], Victor Meriguetti Pinto[3] & Klaus Reichardt[3]

[1] Department of Crop Science, ESALQ/University of São Paulo, Piracicaba, SP, Brazil

[2] Stable Isotopes Laboratory, CENA/University of São Paulo, Piracicaba, SP, Brazil

[3] Soil Physics Laboratory, CENA/University of São Paulo, Piracicaba, SP, Brazil

Correspondence: Victor Meriguetti Pinto, Soil Physics Lab., CENA/University of São Paulo, Av. Centenário 303, CEP 13418-900, Piracicaba, SP, Brazil. E-mail: meriguett@hotmail.com

Abstract

The main factors determining plant growth and productivity are decisive to be understood since they contribute to maximize plant nitrogen use efficiency. Thus, more reviews related to the correlation between the real content of chlorophyll and real carotenoids with the values obtained by chlorophyll (SPAD) in the early development stages of the maize are important to be obtained. The relation between the maize crop responses to the nitrogen fertilization at different development stages is of fundamental importance as well. The primary objective of this study was to investigate the responses of maize to the nitrogen application, urea fertilizer (^{15}N), in side-dress at different development stages. The secondary objective was verifying the correlation between chlorophylls and carotenoids with SPAD index and these with total biomass (BM), harvest index (HI), grain yield (GY) and grain N content in response to the nitrogen side-dress at different development stages. The nitrogen fertilization was carried out in plots, with the application of 30 kg ha^{-1} of N at planting and 140 kg ha^{-1} N as side-dress at vegetative stages V4, V6, V8, V10, and V12, without incorporation into the soil, and control treatment consisted of non-nitrogen side-dress application was also utilized. The 2011/2012 season presented higher precipitation than 2012/2013. Maize crop responded similarly for GY to the nitrogen application in side-dress in both seasons, however, the nitrogen application in the early stages caused higher values for leaf variables, leaf pigments, and SPAD. Higher amount of nitrogen in all parts of the plants was observed in the 2011/2012 season than in 2012/2013, influenced by the adequate weather conditions at the nitrogen application moment. Grain N content from ^{15}N fertilizer and N uptake and efficiency were greater for early N applications. SPAD values correlated positively with most pigment variables at V16 in both seasons, thus proving that SPAD was an efficient instrument of indirect evaluation of chlorophylls and carotenoids in maize leaves at early stages. Chlorophyll b at V16 was positively correlated ($P < 0.05$) with grain N content, GY, and BM, and total chlorophyll at V16 was positively correlated with GY and grain N content. However the chlorophylls a and total, evaluated at V14, were negatively correlated with GY. So, measurement of real chlorophyll and carotenoid pigment contents should be done after V14 stage when studies aim to evaluate crop nutritional conditions and prescribe future grain production practices.

Keywords: chlorophylls, carotenoids, SPAD, early growth stages

1. Introduction

Appropriate mineral nutrition is among the factors which have influence on crop productivity improvement. Nitrogen (N) is an essential nutrient required by plants, especially maize, which is one of the crops that mostly respond to fertilization with high increases in productivity. The N non-availability in soil causes several problems to crops such as reduced leaf area, decrease of photosynthetic rate, developmental delays, and reduced yield. On the other hand, excessive N application to the soil results in high production costs and can cause environmental problems such as contamination of the ground water and contribution to the increase of global warming due to the N volatilization.

Maize N requirements vary considerably in different plant development stages (Arnon, 1975). Although it is known that this crop requires about 20 kg ha^{-1} of N for each ton of produced grain (Fancelli, 2000; Sousa & Lobato, 2004), the best time for N application to this crop is still controversy. Some authors state the best time for N application is at seeding or close to this event, but others report it is ideal to apply at later stages, thus avoiding losses by leaching and volatilization, and increasing the efficiency of absorption and use of the nitrogen fertilizer by the plant (Cantarella, 1993; Pauletti & Costa, 2000; Ceretta et al., 2002; Basso & Ceretta, 2000).

Great advances have been made to improve the nitrogen use efficiency (NUE), defined as the ratio of dry matter production per unit of applied N. The use of indirect measurements to determine the nutritional status of plants has been the object of research for many crops. Research studies have shown the concentration of chlorophyll or the greening of leaves is positively correlated with leaf N concentrations, because 70% of N contained in the leaves is in the chloroplasts, participating in the synthesis and the structure of chlorophyll molecules (Marenco & Lopes, 2005). Therefore, the content of chlorophyll in the late vegetative stage has been related to the N nutritional status of various crops (Argenta, Silva, & Sangoi, 2001).

The traditional methods used to determine the amount of chlorophyll in the leaf require sampling and destruction of plant tissue, and the chlorophyll extraction and quantification processes are time-demanding. The portable chlorophyll meter SPAD (Soil Plant Analysis Development) is a nondestructive device that allows measuring instantaneously the chlorophyll amount of leaves, and it is an alternative to estimate the relative content of leaves pigments (Dwyer, Tollenaar, & Houwing, 1991; Argenta, Silva, & Sangoi, 2001). The N concentration in plant leaves has strong and positive relationship with the SPAD values, being more evident in the later growth stages (Argenta, Silva, & Sangoi, 2001). The leaf chlorophyll content shows also high correlation with SPAD results (Dwyer, Tollenaar, & Houwing, 1991; Ciampitti et al., 2012).

The knowledge on the effective influence of the factors that determine the performance of the plant can contribute decisively to minimize the stress caused by nitrogen deficiency. Thus, it is important to obtain reviews related to the correlation between the real content of chlorophyll and real carotenoids with the values obtained by chlorophyll (SPAD) in the early development stages of maize. The maize crop response to the nitrogen fertilization at different development stages is as well as important to be evaluated. Through these assessments, it is possible to have a greater knowledge about the plant relationship with the environment in which it is grown and may lead to increases in the grain yield.

The primary objective of this study was to investigate the maize response to side-dress N application by urea fertilizer (^{15}N) at different development stages. The secondary objective was verifying the correlation between the chlorophylls and carotenoids with SPAD values and these with, total biomass, harvest index (HI), grain yield (GY) and grain N content in response to the nitrogen side-dress in different development stages.

2. Materials and Methods

2.1 Experimental Site and Treatments

The experiments was carried out at Tanquinho Farm, in Piracicaba, SP, Brazil (22°34'13.9" S, 47°36'14.3" W) under field conditions during the seasons of 2011/2012 and 2012/2013. The first experiment was installed in December 2011 and completed in March 2012, and the second experiment was carried out from December 2012 to March 2013. The soil was classified as "Latossolo Vermelho Distrófico" (EMBRAPA, 2006), or as Rhodic Hapludox (Soil Survey Staff, 2010). According to Köppen's classification, the climate of the region is Cwa type, with annual average temperature 23.9 °C and annual precipitation 1,257 mm.

Chemical and physical characterization of the soil was performed for the 0-20 cm layer (Table 1) previously the implementation of the experiments. For both seasons, weed plants were controlled with glyphosate before planting. Mechanical seeding was performed with approximately 3.3 seeds per meter (already accounted for 10% surplus due to losses) in order to obtain a final stand of 60,000 plants per hectare. The hybrid used in this study was the 30F35HR (PIONEER, 2014). The experimental design consisted of random blocks having four replicates managed under conventional tillage, with maize as the preceding crop. Plots had 10 rows of maize with 10 m length, spaced 0.5 m apart, totaling an area of 50 m^2. In each of the plots, mini plots (0.5 m wide and 1.5 m long) were delimited for ^{15}N-urea application at the same rate and time of the commercial urea application to the rest of the plot.

Table 1. Soil analysis in both maize growing seasons, 2011/2012 and 2012/2013, (inorganic nitrogen [NO_3^--N/NH_4^+-N], soil pH, potassium content [K], and phosphorus Bray-P 1 [P]) in from 0-20 and 20-40 of the soil profile

Profile	pH	P	K	Ca	Mg	H+Al	Al	T	V	OM[1]	Silt	Clay
-------cm-------	($CaCl_2$)	---mg dm^{-3}---	-----------------$mmol_c$ dm^{-3}----------------						--%--	----------g kg^{-1}----------		
2011/12												
0-20	4.9	27	1.9	30	13	42	1	87	52	29	151	529
20-40	4.5	32	0.6	19	8	58	5	86	32	22	102	548
2012/13												
0-20	4.8	29	1.3	16	9	47	2	73	36	30	-	-
20-40	4.6	21	0.6	10	7	52	3	70	25	24	-	-

Note. [1] = Organic matter.

Planting fertilization was performed with the application of full rates of phosphorus (P) and potassium (K). Nitrogen (N) fertilization was carried out in plots with urea as the source, corresponding to 30 kg N ha^{-1} at planting and 140 kg N ha^{-1} as side-dress, without incorporation. The N, P and K rates were applied aiming at high grain yield (10-12 t ha^{-1}) and following the recommendations for areas with high response to N application (Cantarella, Raij, & Camargo, 1997). The treatments consisted of urea fertilizer application five times as side-dressing, corresponding to the vegetative stages V4, V6, V8, V10 and V12 as described by Ritchie, Hanway, and Benson (2003). A control treatment consisted of non-nitrogen side-dress fertilization.

2.2 Sampling and Analysis

The grain yield (GY), the harvest index (HI) and the plant biomass (BM) were determined at the end of the crop cycle. GY was measured by weighing the grain harvested with moisture correction to 13%. BM was measured based on the wet weight of the residuals, followed by the correction of the previously determined moisture.

Plant material from the shoot was separated into stem, leaf + tassel + ear husk, cob and grain. All material was dried at 60 °C with forced air to a constant weight. Subsequently, the dried material was ground in a Wiley grinder, homogenized and subsampled. In all subsamples, the nitrogen content (g kg^{-1}) was determined by Kjeldahl digestion - distillation and the sulfur content by the nitric perchloric digestion methodology followed by turbidity determination.

Plants collected from the mini plots in the 2012/2013 season were oven dried in laboratory and finally ground for later determination of total-N content and ^{15}N abundance by mass spectrometer (Barrie & Prosser, 1996). Grain nitrogen content derived from the fertilizer (GNCF) as well as the nitrogen fertilizer use and efficiency (NFUE, Equation 3) were calculated according to Gava et al. (2006).

Indirect chlorophyll content index was obtained by using the SPAD-502 chlorophyll meter (Minolta, Japan) (Minolta, 1989; Pestana et al., 2001; Markwell, Osterman, & Mitchell, 1995) on the two uppermost fully developed plant leaves before entering senescence and being photosynthetically active. Six samples per leaf were performed in 4 plants per treatment, carried out at the V14 to V16 stages.

The plant height (PH) was measured with a tape (in meters) placed from soil surface to the highest insertion of the last uppermost leaves on 4 plants per treatment at the flowering period (VT).

The content of pigments in the leaves was evaluated in the laboratory. The same leaves of SPAD evaluation were afterwards used for the analysis of chlorophylls a, b, and total, and carotenoids. Following the methodology adapted from Moran and Porath (1980), two leaves per plant were evaluated from a total of 4 plants per treatment at the V14 and V16 stages.

2.3 Statistical Analysis

Data were tested for normality, as well as the homogeneity of variances, and then subjected to analysis of variance at 5% of significance. Having significant effects of treatments by the F test, comparisons were performed using the t test, also at the level of 5% significance. The Pearson correlation test was also performed among the variables in order to verify the existence of positive correlations.

3. Result and Discussion

3.1 Growing Season and Phenology

The weather conditions during both growing seasons can be seen in Figure 1. The 2011/2012 season presented higher precipitation than 2012/2013, however, in the last one there was regular precipitation during the stages of N application. Figure 1 shows also the occurrence date of plant stages for both growing seasons.

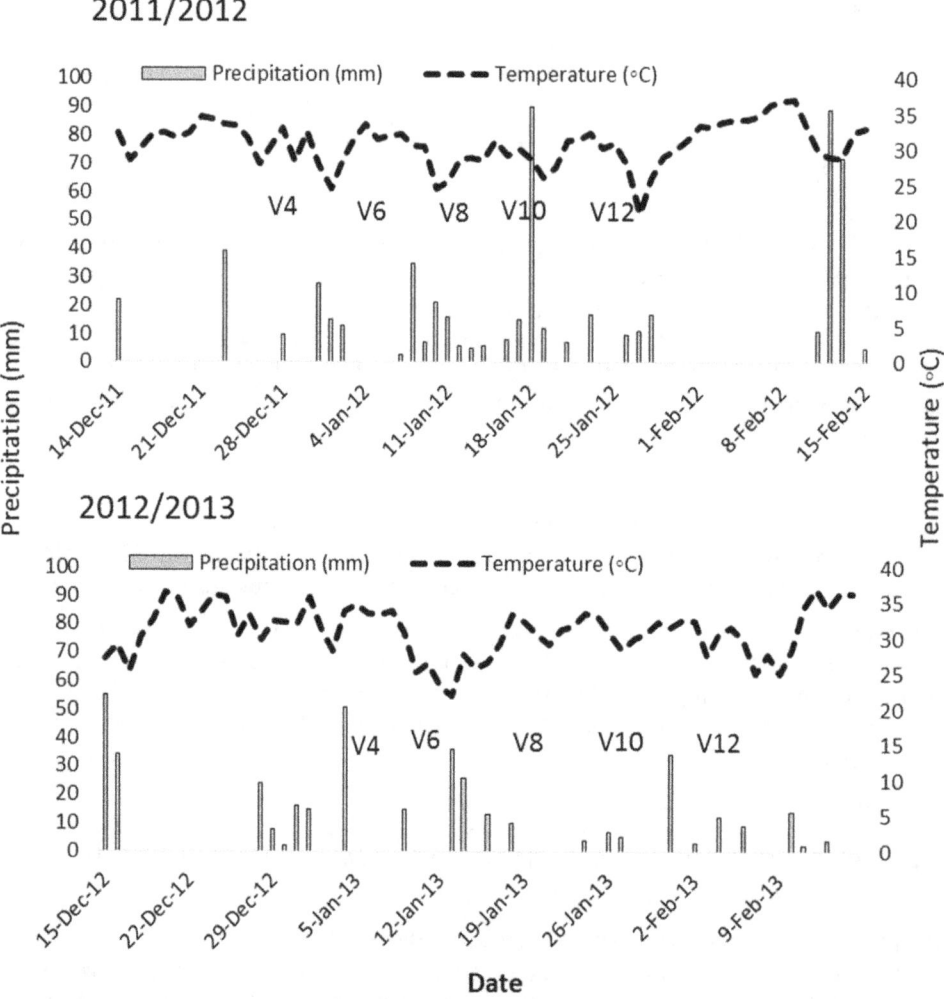

Figure 1. Days of development stages (moment of nitrogen application in side-dress) and weather conditions (maximum air temperature and mean precipitation) for the 2011/2012 and 2012/2013 maize growing seasons at *Tanquinho* Farm, Piracicaba, Sao Paulo, Brazil

3.2 Nitrogen and Sulfur Uptake

During the 2011/2012 season (Table 2), the treatments were significant for most variables, but not for cob N content. For leaf N content there was only difference between the control and all N applications, in all development stages. But for stem N content only the N application in V4 and V6 were different in relation to the control (V4 = V6 > control). Grain N content and total plant N uptake were greater for V6 and V10 in relation to the control. However, the N application at V8 resulted in similar values in relation to the control for both variables. Nonetheless, the total plant sulfur uptake was similar for all treatments with nitrogen application. The N amount in all parts of the plants was higher in the 2011/2012 season than in 2012/2013 (Table 3). The average total N plant uptake value 50 kg ha^{-1} of N was evidently greater in the first season (2011/2012).

Maize responded differently between seasons due to water availability. After the initial stages that presented less variable means of the measured plant parameters, a well-distributed precipitation spell was observed (very close to the nitrogen application in side-dress), while in the other stages there was a lack or small amount of precipitation that could have influenced the N uptake and partitioning. For urea application in dry soil conditions, Black, Sherlock, and Smith (1987) observed 70% of the applied nitrogen remained in the soil in the hydrolyzed form.

Fertilizer N content (GNCF) in grain and N fertilizer use and efficiency (NFUE) calculated from ^{15}N data were greater for the early applications (V4 = V6 > V8, V10, and V12) p < 0.0001.These variables presented lower values for the later N application, the V12 application being the lowest one for GNCF, lower than the N application at V4 stage. This could have happened because plants had a longer period to perform the N uptake from the soil, nevertheless the weather conditions had a greater influence on this.

Table 2. 2011/2012 Season: Nitrogen (N) partitioning into leaf, stem, cob, grain components and total plant; and total plant sulfur (S) uptake at physiological maturity, in response to the nitrogen application as side-dress at the V4, V6, V8, V10, and V12 stages

Treat	Leaf N	Stem N	Cob N	Grain N	Total N	Total S
	--kg ha^{-1}---					
Control	40.9 b	15.2 b	8.4	85.7 bc	150.2 c	8.9 c
V4	50.0 a	24.0 a	10.5	115.7 ab	200.3 ab	12.9 ab
V6	55.5 a	25.9 a	11.6	125.6 a	218.6 a	13.2 ab
V8	55.2 a	23.8 ab	8.7	78.9 c	166.0 bc	11.3 b
V10	53.5 a	21.2 ab	10.9	131.7 a	217.4 a	13.5 a
V12	49.9 a	18.0 ab	11.7	110.5 abc	190.1 abc	12.3 ab
ANOVA						
Treat.	*	*	ns	*	*	**
CV%	11.4	27.3	29.4	21.8	15.4	12.9

Note. ns = not significant; * = P < 0.05; ** = P < 0.01; *** = P < 0.0001.

Table 3. 2012/1013 Season: Nitrogen partitioning into leaf, stem, cob, grain components and total plant; Sulfur total plant uptake; grain nitrogen content from the fertilizer (GNCF), and nitrogen fertilizer use and efficiency (NFUE) at physiological maturity, in response to the nitrogen application as side-dress at the V4, V6, V8, V10, and V12 stages

Treat.	Leaves N	Stem N	Cob N	Grain N	Total N	Total S	GNCF	NFUE
	---kg ha^{-1}---							----%----
Control	35.3	10.6 c	6.6	129.6	182.1	13.6 c	-	-
V4	45.5	15.1 ab	6.8	127.6	195.0	14.7 bc	46.1 a	48.9 a
V6	43.4	16.3 a	7.6	130.7	198.1	16.4 ab	49.7 a	51.8 a
V8	39.0	15.2 ab	8.3	121.9	184.4	15.5 abc	31.9 b	31.2 b
V10	39.3	15.8 ab	9.0	135.6	199.7	17.7 a	27.0 bc	26.7 bc
V12	41.8	13.4 bc	8.8	145.7	209.8	17.4 a	17.7 c	17.8 c
ANOVA								
Treat.	ns	**	ns	ns	ns	*	***	***
CV%	13.7	12.8	15.5	7.7	7.8	11.3	20.56	19.9

Note. ns = not significant; * = P < 0.05; ** = P < 0.01; *** = P < 0.0001.

3.3 Grain Yield, Biomass and Harvest Index

In 2011/2012, total plant biomass was not influenced by treatments (Table 4), only cob weight had significant differences (p < 0.05). On the other hand, in 2012/2013 (Table 5) there was only difference for HI, the V12 application being better than most applications, but not better than V8. França et al. (1994) reported that the splitting of N does not affect the efficiency of nitrogen fertilizer or use of N from the soil, and the results were similar when they applied up to 106 kg of N per ha in a single rate at the stage where the plant had 6 leaves or, when fertilizer is subdivided twice, half at the 6-leaf stage and the other half at the 10 leaf stage. These authors also observed most of the N in the plant was accumulated until flowering, reaching values of up to 93%. They concluded that the nitrogen side-dressing should be made after seeding until early flowering, a period during which the rate of absorption is virtually linear. The N application efficiency prior to maize planting was studied by many authors (Pauletti & Costa, 2000; Ceretta et al., 2002). All of them found little difference among time N applications, but Ceretta et al. (2002) warned that the early application can compromise yield in years of high rainfall, in the early stage of crop development. However, Jokela and Randall (1989) concluded that there was less response of maize to N when it was applied at theV2 stage than at the V8 stage. Maize starts to take up N rapidly at the middle vegetative growth period (V10) and the maximum rate of N uptake occurred near to silking (Hanway, 1963; Settimi & Maranville, 1998). Hence, application of N at V8-V10 stage should be one of the best ways of supplying N to convene this high demand.

Table 4. 2011/2012 Season: Plant biomass of leaf, stem, cob, grain, and total plant components; grain yield (GY); and grain harvest index (HI) at physiological maturity, in response to the nitrogen application as side-dress at the V4, V6, V8, V10, and V12 stages

Treat.	Leaf	Stem	Cob	Total	GY	HI
	--kg ha--					------%------
Control	4.6	3.1	1.2 c	16.7	6.7	43.1
V4	5.3	4.1	1.5 ab	19.2	8.2	42.9
V6	5.3	4.2	1.6 ab	18.7	7.6	40.5
V8	4.8	3.6	1.3 bc	15.4	5.6	36.2
V10	5.1	3.8	1.6 a	20.0	9.4	47
V12	5.2	3.7	1.5 ab	18.9	8.4	44.1
ANOVA						
Treat.	ns	ns	*	ns	ns	ns
CV%	9.5	17.7	12.6	16	15.5	11.6

Note. ns = not significant; * = P < 0.05; ** = P < 0.01; *** = P < 0.0001.

Table 5. 2012/2013 Season: Plant biomass of leaf, stem, cob, grain, and total plant components; grain yield (GY); and grain harvest index (HI) at physiological maturity, in response to the nitrogen application as side-dress at the V4, V6, V8, V10, and V12 stages

Treat.	Leaves	Stem	Cob	Total	GY	HI
	--kg ha^{-1}--					------%------
Control	4.6	3.2	1.5	19.9	11.1	53.1 b
V4	4.7	3.6	1.6	20.9	11.1	52.5 bc
V6	4.8	3.8	1.6	21.2	10.9	51.5 c
V8	4.4	3.2	1.6	20.0	10.1	53.9 ab
V10	4.6	3.4	1.6	20.9	10.6	53.4 b
V12	4.6	3.1	1.5	20.8	11.1	55.1 a
ANOVA						
Treat.	ns	ns	ns	ns	ns	**
CV%	8.2	10.7	7.2	6.9	5.8	2

Note. ns = not significant; * = P < 0.05; ** = P < 0.01; *** = P < 0.0001.

3.4 Leaf Pigment Contents, SPAD, and Plant Height

Early N application triggered the highest value for most variables in both seasons. In 2011/2012 (Table 6), lower values than in 2012/2013 (Table 7) were observed for all variables. The differences were more expressive for the V14 stage than for V16 for most of the variables. However, SPAD had significant differences for these two stages in both seasons, being greater from V4 to V10 than at the V14 stage and having difference only between N fertilized treatments with the control for V16 in 2011/2012. However in 2012/2013 differences for SPAD were larger than in the first season. At the V14 sample stage, V6 was greater than V8, V10, V12, and control. Similar results for V16 were found, however, V6 presented higher values for SPAD only in relation to V4, V12, and the control. PH in 2011/2012 was only different from the control, and in 2012/2013 only V4 presented differences in relation to the control (V4 > Control). Chlorophyll a, b and total were also lower in the non-fertilized treatment than in treatments with N.

Table 6. 2011/2012 Season: Leaf pigment contents (mg g^{-1} fresh leaf mass): chlorophyll a (CA), chlorophyll b (CB), chlorophyll total (CT), carotenoids (Carot), SPAD, evaluated at V14 and V16 stages; and plant height (PH) at VT in response to the nitrogen application as side-dress at the V4, V6, V8, V10, and V12 stages

Treat.	CA V14	CA V16	CB V14	CB V16	CT V14	CT V16	Carot V14	Carot V16	SPAD V14	SPAD V16	PH
	--mg g^{-1}---										---m---
Control	0.770 c	0.77	0.25 c	0.24	1.02 c	1.01	0.16 c	0.17 b	41.1 c	38.0 b	2.59 b
V4	1.130 a	0.88	0.46 a	0.32	1.59 a	1.20	0.23 a	0.21 ab	55.2 a	51.5 a	2.90 a
V6	1.050 ab	0.9	0.37 ab	0.30	1.42 ab	1.19	0.22 ab	0.19 ab	51.0 a	51.2 a	2.83 a
V8	1.04 ab	0.92	0.39 ab	0.31	1.43 ab	1.23	0.22 ab	0.22 a	52.1 a	50.1 a	2.82 a
V10	0.947 abc	0.9	0.31 bc	0.28	1.26 bc	1.19	0.20 abc	0.21 ab	50.2 ab	50.0 a	2.87 a
V12	0.880 bc	0.83	0.29 bc	0.26	1.17 bc	1.09	0.19 bc	0.20 ab	45.0 bc	48.5 a	2.81 a
ANOVA											
Treat.	*	ns	**	ns	*	ns	*	*	**	**	*
CV%	13.9	17.9	20.2	19.6	15.1	18.1	13.2	15.6	7.7	6.7	4.1

Note. ns = not significant; * = P < 0.05; ** = P < 0.01; *** = P < 0.0001.

Table7. 2011/2012 Season: Leaf pigment contents: chlorophyll a (CA), chlorophyll b (CB), chlorophyll total (CT), carotenoids (Carot), SPAD, evaluated at the V14 and V16 stages; and plant height (PH) at VT in response to the nitrogen application as side-dress at the V4, V6, V8, V10, and V12 stages

Treat.	CA V14	CA V16	CB V14	CB V16	CT V14	CT V16	Carot V14	Carot V16	SPAD V14	SPAD V16	PH
	--mg g^{-1}---										----m----
Control	1.26 c	1.18	0.40 b	0.35	1.66 b	1.53	0.29	0.21	50.6 e	45.9 c	2.26 b
V4	1.44 ab	0.98	0.61 a	0.38	2.04 a	1.30	0.32	0.12	58.3 ab	51.8 bc	2.42 a
V6	1.43 ab	1.19	0.52 ab	0.39	1.95 a	1.58	0.32	0.23	58.8 a	61.5 a	2.37 ab
V8	1.53 a	0.98	0.51 ab	0.28	2.04 a	1.26	0.31	0.12	56.3 bc	55.5 ab	2.37 ab
V10	1.43 ab	1.11	0.52 ab	0.31	1.94 a	1.42	0.35	0.14	56.0 c	54.8 ab	2.32 ab
V12	1.36 bc	1.22	0.55 ab	0.36	1.91 a	1.58	0.31	0.23	53.0 d	53.8 b	2.36 ab
ANOVA											
Treat.	**	ns	*	ns	*	ns	ns	ns	***	**	ns
CV%	5	27.8	23.4	30.4	8.5	26.3	15.4	63.7	2.64	9.22	3.7

Note. ns = not significant; * = P < 0.05; ** = P < 0.01; *** = P < 0.0001.

3.5 Person Correlations

SPAD at V16 had significant positive correlations with all pigments at both stages evaluated in 2011/2012 (Table 8), this variable correlated also positively and significantly (both stages, V14 and V16) with total chlorophyll evaluated in laboratory conditions. The strongest SPAD V16 correlation related to pigments was with chlorophyll a in the V14 stage (0.61, p < 0.05). In the same way SPAD at V14 also presented a strong correlation with other pigments, however, this variable only had correlation with the variables evaluated at the same stage. The strongest correlation for SPAD evaluated at V14 with pigments was with total chlorophyll (0.80, p < 0.0001). Nevertheless, in 2012/2013 (Table 9), there were significant correlations for SPAD with pigments only between SPAD/V14 with chlorophyll a and total (0.40, p < 0.05, for both).

Table 8. 2011/2012 Season: Pearson correlation analysis for leaf pigment contents: chlorophyll a (CA), chlorophyll b (CB), chlorophyll total (CT), carotenoids (Carot), SPAD, evaluated at V14 and V16 stages; total plant biomass (BM), grain harvest index (HI), grain yield (GY), and grain nitrogen content (GNC) in response to the nitrogen application as side-dress at the V4, V6, V8, V10, and V12 stages

	GNC	GY	HI	BM	SPAD16	SPAD14	Carot16	Carot14	CT16	CT14	CB16	CB14	CA16	CA14
CA14	0.06	0.2	-0.30	0.44	0.61**	0.76***	0.14	0.95***	0.20	0.99***	0.32	0.92***	0.14	1.00
CA16	0.13	0.17	-0.20	0.11	0.42*	0.23	0.08	0.07	0.99***	0.12	0.93***	0.08	1.00	
CB14	0.02	0.12	-0.22	0.05	0.52**	0.71***	0.90***	0.90***	0.12	0.97***	0.27	1.00		
CB16	0.1	0.14	-0.19	0.07	0.50*	0.40	0.24	0.12	0.97***	0.31	1.00			
CT14	0.05	0.15	-0.25	0.05	0.59**	0.80***	0.12	0.95***	0.20	1.00				
CT16	0.12	0.12	0.15	0.10	0.44*	0.23	0.84***	0.13	1.00					
Carot14	0.01	0.20	-0.30	0.04	0.60**	0.70***	0.13	1.00						
Carot16	0.01	0.05	-0.55	0.14	0.50**	0.4	1.00							
SPAD14	0.33	0.34	-0.05	0.34	0.75***	1.00								
SPAD16	0.50*	0.30	-0	0.45*	1.00									
BM	0.92***	0.70***	0.68	1.00										
HI	0.80***	0.50*	1.00											
GY	0.60**	1.00												
GNC	1.00													

Note. * = P < 0.05; ** = P < 0.01; *** = P < 0.0001.

Table 9. 2012/2013 Season: Pearson correlation analysis for leaf pigment contents: chlorophyll a (CA), chlorophyll b (CB), chlorophyll total (CT), carotenoids (Carot), SPAD, evaluated at V14 and V16 stages; total plant biomass (BM), grain harvest index (HI), grain yield (GY), and grain nitrogen content (GNC) in response to the nitrogen application as side-dress at the V4, V6, V8, V10, and V12 stages

	GNC	GY	HI	BM	SPAD16	SPAD14	Carot16	Carot14	CT16	CT14	CB16	CB14	CA16	CA14
CA14	-0.35	-0.61*	-0.35	-0.22	0.23	0.40*	-0.13	0.29	-0.18	0.83***	-0.22	0.41*	-0.14	1.00
CA16	0.40	0.38	0.22	0.24	0.12	-0.10	0.85***	-0.004	0.97***	0.23	0.59***	-0.25	1.00	
CB14	0.06	-0.22	-0.16	0.13	0.05	0.3	-0.12	0.34	-0.12	0.84***	-0.11	1.00		
CB16	0.61*	0.55**	-0.01	0.53*	0.31	0.12	0.075***	0.21	0.75***	-0.23	1.00			
CT14	-0.16	-0.50*	-0.30	-0.04	0.16	0.40*	-0.15	0.34	-0.24	1.00				
CT16	0.48*	0.50*	0.18	0.34	0.20	-0.05	0.90***	0.05	1.00					
Carot14	0.31	-0.14	-0.31	0.36	0.19	0.31	0.03	1.00						
Carot16	0.50**	0.36	0.08	0.28	0.19	-0.05	1.00							
SPAD14	-0.01	0.01	-0.38	0.34	0.47*	1.00								
SPAD16	0.26	0.01	-0.06	0.36	1.00									
BM	0.82***	0.60*	0.00	1.00										
HI	0.26	0.31	1.00											
GY	0.60**	1.00												
GNC	1.00													

Note. * = P < 0.05; ** = P < 0.01; *** = P < 0.0001.

Even with less correlation between SPAD and pigments in the second season, we can consider the SPAD measuring device as efficient to evaluate the actual amount of these pigments. These results are in accordance with other studies. According to Piekielek et al. (1995) and Dwyer, Tollenaar, and Houwing (1991) this indirect evaluation of chlorophyll content in the leaf can be used to predict the nutritional N level in plants, because the correlation with the amount of pigment was positive in relation to N concentration. There is a strong positive relationship between the SPAD and N concentration in the leaves of the plants, although this is more evident in the later growth stages (Argenta, Silva, & Sangoi, 2001), and there is also a high correlation of SPAD with chlorophyll content (Dwyer, Tollenaar, & Houwing, 1991; Ciampitti et al., 2012). A significant positive correlation between SPAD/V16 and grain N content was found in the first season, but there was no significant correlation between SPAD with grain yield, harvest index, and grain N content in the second season (2012/2013). Chlorophyll b at V16 presented significant positive correlation with grain N content (0.61, $p < 0.05$), GY (0.55, $p < 0.01$), and total biomass (0.53, $p < 0.05$). The chlorophyll total at V16 also presented a positive correlation with GY (0.50, $p < 0.05$) and grain N content (0.48, $p < 0.05$), however, the chlorophylls a and total evaluated at V14 presented a negative significant correlation with GY. Thus, the measurement of pigment contents aiming to study nutritional crop conditions and predict grain production should be performed after the V14 stage.

4. Conclusion

This growth and development experiment for maize (*Zea mays* L.) evidenced a similar response to the treatments for grain yield, harvest index and total plant biomass in both seasons with distinct climatic conditions. However, as a result of the different climatic conditions and the N application moment, a higher amount of nitrogen was observed in all parts of the plants in the 2011/2012 season in relation to the 2012/2013.

Fertilizer N in grain and nitrogen fertilizer use and efficiency were greater for the early applications, at stages V4 and V6.

Chlorophyll contents estimated through SPAD measurements and leaf pigments were largely influenced by the development stage. Evaluations made at V14 presented more differences in treatments than at V16, and the N application in the early stages caused higher values for most of the leaf variables (mainly pigments and SPAD).

SPAD correlated positively and significantly with most pigment variables at V16, for both seasons, mainly in the first season in which this variable was correlated with all chlorophylls and carotenoids, showing that the SPAD-502 chlorophyll meter is an efficient instrument for the indirect evaluation of chlorophylls and carotenoids in maize leaves. Also, SPAD in V16 sample stage had a positive correlation with grain nitrogen content and total plant biomass.

Chlorophyll b at V16 presented a significant and positive correlation with grain N content, grain yield, and total biomass. Total Chlorophyll at the same stage also presented a positive correlation with grain yield and grain N content, however, the chlorophylls a and total, evaluated at V14, presented a negative significant correlation with grain yield. Therefore, we recommend that measurements of real pigment contents aiming to study the nutritional maize crop conditions and predict grain production should be made after the V14 stage.

Acknowledgements

The authors are grateful to the São Paulo Research Foundation (FAPESP/Grant #:2011/23303-3) and to the National Council for Scientific and Technological Development (CNPq/Grant #:302261/2011-7) for the financial support.

References

Argenta, G., Silva, P. R. F., & Sangoi, L. (2001). Arranjo de plantas em milho: Análise do estado da arte. *Ciência Rural, 31*, 1075-1084. http://dx.doi.org/10.1590/S0103-84782001000600027

Arnon, I. (1975). *Mineral nutrition of maize*. Bern: International Potash Institute.

Basso, C. J., & Ceretta, C. A. (2000). Manejo do nitrogênio no milho em sucessão a plantas de cobertura de solo, sob plantio direto. *Revista Brasileira de Ciência do Solo, 24*, 905-915. http://dx.doi.org/10.1590/S0100-06832000000400022

Black, A. S., Sherlock, R. R., & Smith, N. P. (1987). Effect of urea granule size on ammonia volatilization from surface-applied urea. *Fertilizer Research, 11*, 87-96. http://dx.doi.org/10.1007/BF01049567

Cantarella, H. (1993). Calagem e adubação do milho. In L. T. Büll, & H. Cantarella (Eds.), *Cultura do milho: Fatores que afetam a produtividade* (pp. 148-196). Piracicaba: POTAFOS.

Cantarella, H., Raij, B. van, & Camargo, C. E. O. (1997). Cereais. In B. van Raij, H. Cantarella, J. A. Quaggio, & A. M. C. Furlani (Eds.), *Recomendações de adubação e calagem para o Estado de São Paulo* (2nd ed., pp. 45-71). Campinas: IAC. (Boletim Técnico, 100).

Ceretta, C. A., Basso, C. J., Flecha, A. M. T., Pavinato, P. S., Vieira, F. C. B., & Mai, M. E. M. (2002). Manejo da adubação nitrogenada na sucessão aveia preta/milho, no sistema plantio direto. *Revista Brasileira de Ciência do Solo, 26*, 163-171. http://dx.doi.org/10.1590/S0100-06832002000100017

Ciampitti, I. A., Zhang, H., Friedemann, P., & Vyn, T. J. (2012). Potential physiological frameworks for mid-season field phenotyping of final plant nitrogen uptake, nitrogen use efficiency, and grain yield in maize. *Crop Science, 52*, 2728-2742. http://dx.doi.org/10.2135/cropsci2012.05.0305

Dwyer, L. M., Tollenaar, M., & Houwing, L. (1991). A nondestructive method to monitor leaf greenness in corn. *Canadian Journal of Plant Science, 71*, 505-509. http://dx.doi.org/10.4141/cjps91-070

Empresa Brasileira de Pesquisa Agropecuária (EMBRAPA). (2006). *Brazilian Soil Classification System* (2nd ed.). Rio de Janeiro.

Fancelli, A. L. (2000). *Nutrição e adubação do milho*. Piracicaba: ESALQ.

França, G. E., Coelho, A. M., Resende, M., & Bahia Filho, A. F. C. (1994). *Parcelamento da adubação nitrogenada em cobertura na cultura do milho irrigado* (pp. 28-29). EMBRAPA, Centro Nacional de Pesquisa de Milho e Sorgo. Relatório técnico anual do Centro Nacional de Pesquisa de Milho e Sorgo: 1992-1993, Sete Lagoas.

Gava, G. J. C., Trivelin, P. C. O., Oliveira, M. W., Heinrichs, R., & Silva, M. A. (2006). Balanço do nitrogênio da uréia (^{15}N) no sistema solo-planta na implantação da semeadura direta na cultura do milho. *Bragantia, 65*, 477-486. http://dx.doi.org/10.1590/S0006-87052006000300014

Hanway, J. J. (1963). Growth stages of corn (*Zea mays* L.). *Agronomy Journal, 55*, 487-492. http://dx.doi.org/10.2134/agronj1963.00021962005500050024x

Jokela, W. E., & Randall, G. W. (1989). Corn yield and residual soil nitrate as affected by time and rate of nitrogen application. *Agronomy Journal, 81*, 720-726. http://dx.doi.org/10.2134/agronj1989.00021962008100050004x

Marenco, R. A., & Lopes, N. F. (2005). *Fisiologia vegetal: Fotossíntese, respiração, relações hídricas e nutrição mineral* (2nd ed.). Viçosa: UFV.

Markwell, J., Osterman, J. C., & Mitchell, J. L. (1995). Calibration of the Minolta SPAD-502 leaf chlorophyll meter. *Photosynthesis Research, 46*, 467-472. http://dx.doi.org/10.1007/BF00032301

Minolta C. (1989). *Manual for chlorophyll meter SPAD 502*. Osaka: Radiometric Instruments Divisions.

Moran, R., & Porath, D. (1980). Chlorophyll determination in intact tissues using n,n dimethylformamide. *Plant Physiology, 65*, 478-479. http://dx.doi.org/10.1104/pp.65.3.478

Pauletti, V., & Costa, L. C. (2000). Época de aplicação de nitrogênio no milho cultivado em sucessão à aveia preta no sistema plantio direto. *Ciência Rural, 30*, 599-603. http://dx.doi.org/10.1590/S0103-84782000000400007

Pestana, M., David, M., Varennes, A., Abadia, J., & Faria, E. A. (2001). Responses of "Newhall" oranges trees to iron deficiency in hidroponics: effects on leaf chlorophyll, photosynthetic efficiency, and root ferric chelate reductase activity. *Journal of Plant Nutrition, 24*, 1609-1620. http://dx.doi.org/10.1081/PLN-100106024

Piekielek, W. P., Fox, R. H., Toth, J. D., & Macneal, K. E. (1995). Use of a chlorophyll meter at the early dent stage of corn to evaluate N sufficiency. *Agronomy Journal, 87*, 403-408. http://dx.doi.org/10.2134/agronj1995.00021962008700030003x

Pionner. (2014). *Hibridos de milho: 30F35HR*. Santa Cruz do Sul. Retrieved from http://www.pioneersementes.com.br/DownloadCenter/Catalogo-De-Produtos-Milho-Safrinha-2014.pdf

Ritchie, S. W., Hanway, J. J., & Benson, G. O. (2003). Como a planta de milho se desenvolve. *Informações Agronômicas, 103*, 1-11.

Settimi, J. R., & Maranville, J. W. (1998). Carbon dioxide assimilation efficiency of maize leaves under nitrogen stress at different stages of plant development. *Soil Science and Plant Analysis, 29*, 777-792. http://dx.doi.org/10.1080/00103629809369985

Soil Survey Staff. (2010). *Keys to Soil Taxonomy* (11th ed.). USDA-Natural Resources Conservation Service, Washington.

Sousa, D. M. G. de, & Lobato, E. (2004). Calagem e adubação para culturas anuais e semiperenes. In D. M. G. de Sousa, & E. Lobato, (Eds.), *Cerrado: Correção do solo e adubação* (pp. 283-315). Planaltina: Embrapa Cerrados.

Trivelin, P. C. O., Oliveira, M. W., Vitti, A. C., Gava, G. J. C., & Bendassolli, J. A. (2002). Perdas do nitrogênio da uréia no sistema solo-planta em dois ciclos de cana-de-açúcar. *Pesquisa Agropecuária Brasileira, Brasília, 37*, 193-201. http://dx.doi.org/10.1590/S0100-204X2002000200011

8

Searching about Resistance of Common Cultivated Varieties in Varamin to Separated Fungal *Phytophthora drechsleri* from the Same Place

Majid Shahi-Bajestani[1] & Kheyzaran Dolatabadi[2]

[1] Research and Innovation Center of Etka Organization, Tehran, Iran

[2] Department of Horticulture, Ferdowsi University of Mashhad, Mashhad, Iran

Correspondence: Majid Shahi-Bajestani, Research and Innovation Center of Etka Organization, Tehran, Iran.
E-mail: majid.shahi.1364@gmail.com

The research is financed by Research and Innovation Center of Etka Organization, Tehran, Iran

Abstract

Dieback plant disease caused by fungi species Phytophthora is one of the most important soil borne disease in Iran. During studies in Varamin, one of the most harmful factors on crops is *Phtophtora* species in that place. So that in almost all different villages of this city, limitation of cultivation of crops such as cantaloupe, that is one of the most important product in this region, has arisen. This study had been done for searching about reaction of common cultivars in some plants in Varamin to the separated *Phtophthora drechsleri* from the same place by measuring plant growth factors. Seeds of melon in Ivanaki cultivar, cantaloupe in Samsoori cultivar, tomato in Urbana cultivar, red bean in Mahalli cultivar, were planted in pots containing sterile soil, then mentioned fungal were infected by Zoospore suspension and kept in greenhouse condition. In addition, the reaction of safflower seedlings of species *Phytpophthora melonis* was used to differentiate species. Percentage of disease as well as growth factors such as stem fresh and dry weight, stem length and root length during the time of 1, 2 and 3 weeks after inoculation had been measured. Symptoms in different hosts were seen such as reducing growth, root, and crown rot, yellowing and wilting of aerial organ and ultimately these symptoms led to death in susceptible hosts and destroyed them. Due to the discussed factors, cantaloupe and, melons were very sensitive hosts, tomatoes were sensitive hosts and beans were relatively resistant hosts, the results indicated sensitivity of used cultivars to the phytophthora in this area. Also in checking the germination percentage of seeds, separated *Phtophthora drechsleri* could affect the germination of melon and cantaloupe seeds, their germination percentage is drastically reducing. In a parallel study, the same research had been done to measure the effect of biofertilizers Trichodermin B and Subtilin for *Phtophthora drechsleri* control in mentioned cultivars, the results indicated positive effects of these two biofertilizers in pathogenic phytophthora control.

Keywords: *Phtophthora drechsleri*, varamin, trichodermin B, subtilin

1. Introduction

Phytophthora has more than 60 species, they were known as the main factors for different diseases in lots of crops and vegetables (Shekari et al., 2006). *Phtophthora drechsleri*, is known as the most important factor for root and crown rot in different places in Iran, sometimes its damage had been reported about 80 percent in some farms in Fars province of Iran (Ghaderi et al., 2011).

Crown and root rot diseases of cucurbits caused by the Oomycete pathogen, *Phytophthora drechsleri* Tucker, have been reported from many cucurbits in Iran and other countries (Ershad, 1992).

in original description of *Phtophthora drechsleri* from Iran, separated the putative *Phtophthora drechsleri* isolates into two cucurbit and non-cucurbit groups (Mostowfizadeh-Ghalamfarsa et al., 2015). *Phytophthora drechsleri* rotting can damage other economically important plants like tomato and alfalfa as well as safflower (Mostowfizadeh-Ghalamfarsa et al., 2015). For the first time *Phtophthora drechsleri* was reported from sugar beet root (Tompkinst et al., 1936). Also this species was reported from cucumber and melon in Varamin, Karaj,

Hamedan, Ghazvin, Isfehan, Yazd, Mahhad, Saveh in 1969, from beet root in Ghazvin in 1971 and 1998 in Kermamshah.

In other searching about root and crown rot of beet in Khorasan in 1998 *Phtophthora drechsleri* were isolated and reported (Shekari et al., 2006). In other researches *Phtophthora drechsleri* Tucker was reported as one of the most important pathogens in Iran that causes the disease in cucurbits, sugar beet and sunflower. Also this pathogen is the cause of death in planted pine in Australia and root rot in safflower and fruit in United States (Mansoori & Banihashemi, 1982).

Mansoori and Banihashemi (1981) studied resistance of different varieties of squash to the *Phtophthora drechsleri*, then they announced that the most sensitive variety was Cucumis melon and Cucurbita pepo variety was the most resistant variety to the *Phytophthora*. El-Helaly et al. (1968) isolated this species of cucumber plants in Egypt with root rot signs in 1981.

Ho et al. (1984) tested on separated *Phytophthora* species of cucumber and they had been identified based on morphological characteristics called *Phtophthora drechsleri*.

In other research *Phtophthora drechsleri*, *Phtophthora cryptogea*, *Phtophthora nicotiana* and *Phtophthora capsici* as productive root rot of cucurbits have been mentioned in Khuzestan province (Shekari et al., 2006). During the study, Shekari et al. (2006) had isolated *Phtophthora drechsleri*, *Phtophthora cryptogea*, *Phtophthora nicotiana* and *Phtophthora capsici* species from a number of crops and vegetables in East Azerbaijan province.

Due to huge losses of this species to the farms and greenhouses, its controlling by chemical pesticides requires spending too much expense (Heidari Faroughi et al., 2005). The alternatives of biological control methods are appropriate, because the continued usage of these pesticides cause environmental pollution and pathogen resistance. Among the types of bio-control methods, the use of *Trichoderma* spp. has been attracted so many researchers due to control a wide variety of plant diseases and enhance the growth of some products (Subash et al., 2014).

Trichoderma spp. is an imperfect fungi in an order of *Hypocreales* that can do all three biological decomposition mechanisms of reproductive organ, survival, reproduction, expelling pathogens from chaff, and preventing of residue contamination source, as a result it can be used as a biological fungicide, plant growth promoter and activity of beneficial micro organisms (Behboudi et al., 2005). As an example, one of the bio-control agents is *T. harzianum* that is used to control lots of *Phytophthora* species diseases in commercial plants (Subash et al., 2014).

Determining the host range of pathogens often is essential to control and reduce the damages that caused by them. Damping off in cucurbit plant that caused by *Phtophthora drechsleri* is an important disease, it has extensive host range, and it causes massive damages on its host that many searching have not been done on the host range of that in Iran.

The aims of this study are determining the potential host of this fungus, comparing with sensitive and resistant host and studying about the effect of biological fertilizers Trichodermin B and Subtilin to control it.

2. Materials and Methods

2.1 Separation and Identification of Phytophtora Species

Studied Species were separated from Aliabad and Ghareh tape in Varamin by method of baiting (Grimm & Alexander, 1973) and it was identified in morphological method by valid keys.

2.2 Preparing of Phytophthora Zoospore

Some sterilized cannabis seed were put on the seven-day pure culture of phytophthora, after 24 hours, cannabis seeds were put in other sterile Petri dishes containing 20 ml of distilled water. Dishes were kept in 15 degree of centigrade under fluorescent light in 30 centimeter distance for 24 hours. Petri dishes were put in 20-22 degree of centigrade for hours to exit zoospore in a same time.

2.3 Reaction of Safflower Seedling

Reaction of safflower seedling method had been used to distinguish studied species of *Phytpophthora melonis* (Banihashemi & Mitchell, 1975).

2.4 Planting, Inoculation, and Reaction of Common Cultivars in Varamin

In this study, Urbana tomato cultivar, Samsoori cantaloupe cultivar, Ivanaki melon cultivar, Mahalli red been cultivar were used to determine host range of separated *Phtophthora drecsleri* fungi. Seeds of mentioned plants were prepared and they were disinfected by one percent sodium hypochlorite. Then the seeds had been planted in mixture of sterilized soil of farm, sand and peat soil in ratio of 1:1:2 in pots.

After the tow leaves stages, three seedlings had been selected in each pot and others had been omitted. After that 50 ml of a zoospore suspension of separated *Phtophthora drechsleri* to concentration of 1×10^5 per ml, was added to the soil around the crown. Distilled water had been added to control pots and then adapted plants had been irrigated with water requirement. To determine the percentage of germination of the plants, potting soil contaminated with prepared suspension then 4 seeds were planted in each pot. All the pots were kept in a greenhouse at temperature of 28±1 in days and 22±1 at nights.

In the period of 1, 2 and 3 weeks after inoculation, the plants had been sampled and the samples to measure specific factors had been transported to the laboratory. In order to determine the authenticity of disease percentage, fragments of their root and crown had been cultured in specific mediums as well as the percentage of each plant infection was determined by the level meter device. Fresh and dry weight of aerial organ and root and their length were measured. After reviewing the degree of contamination in the samples, the degree had been awarded to them that was 0 to 4 based on the severity of the disease. Number 0 means no disease; one means 1 to 25 percent of disease in host organ, 2 shows 25 to 50 percent of disease in host organ, 3 indicates 50 to 75 percent of disease in host organ and 4 shows 75 to 100 percent of disease in host organ.

For each treatment, 4 times of repetition were considered for both tests of determining host range and germination percentage. This test had been done completely random.

2.5 To Study Biological Control of Fertilizers, Subtilin and Trichodermin B

The effect of biological fungicide Trichodermin B and Subtilin had been evaluated with fungicidal properties against separated *Phytophthora drechsleri* in green house condition. The test was completely random in 4 treatments and 4 repetitions. Treatments had been done as said in below:

One: Phytophthora zoospore without any biological fertilizers (control);

Two: Phytophthora zoospore in a pot contains Subtilin;

Three: Phytophthora zoospore in a pot contains Trichodermin B;

Four: Phytophthora zoospore in a pot contains both fertilizers.

Seeds should be a little wet for mixing them with biological fertilizers, after adding Trichodermin B (powder in particles of 120 microns) and the required amount of Subtilin (nearly 10 to 15 kg for each kilogram of seed) mixed them so that fertilizer covered seeds then seeds were planted. Also by preparing suspension 2% and irrigation plants by it during the period of growth the effect of fertilizer in pathogenic control was studied after a week of inoculation by 50 ml of zoospore suspension of separated *Phytophthora drechsleri* with the concentration of 1×10^5 per ml. Infection of each treatment had been studied by using scoring system of 0-4. Evaluation of treatment effects with determining infection percentage and disease severity and performance calculation had been done.

Statistical analysis of data was done by Mini-Tab and JMP-9 software and comparison of data were performed with LSD test. Excel 2013 software was used for drawing diagrams of the results.

3. Results

3.1 Reaction of Safflower Seedling

The results of infection of safflower seedlings indicated that separated relevance was *Phtophthora drechsleri* because it had ability to infect seedlings and all seedlings showed dieback signs. Mirtalebi and Banihashemi (2006) announced the resistance of all safflower cultivars to the *Phytpophthora melonis* so that mentioned separated, *Phtophthora drechsleri* was confirmed.

3.2 Germination

In terms of effect on germination of seeds, *Phtophthora drechsleri* could reduce the amount of 31.25% in generation percentage compared to the control in Samsoori cultivar cantaloupe, the lowest amount of germination percentage reducing had been seen in Mahalli cultivar red bean, this ratio was 6.25%. Germination percentage reducing, in Ivanaki cultivar melon was 25% compared to the control that showed nearly extreme reducing of seeds germination in cucurbits compared to the other plants (Figure 1).

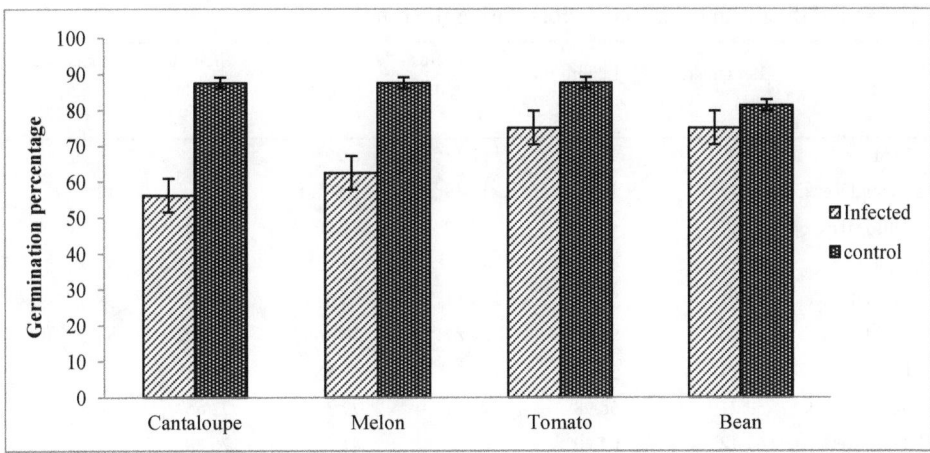

Figure 1. Comparing seeds germination percentage of different plants in control and infected soil to the *Phtophthora drechsleri*

3.3 Host Rang

In host range studied, plants showed many differences to each other and data had the significant difference in 1% level (Table 1). So that disease progressed so fast in Samsoori cultivar cantaloupe, disease percentage was 79.80% after passing 3 weeks of inoculation but results showed the low speed of disease progress in Urbana cultivar tomato and the lowest speed of disease progress was seen in Mahalli cultivar bean after 3 weeks.

For disease percentage, level meter device reported number 8.05% of disease compared with the control for been (Figure 2), in cantaloupe and melon plant growth was so poor and slow even in some samples no growth in root and stem. In some samples, roots were atrophied because of increasing disease so that they could not continue to their growth. Weight and percent of dry root and stem had reduced (Table 2) in plants with high disease percentage.

Table 1. Analysis of variance of measured mean square factors during the time

Sources changes	DF	Mean squares						
		Length stem	Length root	Stem fresh weight	Root fresh weight	Stem dry weight	Root dry weight	%Diseases
Plant	3	4171.667**	4328.031**	7312.071**	8596.564**	7147.294**	7366.099**	4455.115**
Error (main factor)	8	1.209	2.088	1.459	1.535	1.025	1.102	0.546
Time	2	249.357**	175.976**	305.775**	440.434**	232.098**	333.152**	2603.222**
Plant × Time	6	20.147**	6.477**	32.433**	49.148**	19.771**	36.812**	632.952**
Error (Sub)	16	0.487	0.498	0.57	0.46	0.68	0.46	0.63

Table 2. Comparison of the mean measured factors during the time

Plant	Time	Length stem (%)	Length root (%)	Stem fresh weight (%)	Root fresh weight (%)	Stem dry weight (%)	Root dry weight (%)	%Diseases
Bean	First week	9.69[l]	9.84[k]	11.42[k]	8.20[k]	7.97[j]	11.96[k]	6.27[i]
	Second week	11.49[k]	11.13[j]	12.49j[k]	8.89[k]	9.95[i]	13.37[j]	9.05[h]
	Third week	13.17[j]	13.82[i]	13.77[j]	11.30[j]	11.20[i]	14.70[i]	8.05[h]
Melon	First week	33.19[f]	46.16[e]	47.51[f]	53.42[f]	50.92[e]	51.98[e]	20.70[e]
	Second week	35.91[e]	50.01[d]	54.51[e]	66.51[d]	55.74[d]	58.00[d]	44.17[d]
	Third week	47.06[d]	54.60[c]	64.87[d]	70.16[c]	63.68[c]	64.20[c]	70.08[b]
Cantaloupe	First week	55.74[c]	53.40[c]	67.40[c]	62.48[e]	64.01[c]	63.77[c]	20.78[e]
	Second week	59.21[b]	58.98[b]	73.31[b]	77.38[b]	72.65[b]	78.56[b]	53.17[c]
	Third week	67.12[a]	64.40[a]	80.27[a]	81.19[a]	76.88[a]	82.02[a]	79.80[a]
Tomato	First week	18.26[i]	21.65[h]	18.27[i]	18.86[i]	22.10[h]	20.31[h]	7.90[h]
	Second week	21.73[h]	24.79[g]	22.79[h]	23.86[h]	25.69[g]	24.73[g]	13.08[g]
	Third week	25.24[g]	28.82[f]	26.01[g]	27.35[g]	28.38[f]	28.70[f]	15.41[f]

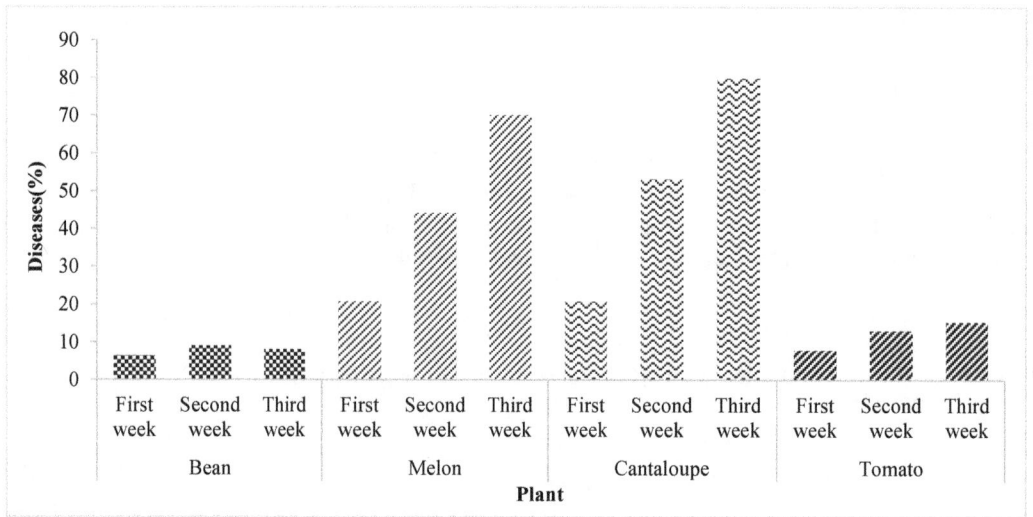

Figure 2. The percentage of disease changes in different plants during 3 weeks

3.4 The Effect of Fertilizers on Control of Separation

The study factors in the biological control test had shown significant difference at 1% level in all of the plants (Table 3). Collection and final evaluation of the results had done with determining of infection percent and evaluation of performance.

The results showed (Table 4) there was a meaningful difference in treatments in terms of infection amount and performance compared with the control and fertilizers was the reason for reducing infection and decreasing product performance compared with the control. The results indicated positive effect of biological fertilizers on disease control of *Phytophthora drechsleri* and the most effective one was treatment of using both fertilizers together.

Table 3. Analyzing variance of mean squares results of biological control by fertilizers

Sources changes	DF	Mean squares			
		Length stem	Stem fresh weight	Length root	Root fresh weight
Fertilizer treatment (safflower)	3	46.19000**	0.0284750*	8.110000**	0.2510570**
Error	8	0.0625	0.005025	0.08000	0.007408
Fertilizer treatment (melon)	3	136.1475**	1.471453**	20.22000**	0.7118750**
Error	8	0.092	0.00404	0.0600	0.002250
Fertilizer treatment (tomato)	3	7.347500**	0.6962000**	8.827500**	0.0789667**
Error	8	0.09750	0.005783	0.06500	0.008208
Fertilizer treatment (bean)	3	34.74750**	8.106600**	109.3875**	2.098608**
Error	8	0.1050	0.00543	0.088	0.01198
Fertilizer treatment (cantaloupe)	3	54.32750**	3.947800**	9.587500**	1.878608**
Error	8	0.0925	0.00930	0.09000	0.00702

Table 4. Comparing the mean results of biological control of the pathogen

plant	Fertilizer treatment	Length stem (cm)	Stem fresh weight (gr)	Length root (cm)	Root fresh weight (gr)	Stem disease degree*	Root and crown disease degree*
Safflower	control	18.2^d	2.22^b	6.4^c	0.92^c	3	3
	subtilin	19.6^c	2.34^{ab}	7.2^b	1.14^b	2	2
	Trichodermin	20.2^b	2.40^a	7.6^b	1.23^b	1	1
	Trichodermin+subtilin	27.0^a	2.44^a	10.2^a	1.61^a	1	1
Melon	control	12.5^d	1.14^b	3.7^d	0.16^c	4	4
	subtilin	13.4^c	1.16^b	5.2^c	$0.24b^c$	3	3
	Trichodermin	18.4^b	1.22^b	6.1^b	0.30^b	3	3
	Trichodermin+subtilin	27.2^a	2.57^a	9.8^a	1.20^a	1	1
Tomato	control	5.9^c	0.82^c	3.8^d	0.73^b	3	3
	subtilin	6.3^c	1.17^b	4.4^c	0.75^b	2	2
	Trichodermin	7.3^b	1.28^b	5.4^b	0.98^a	2	2
	Trichodermin+subtilin	9.4^a	1.97^a	7.7^a	1.05^a	1	2
Bean	control	23.5^d	2.44^d	15.7^d	1.40^d	2	2
	subtilin	26.9^c	3.11^c	19.6^c	1.91^c	2	1
	Trichodermin	28.4^b	4.09^b	21.4^b	2.86^b	1	1
	Trichodermin+subtilin	31.7^a	6.21^a	30.0^a	3.21^a	0	0
Cantaloupe	control	7.9^d	0.87^d	5.7^d	0.52^c	4	4
	subtilin	13.8^c	2.53^c	7.3^c	1.08^b	2	2
	Trichodermin	14.4^b	2.80^b	8.1^b	1.16^b	2	3
	Trichodermin+subtilin	18.2^a	3.60^a	10.0^a	2.40^a	2	2

Note. *Grading the severity of infection: 0 = no disease, 1 = 1 to 20 percent of disease in host organ, 2 = 25 to 50 percent of disease in host organ, 3 = 50 to 75 percent of disease in host organ and 4 = 75 to 100 percent of disease in host organ.

4. Discussion

Based on measured factors and the percentage of virulence in studied plants, it could be said that all cultivated cultivars in Varamin are sensitive to the separated *Phytophthora drechsleri* in the same place, that distribution area of the region is high. Mehrabi Koshki et al. (2007) had done evaluation of the effect of two biological products, Trichodermin B and Subtilin, on the wheat take-all disease control. The searches showed Terchodermin B had the most biocontrol effects on separated pathogen. Since in Mehrabi Koshki et al. (2007) research there was no combination of two products, therefore in this study have been seen the effect of both products on biological control of separated *Phytophthora drechsleri* had the most effects and it could be good option for usage, however using resistant cultivars seem the best way to control. For this reason, it is recommended that studying about the amount of intended separated effects on different cultivars of each plants are planted in the area could be the next research.

5. Conclusion

Achievements of these study shows that not only searching about the resistance of different cultivated varieties of a same place and selection among them can be a good solution in building combat with pathogen, but also biological fertilizers could be safe and secure alternative to chemical fertilizers and pesticides. In addition, we can help to increase level of soil fertility by continuous and timely usage of these biological products.

References

Banihashemi, Z., & Mitchell, J. E. (1975). Use of safflower seedlings for the detection and isolation of *Phytophthora cactorum* from soil and its application to population studies. *Phytopathology, 65*(12), 1424-1430. http://dx.doi.org/10.1094/Phyto-65-1424

Behboudi, K., Sharifi Tehrani, A., Hejaroud, G. A., & Zad, J. (2005). Antagonistic effects of Trichoderma species on *Phytophthora capsici*, the causal agent of pepper root and crown rot. *Iranian Journal of Plant Pathology, 41*(3), 345-362.

El-Helaly, A. F., Assawah, M. W., Elarosi, H. M., & Wasfy, E. E. H. (1968). Fruit rots of vegetable marrow in Egypt (United Arab Republic). *Phytopathologia Mediterranea, 7*(2/3), 107-115.

Ershad, J. (1992). *Phytophthora species in Iran* (p. 217). Agricultural Research Organization, Tehran, Iran.

Ghaderi, F., Askari, S., & Abdollahi, M. (2011). *Isolation and identification of phytophthora species on greenhouse cucumber in southern kohgiluyeh and boyerahmad province and evaluation of relative resistance of greenhouse cucumber cultivars to them.*

Grimm, G. R., & Alexander, A. F. (1973). Citrus leaf pieces as traps for *Phytophthora parasitica* from soil slurries. *Phytopathology.* http://dx.doi.org/10.1094/Phyto-63-540

Heidari Faroughi, S., Etebarian, H. R., & Zamanizadeh, H. R. (2005). Evaluation of Trichoderma isolates for biological control of *Phytophthora drechsleri* in glasshouse. *Applied Entomology and Phytopathology, 72*(2), 113-134.

Ho, H. H., Jiayun, L., & Longyin, G. (1984). *Phytophthora drechsleri* causing blight of Cucumis species in China. *Mycologia, 76*(1), 115-121. http://dx.doi.org/10.2307/3792842

Mansoori, B., & Banihashemi, Z. (1982). Evaluating cucurbit seedling resistance to *Phytophthora drechsleri*. *Plant Diseases (USA), 66*(1), 373-376. http://dx.doi.org/10.1094/PD-66-373

Mehrabi Koushki, A., Zafari, D. M., Rouhani, H., & Ghalandar, M. (2007). Effectiveness of trichoderma isolates, mustard flour and two biologic commercial products for biocontrol of wheat take-all. *Journal of Agricultural Science (University of Tabriz), 17*(3), 197-208.

Mirzaiian, A., Pahlevani, M., Soltanloo, H., & Razavi, S. E. (2015). Improving field establishment of safflower in soils infected by *Phytophthora drechsleri* and *Pythium ultimum*. *International Journal of Plant Production, 9*(1).

Mostowfizadeh-Ghalamfarsa, R., & Banihashemi, Z. (2015). A revision of Iranian *Phytophthora drechsleri* isolates from Cucurbits based on multiple gene genealogy analysis. *Journal of Agricultural Science and Technology, 17*(5), 1347-1363.

Shekari, A., Mirabolfathi, M., Mohammadi-Pour, M., Zad, J., & Okhovvat, S. M. (2006). Phytophthora root and crown rot of several field and vegetable crops in East Azarbaijan Province. *Iranian Journal of Plant Pathology, 42*(2), 293-308.

Subash, N., Meenakshisundaram, M., Sasikumar, C., & Unnamalai, N. (2014). Mass cultivation of *trichoderma harzianum* using agricultural waste as a substrate for the management of damping off disease and growth promotion in chilli plants (*Capsicum annuum* L.). *Inter. J. Phar. and Phar. Sci., 6*(5), 188-192.

Sharma, P. D. (2003). Epidmiology assessment and forecasting of plant disease. *Plant Pathology* (2nd ed., pp. 91-106). Rastogi Publications, India.

Tompkins, C. M., Tucker, C. M., & Gardner, M. W. (1936). Phytophthora root rot of cauliflower. *Journal of Agricultural Research, 53*(9), 685-692.

9

Antixenosis Studies in Different Genotypes of Bitter Gourd Fruits against the Infestation of Melon Fruit Fly

Paras Nath[1], A. K. Panday[2], Akhilesh Kumar[2] & Hemalatha Palanivel[1]

[1] College of Agriculture, Fisheries and Forestry, Fiji National University, Koronivia, Fiji

[2] Department of Entomology and Agricultural Zoology, Institute of Agricultural Sciences, Banaras Hindu University, Varanasi, India

Correspondence: Hemalatha Palanivel, College of Agriculture, Fisheries and Forestry, Fiji National University, Koronivia, Fiji. E-mail: hemalatha.palanivel@fnu.ac.fj

Abstract

In crop plants, three principle mechanism viz., non-preference (antixenosis), antibiosis and tolerance are responsible for imparting resistance to insects. Non preference denotes a group of plants character and insect response that keep away the insect from using a particular plant or variety for oviposit ion, food, shelter, or combination of both. Keeping in this view, seventy four promising genotypes of bitter gourd were screened against fruit fly infestation to identify the antixenosis traits involved in host plant selection by the melon fruit fly. On the basis of percent fruit infestation and the average number of larvae per damage fruit, the genotypes were categorized in to different groups i.e. (Highly resistant, resistant, moderately resistant, susceptible, and highly susceptible). The fruit infestation during the 2006 and 2007 summer season (Average of two years) ranged from 13.23% to 83.75%. Larval density per fruit ranged from 2.59 to 8.13 larvae per fruit. The larval density increased with the increase in percent fruit infestation and showed significant positive correlation (r = 0.98).The depth of ribs in resistant genotypes was higher as compared to susceptible genotypes. The number of seed was recorded maximum in the genotype VRBT-96 (31.8) and minimum was recorded in the genotype IC 68314(14.8). The fruit toughness had significant negative effect on fruit infestation(r = -0.52) and larval density (r = -0.57) of fruit fly. There was a strong correlation between number of ribs and fruit toughness, and these traits can be used as markers to select bitter gourd genotype resistant to melon fruit fly.

Keywords: genotypes, bitter gourd fruits, infestation, melon fruit fly

1. Introduction

The vegetables form an essential component of the human diet especially in the case of India and some Southeast Asian countries where sizable population basically consists of vegetarians. In the vegetable kingdom, cucurbitaceous crops occupy the major share in terms of kind of total crops grown, area covered, crops produced and consumed worldwide (Nath, 1965, 2008). In terms of nutritive value, bitter gourd (*Momordica charentia* L.) ranks first among cucurbits, being rich in iron, phosphorus and ascorbic acid (Awasthi & Jaiswal, 1986). It is a popular vegetable cultivated all over the world especially in India, Pakistan, Srilanka and China (Panday et al., 2008).

In India, it is one the most important and round the year cultivated popular vegetable crops grown for its immature fruits which are consumed as stuffed-fried and in many other ways. It has immense medicinal properties due to the presence of beneficial phytochemicals which are known to have antibiotic, antimutagenic, antioxidant, antiviral, antidiabetic and immune enhancing properties (Grover & Yadav, 2004). A compound known as charantin, present in the bittergourd is used in the treatment of diabetes in reducing blood sugar level (Lotlikar et al., 1966). Bitter gourd has good export potential and its share in export of green vegetable is to the extent of 20 per cent (Anonymous, 1992). Insect pests are a major constraint for increasing the production and productivity of this crop. Among them the fruit fly is one of the most destructive insect-pest (Panday et al., 2008). The melon fruit fly has been observed on 81 host plants but bitter gourd is one of the most preferred hosts and has been a major limiting factor in attaining good quality fruits and high yield (Panday et al., 2009; Lall & Singh, 1969). Depending on the environmental conditions and susceptibility of the crop species, the extent of losses varies between 30 to 100% (Dhillon et al., 2005a, 2005b, 2005c; Panday et al., 2009).

The melon fruit fly is active throughout the year except for a short period from December to mid February (due to excessive cold). It prefers young, green and tender fruits as compared to bigger ones with hard rind for egg laying (Narayan, 1953). When eggs are hatched into maggots, they tunnel deep into fruit pulp and the entire fruits get spoiled.

There is no satisfactory chemical control of melon fruit fly, however, some pesticides such as malathion, dichlorovos, phosphamidon and endosulfan partially control its attack (Agarwal et al., 1987). The control measures adopted rely mainly on contact poison or baits (Gupta & Verma, 1979; Lee, 1988). Contact poisons may have serious deleterious effect on health as fruits in India and other developing countries are consumed raw, often unwashed. The fruits of bitter gourd of which the melon fruit fly is a serious pest, are picked up at short interval for marketing and self consumption. Therefore, it is difficult to rely on insecticides as a means of controlling this pest (Panday et al., 2008). Hence development of resistant varieties is the most effective and cheapest method of controlling this pest. To identify the source of resistance, screening of bitter gourd germplasms against this pest is a pre-requisite to achieve this objective. The identification of morphological factors governing resistance is helpful in the development of rapid screening technique. The mechanism of resistance may be antixenosis, antibiosis and tolerance. The bitter gourd varieties having such inhibitory mechanism of resistance to melon fruit fly can be used in transferring the resistance in to the commercially acceptable varieties. Even partially resistant cultivars may also provide adequate control even with minimum usage of insecticides. It will help to prolong the useful commercial life of existing insecticides by discouraging the development of insecticides resistance strains of the insect. More recent research has found, for instance, that higher concentrations of glucosinolates and greater densities of trichomes in Arabidopsis thaliana reduced herbivory by two flea beetle species (Mauricio, 1998). These traits can also evolve as adaptive defences since there exists heritable variation for glucosinolate and trichome levels, and herbivores selected for an increase in these levels (Mauricio & Rausher, 1997). We lack an understanding of the traits that are most strongly associated with resistance against this pest, and the relative importance of different types of traits involved in defense (Stamp, 2003). Keeping in view the above facts, the present work entitled "Antixenosis studies in different genotypes of bitter gourd against the infestation of melon fruit fly" was under taken.

2. Materials and Methods

Morphological resistance factors interfere physically with mechanisms of feeding and oviposition. Size and shape of bitter gourd fruits may affect the egg laying and developmental behaviour of fruit fly (Boller & Prokopy, 1976). Such morphological factors which may provide resistance to melon fruit fly were investigated so that it could be possible to develop rapid screening techniques based on that factors. The experiments were conducted in the summer seasons of the year 2009 and 2010, to screen different genotypes of bitter gourd against the infestation of fruit fly. The fruit infestation was calculated based on the fly ovipuncture is visible. Sometimes it is difficult to view the ovipuncture, then the fruit rearing was done for a week to check the presence of larvae in the harvested fruit. All the damaged fruits were cut opened and count the number of larvae per fruit and averaged. The observations on fruit damage were recorded at weekly intervals from one week of fruit initiation stage till harvest the crop. All the genotypes were grouped in to different categories on the basis of percent fruit damage and number of larvae per fruit. Plant herbivore interactions are influenced by several morphological and biochemical plant traits, environmental conditions and physiological status of the tests insects (De- Ponti, 1977).

In the present study, morphological factors such as fruit length, fruit diameter, fruit flesh thickness, fruit skin thickness, ridge depth, number of ridge (cm^{-2}) area, number of seeds fruit^{-1}, and fruit toughness, of different genotypes of bitter gourd were studied with a view to understand their contribution in imparting resistance or susceptibilities nature of bitter gourd plant. The fresh fruits were harvested from the experimental plots. Out of the harvested fruits of various germplasms, five fresh, healthy and average size fruits were selected to record the morphological characters. The observations on fruit diameter, were measured with the help of Verniercaliper. Depth of ribs were measured by cut the fruits longitudinally and depth of ribs was measured on two opposite sides and was averaged. All the fruits used for observation on depth of ribs were also used for measuring the skin thickness. Skin thickness was measured at two opposite points with the help of Verniercaliper in mm and averaged for five fruits. Intensity of small ribs in number was measured by counting the number of ribs in one cm^2 area and average of five fruits was taken for study. The fruits were cut longitudinally and the number of seeds per fruit was counted. Toughness of the fruit was tested by penetrometer/pressure tester (Ogawa Seiki Co. Ltd. Tokyo, Japan) and was expressed in terms of kg per cm^2. Pressure tester was placed on surface of the fruit and was pressed till the fruit surface ruptured and at this point pressure was noted and as such observation on five randomly selected fruits was averaged for recording the toughness/firmness. The percent fruit damage and number of larvae were correlated with different morphological factors to know the role of different morphological traits in imparting resistant or susceptibility in bitter gourd fruits. The data on percentage fruit infestation and larval density per fruit and biochemical fruit traits were analyzed through one-way ANOVA using SPSS software (O'Connor, 2000).

3. Results and Discussion

3.1 Fruit Infestataion

In crop plants, three principle mechanism viz., non-preference (antixenosis), antibiosis and tolerance are responsible for imparting resistance to insects (Painter, 1968). Non preference denotes a group of plants character and insect response that keep away the insect from using a particular plant or variety for oviposition, food, shelter, or combination of both. Keeping in this view, seventy four promising genotypes of bitter gourd were screened against fruit fly infestation to identify the antixenosis traits involved in host plant selection by the melon fruit fly. On the basis of percent fruit infestation and the average number of larvae per damage fruit, the genotypes were categorized in to different groups *i.e.* (Highly resistant, resistant, moderately resistant, susceptible, and highly susceptible). There were significant differences in percent fruit infestation and larval density per fruit among the genotypes tested across the years. The fruit infestation during the 2009 and 2010 summer season (Average of two years) ranged from 13.23% to 83.75% while larval density per fruit ranged from 2.59 to 8.13 larvae per fruit.

Table 1. Grouping of bitter gourd germplasms/lines/genotypes in to different categories on the basis of percent fruit infestation in summer season (average of two years)

Sr. No.	Percent fruit infestation	No. of larvae/fruit	Reaction	No. of genotypes	Genotypes
1	1-10	4.3	Highly resistant	0	-
2	11-20	4.9	Resistant	5	IC-248282, Kerala collection-1, VRBT-4, DRAR-1, IC-68314
3	21-50	5.0	Moderately resistant	61	VRBT-6, VRBT-7, VRBT-11, VRBT-14, VRBT-28, VRBT-29, VRBT-31, VRBT-32, VRBT-35, VRBT-37, VRBT-39, VRBT-41, VRBT-46, VRBT-54, VRBT-63, VRBT-68, VRBT-69, VRBT-71, VRBT-72, VRBT-73, VRBT-75, VRBT-76, VRBT-77, VRBT-83, VRBT-85, VRBT-86, VRBT-87, VRBT-90, VRBT-91, VRBT-92, VRBT-94, VRBT-95, VRBT-96, VRBT-98, VRBT-99, VRBT-100, VRBT-103, VRBT-107, VRBT-113, VRBT-115, VRBT-128, VRBT-135, VRBT-139, VRBT-145, VRBT-147, VRBT-178, VRBT-179, VRBT-187, VRBT-188, DVBTG-1, DVBTG-2, DVBTG-5, DVBTG-7, NDBT-1, Long green, Pusa vishesh, Hirkani, US-6201, Jhalri baramasi, BL-237, Konkan Tara
4	51-75	6.1	Susceptible	5	VRBT-21, VRBT-22, VRBT-38, VRBT-93, VRBT-175
5	76-100	8.1	Highly Susceptible	3	Jaunpuri, Arka harit, Pusa Do Mausmi

The lowest fruit infestation was observed in the genotypes IC 248282, Kerala collection-1, VRBT-4, DRAR-1 and IC 68314 these were grouped as resistant genotypes while 61 genotypes as moderately resistant, 5 genotypes as susceptible and 3 genotypes highly susceptible. Painter (1951) emphasized the need to identify sources of resistance to target pests, followed by identification of physico-chemical factors involved in host plant selection by the insects, both for oviposition and feeding (Maxwell & Jennings, 1980). The variety Pusa Do Mausami was most susceptible (having 81.57% fruit infestation) to the attack of this pest followed by Arka harit (78.17%) and Jaunpuri (76.21%). None of the genotypes were found highly resistance, out of 74 genotypes.

3.2 Morphological Traits

The difference in the length of bitter gourd fruits of various genotypes were observed significant. However maximum fruit length was observed in the genotype VRBT-11 (19.4 cm) followed by VRBT-6 (18.2 cm) while minimum fruit length was observed in the genotype VRBT-94 (6.18 cm) and VRBT-73 (6.22 cm). The effect of fruit length was positive on fruit infestation and larval density of fruit fly but the correlation coefficient was not found significant. The fruit diameter was recorded maximum in case of genotype VRBT-187 (4.28 cm) while the minimum fruit diameter was recorded in genotype VRBT-4 (2.2 cm) followed by IC 248282 (2.56 cm). The fruit diameter showed significant positive effect on fruit infestation and larval density of fruit fly. The fruit flesh thickness was recorded maximum in the genotype VRBT-187 (3.72 cm) and minimum flesh thickness was recorded in the genotype VRBT-4 (1.86 cm). Fruit skin thickness was maximum in case of the genotype PDM (0.62 mm) and minimum in the genotypes IC 248282 followed by IC 68314 (0.26 mm). The fruit diameter, fruit flesh thickness and skin thickness of resistant genotype was lesser as compared to susceptible and highly susceptible genotypes. Shape of the fruit influences the orientation of fruit flies to a potential ovipositional site (Boller & Prokopy, 1976). Pal et al. (1984) also found thick and tough rind fruits of IHR 89 and IHR 213 genotypes resistant to melon fruit fly. The fruit infestation is reported to increase with an increase in fruit length and diameter (Jaiswal et al., 1990; Tewatia et al., 1997).

The number of ridge per cm^{-2} area was maximum in the genotype Harikani (31.2 cm^{-2}) followed by IC 248282 (31.0 cm^{-2}) while it was minimum in the genotype VRBT-37 (7.6 cm^{-2}). The number of ridge per cm^{-2} area had negative effect on fruit infestation and larval density of fruit fly but the correlation coefficient was found non-significant. The ridge area of resistant genotypes was higher as compared to susceptible genotypes. The highest depth of fruit rib was recorded in the genotypes Jaunpuri (0.82 mm) and lowest was recorded (0.33 mm) in the genotypes IC 248282 and DRAR 1. The depth of ribs in resistant genotypes was higher as compared to susceptible genotypes. The number of seed was recorded maximum in the genotype VRBT-96 (31.8) while the number of seed was recorded minimum in the genotype IC 68314 (14.8) followed by IC 248282 (15.0). The number of seeds had positive effect on fruit infestation and larvae density of fruit fly. The fruit toughness varied significantly from each other. The highest fruit toughness was recorded in the genotype IC 248282 (9.42 kg/cm^2) followed by Kerala collection-1 (8.81 kg/cm^2). While minimum fruit toughness was recorded in case of genotype VRBT-93 (7.19 kg/cm^2) followed by VRBT-38 (7.2 kg/cm^2). The fruit toughness had significant negative effect on fruit infestation and larval density of fruit fly. Fruit toughness of resistant genotypes was higher as compared to susceptible and highly susceptible genotypes. Similar result is reported by Chelliah and Sambandam (1971) they have reported that egg laying by the melon fruit fly is only 11.77 percent of the fruits having tough rind in *C. callosus*, while egg laying by the fruit fly was recorded as high as 87.33 percent fruits in the susceptible variety Delta gold. Similarly, resistance to squash vine borer in *Cucurbita* spp. has also been reported dew to tough vascular bundles (Howe, 1949). Pal et al. (1984) also found thick and tough rind fruits of IHR 89 and IHR 213 genotypes resistant to melon fruit fly.

Table 2. Morphological characters of the fruits of bitter gourd genotypes screened against fruit fly infestation during summer sseason of the year 2009 and 20010 (Average of two years)

Genotypes	Fruit damage (%)	No. of larvae/ fruit	Fruit length (cm)	Fruit diameter (cm)	Fruit flesh thickness (cm)	Fruit skin thickness (mm)	Ridge area (cm^2)	Rib depth (mm)	No. of seeds	Fruit toughness (kg/cm^2)
VRBT-4	18.76	3.36	8.34	2.24	1.90	0.34	17.4	0.5	18.4	8.47
VRBT-6	31.89	4.35	17.44	3.53	3.01	0.52	15.6	0.52	29.5	7.32
VRBT-7	22.78	3.42	13.88	3.29	2.92	0.37	15.6	0.49	25.9	8.425
VRBT-11	29.01	4.10	19.01	3.54	3.03	0.51	12.6	0.56	31.5	7.5
VRBT-14	31.34	4.69	7.63	2.65	2.16	0.49	10.2	0.51	18.0	7.7
VRBT-21	67.01	6.74	8.26	3.65	3.19	0.46	18.7	0.52	18.8	7.305
VRBT-22	56.29	6.14	10.38	3.17	2.62	0.55	15.1	0.59	20.4	7.415
VRBT-28	31.15	4.68	14.12	3.29	2.78	0.51	13.1	0.57	22.8	7.495
VRBT-29	36.71	4.91	16.88	4.17	3.62	0.55	12.9	0.54	28.8	7.485
VRBT-31	26.85	4.25	8.33	3.6	3.11	0.49	19.1	0.53	18.5	8.28
VRBT-32	31.19	4.63	12.79	3.59	3.06	0.53	11.9	0.56	24.1	7.585
VRBT-35	46.49	5.65	7.25	4.1	3.54	0.56	10.3	0.53	19.1	7.36
VRBT-37	36.12	5.08	10.61	3.52	3.03	0.49	7.2	0.50	21.3	7.465
VRBT-38	57.33	6.07	16.95	4.09	3.59	0.50	10.3	0.54	30.2	7.20
VRBT-39	32.43	4.71	10.83	3.67	3.23	0.44	13.8	0.51	21	7.92
VRBT-41	26.16	4.27	10.65	3.30	2.77	0.53	13.7	0.49	18.6	8.225
VRBT-46	30.31	4.53	8.05	3.56	3.09	0.47	11.8	0.55	17.4	7.84
VRBT-54	26.28	4.09	13.54	3.36	2.88	0.48	13.7	0.57	23.5	8.275
VRBT-63	31.54	4.67	10.16	2.74	2.30	0.44	10.8	0.50	18.9	8.11
VRBT-68	30.26	4.48	11.00	3.68	3.19	0.49	12.1	0.53	21.3	7.735
VRBT-69	30.64	4.57	9.38	3.12	2.76	0.36	17.2	0.44	18.9	7.64
VRBT-71	25.31	3.97	7.89	3.55	3.07	0.48	14.2	0.53	18.5	8.225
VRBT-72	30.24	4.57	6.44	3.20	2.74	0.46	10.0	0.51	17.2	8.175

VRBT-73	28.74	4.37	6.82	2.57	2.20	0.37	17.7	0.47	15.9	8.435
VRBT-75	29.74	4.56	13.06	3.47	2.94	0.53	12.3	0.57	22.7	7.545
VRBT-76	28.51	4.46	8.63	3.14	2.65	0.49	17.3	0.47	19.5	8.335
VRBT-77	26.48	4.15	7.34	3.43	3.00	0.43	13.9	0.50	18.2	8.39
VRBT-83	29.37	4.44	8.91	3.07	2.54	0.53	15.9	0.55	18.6	8.415
VRBT-85	37.01	5.03	15.3	3.71	3.14	0.57	16.5	0.54	26.9	7.425
VRBT-86	29.42	4.42	10.7	3.62	3.12	0.5	10.0	0.53	21.3	7.515
VRBT-87	34.67	4.83	17.24	3.45	2.92	0.53	9.40	0.58	28.7	7.39
VRBT-90	32.85	4.80	7.36	3.35	2.81	0.54	16.8	0.51	18.0	7.51
VRBT-91	26.29	3.54	16.51	3.57	3.14	0.43	19.2	0.52	30.4	8.19
VRBT-92	28.61	4.32	6.30	3.24	2.76	0.48	10.6	0.53	17.7	7.82
VRBT-93	55.11	6.16	7.10	3.34	2.89	0.45	14.8	0.48	16.0	7.195
VRBT-94	28.47	4.34	5.98	2.76	2.31	0.45	16.7	0.45	16.9	8.125
VRBT-95	24.53	3.96	13.72	3.56	3.01	0.55	13.0	0.46	24.0	7.45
VRBT-96	31.80	4.54	16.63	3.76	3.25	0.51	13.7	0.52	31.3	7.465
VRBT-98	27.41	4.27	8.71	3.17	2.76	0.41	16	0.43	19.3	8.135
VRBT-99	28.90	4.44	14.52	3.64	3.09	0.55	15.3	0.50	25.5	7.31
VRBT-100	34.68	4.84	8.55	3.63	3.14	0.49	12.7	0.49	17.2	7.5
VRBT-103	27.63	4.31	15.96	3.43	2.98	0.45	13.4	0.45	29.0	8.18
VRBT-107	25.12	4.05	13.68	3.46	2.97	0.49	11.7	0.59	23.8	7.85
VRBT-113	29.13	4.44	10.70	3.01	2.55	0.46	14.0	0.50	18.4	8.11
VRBT-115	34.93	4.92	17.67	3.27	2.76	0.51	12.7	0.53	29.2	7.325
VRBT-128	28.86	4.34	14.40	3.40	2.91	0.49	14.3	0.57	24.2	7.605
VRBT-135	34.33	4.79	10.12	3.40	2.87	0.53	14.5	0.56	19.7	7.305
VRBT-139	29.96	4.58	10.21	3.67	3.15	0.52	14.4	0.52	21.5	7.78
VRBT-145	31.57	4.71	9.53	3.47	2.95	0.52	14.2	0.54	17.5	7.645
VRBT-147	37.27	5.10	17.32	4.15	3.58	0.57	11.5	0.57	30.7	7.305
VRBT-175	65.54	6.68	9.90	3.22	2.73	0.49	11.6	0.55	20.4	7.31
VRBT-178	28.91	4.37	7.95	3.69	3.14	0.55	14.3	0.46	19.7	7.62
VRBT-179	27.92	4.29	10.80	3.42	2.95	0.47	12.0	0.57	20.7	8.3
VRBT-187	30.67	4.65	13.16	4.36	3.81	0.55	11.5	0.60	26.2	7.385
VRBT-188	34.16	4.73	16.34	4.0	3.44	0.56	8.3	0.70	31.4	7.3
DVBTG-1	44.16	5.60	15.52	2.94	2.53	0.41	12.0	0.44	26.0	7.895
DVBTG-2	33.74	4.61	11.03	3.11	2.62	0.49	11.4	0.55	20.4	8.105
DVBTG-5	27.62	4.23	12.81	3.05	2.61	0.44	12.0	0.47	23.5	8.19
DVBTG-7	27.43	4.28	13.57	3.39	2.97	0.42	12.2	0.44	23.2	8.195
DRAR-1	18.12	3.18	10.96	2.79	2.51	0.28	19.3	0.35	17.4	8.455
NDBT-1	40.11	5.20	8.27	3.24	2.81	0.43	14.8	0.60	16.8	7.875
Long green	26.42	4.17	7.63	3.41	3.15	0.26	23.1	0.54	17.1	8.475
Jaunpuri	76.21	7.42	14.48	3.97	3.67	0.30	18.3	0.80	26.5	7.35
Arkaharit	78.17	7.73	9.79	3.67	3.17	0.50	19.9	0.60	18.6	7.6
PusaVishesh	39.48	5.19	10.97	3.61	3.12	0.49	28.8	0.58	21.3	8.34
IC-68314	18.11	3.20	7.40	2.71	2.45	0.26	25.3	0.50	14.8	8.475
IC-248282	13.64	2.74	6.29	2.60	2.32	0.28	31.1	0.38	13.1	9.4
Hirkani	30.91	4.63	10.83	3.81	3.48	0.33	32.1	0.67	24.5	8.565
PDM	81.57	7.84	9.58	3.53	2.92	0.61	18.2	0.64	21	7.525
US-6201	33.87	4.77	7.32	3.18	2.69	0.49	10.8	0.55	17.9	8.4
Jhalribaramasi	26.11	3.91	13.2	3.74	3.45	0.29	18.7	0.64	26.6	7.555
BL-237	26.25	4.12	8.33	3.64	3.34	0.3	27.8	0.52	17.2	7.625
Konkan Tara	24.80	4.09	7.42	3.54	3.25	0.29	29.8	0.56	19.00	7.475
Kerala collection-1	15.68	2.84	10.9	3.21	2.83	0.38	23.2	0.53	26.10	8.8
C.D. (P = 0.05%)			1.06	0.31	0.30	0.09	2.42	0.09	3.49	0.04
SEM ±			0.54	0.155	0.15	0.04	1.23	0.045	1.77	0.015

3.3 Fruit Damage and Larval Density

The larval density increased with the increase in percent fruit infestation and showed significant positive correlation (r = 0.98). Dhillon et al. (2005) was also reported that the genotypes with low fruit fly infestation had low larval numbers in the fruits and there was positive correlation (r = 0.96) between percentage fruit infestation and number of larvae per fruit. The average of two years data on fruit infestation (%) and larval density (fruit^{-1}) showed significant positive correlation with fruit diameter (r = 0.32 and 0.35), flesh thickness (r = 0.27 and 0.28), skin thickness (r = 0.29 and 0.37) and depth of ribs (r = 0.46 and 0.46) and negative correlation with fruit toughness (r = -0.52 and -0.57). The toughness of fruits appeared to play significant role in fruit fly infestation. The length of fruits and number of seeds had no significant impact on fruit fly infestation though they had positive correlation with fruit fly infestation (r = 0.05 and 0.04, 0.07 and 0.05) respectively. The number of ridge showed non-significant negative correlation with fruit fly infestation (r = -0.11 and -0.18). There was a strong correlation between number of ribs and fruit toughness, and these traits can be used as markers to select bitter gourd genotype resistant to melon fruit fly. In the present study there was a significant and positive correlation (r =) between percent fruit infestation and larval density (fruit^{-1}). The fruit infestation (%) and larval density (fruit^{-1}) were positively correlated with flesh thickness, fruit diameter, fruit length, skin thickness and depth of ribs and negatively correlated with fruit toughness. There was a strong correlation between number of ribs and fruit toughness, and similar results have been reported by Dhillon et al. (2005).

Table 3. The correlation coefficients of the bitter gourd fruit damage, larval density and morphological characters of the fruits of different germplasms/lines/genotypes of bitter gourd during summer season of the year 2009 and 2010

Morphological Traits	Fruit damage (%)	Larval density (fruit^{-1})	Fruit length (cm)	Fruit diameter (cm)	Fruit flesh thickness (cm)	Fruit skin thickness (mm)	No of ridge (cm^{-2}) area of fruit	Ridge depth (mm)	No of Seeds (fruit^{-1})	Fruit toughness (kg/cm^2) of fruits
Fruit damage (%)	1.00									
Larval density (fruit^{-1})	0.98**	1.00								
Fruit length (cm)	0.05	0.04	1.00							
Fruit diameter (cm)	0.32**	0.35**	0.42**	1.00						
Fruit flesh thickness (cm)	0.27*	0.28*	0.39**	0.98**	1.00					
Fruit skin thickness (mm)	0.29*	0.37**	0.29**	0.40**	0.20	1.00				
No of ridge(cm^{-2}) area of fruit	-0.11	-0.18	-0.29*	-0.14	0.00	-0.62**	1.00			
Ridge depth (mm)	0.46**	0.46**	0.25*	0.51**	0.51**	0.17	0.00	1.00		
No. of seed (fruit^{-1})	0.07	0.05	0.94**	0.53**	0.50**	0.30**	-0.26*	0.34**	1.00	
Fruit toughness (kg/cm^2) of fruits	-0.52**	-0.57**	-0.38**	-0.57**	-0.49**	-0.53**	0.43**	-0.36**	-0.39**	1.00

4. Conclusion

Development of muskmelon varieties/genotypes resistant to fruit fly has been restricted in India due to inadequate information on the sources of plant traits associated with resistance to pest infestations. Our study was proved that the various morphological traits of bitter gourd of varieties/genotypes fruit traits allied with resistance against melon fruit fly in terms of fruit infestation and larval density under field conditions. The resistance varieties can play a vital role integrated pest management in bitter gourd.

References

Agarwal, M. L., Sharma, D. O., & Rahman, O. (1987). Melon fruit fly and its control. *Hort, 32*(2), 10-11.

Anonymous. (1992). *Report on infrastructure for export of agricultural commodities and processed food* (p. 44). Govt. of India, Planning commission (Agriculture Division), Yojana Bhawan, New Delhi.

Aslam, M., & Stockley, I. H. (1979). Interaction between curry ingredient (Karela) and drug (chlorpropamide). *Lancet, 1*(8116), 607. https://doi.org/10.1016/S0140-6736(79)91028-6

Awasthi, C. P., & Jaiswal, R. C. (1986). Biochemical composition and nutritional quality of fruits of bitter gourd grown in Uttar Pradesh. *Prog. Hort., 18*, 265-269.

Baldwa, V. S., Bhandari, C. M., Pangaria, A., & Goyal, R. K. (1977). Clinical trial in patients with diabetes mellitus of an insulin like compounds obtained from plant source. *Uppasal J. Med. Sci., 82*, 39. https://doi.org/10.3109/03009737709179057

Boller, E. F., & Prokopy, R. J. (1976). Bionomics and management of Rhagoletis. *Annu. Rev. Ent., 21*, 223-246. https://doi.org/10.1146/annurev.en.21.010176.001255

Chelliah, S., & Sambandam, C. N. (1971). Role of certain mechanical factors in *Cucumis callosus* (Rottl.) Cogn. in imparting resistance to *Dacus cucurbitae. AUARA, 3*, 48-53.

De-Ponti, O. M. B. (1977). Resistance in *Cucumis sativus* L. to *Tetranychus urticae* Koch. I. The role of plant breeding in integrated control. *Euphytica, 26*, 633. https://doi.org/10.1007/BF00021688

Dhillon, M. K., Singh, R., Naresh, J. S., & Sharma, N. K. (2005). Influence of physio-chemical traits of bitter gourd, *Momordica charantia* L. on larval density and resistance to melon fruit fly, *Bactrocera cucurbitae* (coquillent). *J. Applied Entomology, 129*(7), 395-399. https://doi.org/10.1111/j.1439-0418.2005.00911.x

Fletcher, B. S. (1987). The biology of Dacine fruit flies. *Annu. Rev. Entomol., 32*, 115-144. https://doi.org/10.1146/annurev.en.32.010187.000555

Grover, J. K., & Yadav, S. P. (2004). Pharmacological action and potential uses of *Momordica charentia* L. *J. Ethnopharmacol, 93*(1), 123-132. https://doi.org/10.1016/j.jep.2004.03.035

Gupta, J. N., & Verma, A. N. (1979). Relative efficacy of insecticides as contact poison to the adults of melon fruit fly, *Dacus cucurbitae* (Coq). *Indian J. Ent., 41*, 117-120.

Howe, W. L. (1949). Factors affecting the resistance of certain cucurbits to squash borer. *J. Econ. Ent., 42*, 321. https://doi.org/10.1093/jee/42.2.321

Jaiswal, R. C., Kumar, S., Raghav, M., & Singh D. K. (1990). Variation in quality traits of bitter gourd (*Momordica charantia* L.) cultivars. *Veg. Sci., 17*, 186-190.

Kapoor, V. C. (1993). *Indian fruit flies* (p. 228). Oxford and IBH Publication, India.

Kedar, P., & Chakraborti, C. H. (1982). Effect of bitter gourd and glibenclamide in streptozotocin induced diabetes Mellitus. *Indian J. Exp. Biol., 28*, 232-235.

Kushwaha, K. S., Pareek, B. L., & Noor, A. (1973). Fruit fly damage in cucurbits at Udaipur. *Univ. Udaipur Res. J., 11*, 22-23.

Lall, B. S. (1964). Vegetable pests. *Entomology in India* (pp. 199-202). Ent. Soc., India, New Delhi.

Lall, B. S., & Singh, B. N. (1969). Studies on the biology and control of melon fly, *D. cucurbitae* (Dip.: Teph.). *Labdev J. Sci. and Tech., 7B*(2), 148-153.

Lall, B. S., & Sinha, S. N. (1959). Biology of melon fruit fly, *Dacus cucurbitae* Coquillet. *Sci. & Cult., 25*, 25.

Lee, W. Y. (1988). The control programme of oriental fruit fly in Taiwan. *Spec. Publ. No. 2, Entomol. Soc., Rep. China* (pp. 51-60).

Lefroy, H. M. (1909). *Indian Insect Life* (pp. 303-362). Thacker, Spink and Co., Calcutta.

Lotlikar, M. M., & Rajaramrao, M. R. (1966). Pharmacology of a hypoglycaemic principal isolated from the fruits of *Momordica charentia* L. *Indian. J. Pharm., 28*, 129.

Mauricio, R. (1998). Costs of resistance to natural enemies in field populations of the annual plant *Arabidopsis thaliana. American Naturalist, 151*, 20-28. https://doi.org/10.1086/286099

Mauricio, R., & Rausher, M. D. (1997). Experimental manipulation of putative selective agents provides evidence for the role of natural enemies in the evolution of plant defense. *Evolution, 51*, 1435-1444. https://doi.org/10.2307/2411196

Maxwell, P. G., & Jennings, P. R. (1980). *Breeding plants resistance to insects* (p. 124). A Wiley Interscience Publication, New York.

Mishra, V. K., & Bhatnagar, K. N. (1978). Studies on population and control of some pests of cucurbits, *Univ. Udaipur Res. J., 16*, 143-144.

Morton, J. F. (1967). The balsam pear–An edible, medicinal and toxic plant. *Econ. Bot., 21*, 57-68. https://doi.org/10.1007/BF02897176

Munro, K. H. (1984). A taxonomic treatise on the Decidae (Tephritoidea: Diptera) of Africa. *Memories of the*

Department of Agriculture, Republic of South Africa, 61, 1-313.

Narayan, E. S. (1953). Fruit fly pest of orchards and Kitchen garden. *Indian Farming, 3*(4), 29-31.

Narayan, E. S., & Batra, H. N. (1960). *Fruit flies and their control* (pp. 28-29). 'Monograph' I.C.A.R., New Delhi.

Nath, P. (1966). Varietal resistance of gourds to the fruit flies. *Indian J. Hort., 23*(2), 69-79.

Nath, P. (2008). Increased trend in production and consumption. *Survey of Indian Agriculture* (p. 66).

O'Connor, B. P. (2000). SPSS and SAS programs for determining the number of components using parallel analysis and Velicer's MAP test. *Behavior Research Methods, Instruments & Computers, 32*, 396-402. https://doi.org/10.3758/BF03200807

Painter, R. H. (1951). *Insects resistance in crop plants* (p. 520). The Macmillan Co., New York. https://doi.org/10.1097/00010694-195112000-00015

Painter, R. H. (1968). *Insects resistance in crop plants* (p. 520). The University Press of Kansas, Lawrence and London.

Pal, A. B., Srinivasan, K., & Douode, S. D. (1984). Sources of resistance to melon fruit fly in bitter gourd and possible mechanism of resistance. *SABARAO J., 16*, 57-69.

Panday, A. K., Nath, P., & Rai, A. B. (2008). Efficacy of some ecofriendly insecticides, poisons baits and their combinations against bitter gourd infestation by melon fruit fly (*Bactrocera cucurbitae* Coquillet). *Veg. Sci., 35*(2), 152-155.

Panday, A. K., Nath, P., Rai, A. B., & Kumar, A. (2009). Screening of some bitter gourd varieties/germplasms on the basis of some biological and biometrical parameters of melon fruit fly (*Bactrocera cucurbitae* Coquillet). *Veg. Sci., 36*(Suppl. 3), 399-400.

Pareek, B. L., & Kavadia, V. S. (1986). Seasonal incidence of insect pests of cucurbits in Rajasthan. *Ann. Arid Zone, 25*(4), 300-311.

Singh, S., & Singh, S. (1961). Field key for the determination of insects infesting vegetables. *Punjab Horti. J., 1*(1), 57-59.

Stamp, N. (2003). Out of the quagmire of plant defense hypotheses. *Quarterly Review of Biology, 78*, 23-55. https://doi.org/10.1086/367580

Tewatia, A. S. (1997). Melon fruit fly resistance in cucurbits–A Review. *Veg. Sci., 24*(2), 79-82.

10

Mycorrhizal Inoculation Increases Growth and Induces Changes in Specific Polyphenol Levels in Olive Saplings

Nasir S. A. Malik[1], Alberto Nuñez[1] & Lindsay C. McKeever[1]

[1] USDA-ARS, ERRC, Wyndmoor, PA, USA

Correspondence: Nasir S. A. Malik, USDA-ARS, ERRC, 600 E Mermaid Lane, Wyndmoor, PA 19038-8598, USA. E-mail: nasir.malik@ars.usda.gov

Abstract

This study was conducted to investigate the effect of mycorrhizal symbiosis on the levels of polyphenols in olive saplings. Rooted stem cuttings of olive cultivar, 'Arbequina', were inoculated with AM fungus *Rhizophagus intraradices*. The inoculated plants showed more robust growth after six months, and after nine months the increase in the mycorrhizal plant's height was 146%, and the increase in number of leaves was 117% when compared to uninoculated controls. Polyphenols in the methanol extracts of leaves were separated by HPLC and the peaks identified by using commercially available standard compounds and comparing retention time and the mass obtained with the mass spectrometer. Oleuropein, which is a major component of the olive leaf polyphenols, increased in mycorrhizal plants compared to uninoculated plants by 42%, and its derivatives, oleuroside and ligstroside, increased by 68% and 48%, respectively. The highest increase was found in the levels of luteolin-7'-O-glucoside (107% increase), while its sister compound luteolin-4'-O-glucoside increased by 43%. Only verbascoside levels were lower in mycorrhizal plants versus non-mycorrhizal plants declining to below detectable limits. Thus, inoculation of olive saplings with mycorrhizal fungi produces very positive effects on the levels of olive leaf polyphenols. Higher levels polyphenols mean better quality of leaf material for use as herbal medicine.

Keywords: ligstroside, luteolin-7'-O-glucoside, oleuropein, rooting of olive cuttings, verbascoside, oleuroside

1. Introduction

Olive (*Olea europaea*) trees are perhaps the original fruit trees that were cultivated (Zohary & Spiegel, 1975). To many, they are considered holy as these trees have been mentioned in Jewish, Christian and Muslim scriptures (Malik, 2014). However, their present day fame has more to do with the presence of specific health benefitting polyphenols in their fruits and leaves (Soler-Rivas et al., 2000; Tripoli et al., 2005; Uccella, 2001). For example, oleuropein and its derivatives act as antioxidants (Baldioli et al., 1996), and are known to reduce the risks of cancer (Owen et al., 2000; Soler-Rivas et al., 2000; Tripoli et al., 2005) and cardiovascular (Covas, 2007; Manna et al., 2002; Visioli et al., 1998; Wiseman et al., 1996), microbial and even viral diseases (Bisignano et al., 1999; Federici & Bongi, 1983; Fleming et al., 1973; Lee-Huang et al., 2003). Because of such profound benefits of olive polyphenols, consumption of olive oil and other olive products (including leaf extracts) has been steadily increasing and so is the cultivation of olive on new lands (Connor, 2005; Malik & Bradford, 2004; Malik, 2011; Sebestiani et al., 2006). It is therefore logical to devise better methods for production or to be able to grow olives in poor or saline soils. Mycorrhizal symbiosis is considered an important method to aid the cultivation of field crops, including olives, even under high salinity or in areas prone to water or nutrient deficiencies (Artursson et al., 2006; Auge, 2004; Dag et al., 2009; Farzaneh et al., 2011; Menge, 1983; Porras-Soriano et al., 2009).

Mycorrhizal fungi form symbiotic associations with most terrestrial plants (Simon et al., 1993; Smith & Read, 1997). The plant provides photosynthates to the fungi and in turn the hyphae of arbuscular mycorrhizal (AM) fungus extend into the soil and provide water and nutrients, especially the immobile nutrients, to plants (Koide, 1991; Marschner & Dell, 1994). Therefore, increased tolerance to drought (Allen & Boosalis, 1983; Auge, 2004; Nelsen & Safir, 1982; Ruiz-Lozano et al., 1995) and significant improvement in growth and productivity of several plant species have been reported in mycorrhizal plants under water and nutrient deficiencies (Baslam et al., 2011; Citernesi et al., 1998; Estaun et al., 2003; Gerdemann, 1968; Mosse et al., 1975). In addition, protection against pathogens in mycorrhizal plants has also been shown in different cultivars (Castillo et al.,

2006; Espinosa et al., 2014; Liu et al., 2007; Pozo & Azcon-Aguilar, 2007; Rabie, 1998; Watanarojanaporn et al., 2011). Since polyphenols play an important role in plant resistance to pests (Corcuera, 1993; Feeny, 1976; Jones & Klocke, 1987; Lattanzio, 2006; Nicholson & Hammerschmidt, 1992) and are of immense benefit to human health, we started to study the role of mycorrhizae in changing the levels of polyphenols in different plants (Malik et al., 2015a, 2015b). Our initial studies showed that while some polyphenols increased in mycorrhizal plants compared to uninoculated control plants, some other polyphenol species actually decreased (Malik et al., 2015a, 2015b). This study was therefore conducted to investigate whether or not the polyphenols of olive leaves (including the prized polyphenol 'oleuropein') increase in the leaves of mycorrhizal vs non-mycorrhizal olive saplings.

2. Materials and Methods

2.1 Plant Growth and Propagation

Thirty freshly rooted saplings of olive cultivar 'Arbequina' were gifted to us by Mr. Jim Henry, owner of 'Texas Olive Ranch' Carrizo Springs, TX. Samples of roots taken from several saplings indicated that they had established symbiosis with arbuscular mycorrhizal [AM] fungi and therefore, could not be used directly for this study due to lack of uninoculated controls. Therefore, these saplings were grown for 1 year (so that the these plants become big enough so we could take fresh cuttings from these plants to induce new clean roots) in 22x38 cm plastic pots filled with commercial growth media (Premier 'Pro-mix-BX', supplied by Premier Horticulture Inc., Quakertown, PA). The plants were supplied with the commercial nutrient 'Miracle Grow' as per manufacturer's instruction.

Fifteen centimeter long branches from one year old 'Arbequina' olive trees (growing in the greenhouse at USDA-ERRC facility in Wyndmoor, PA) were used to induce rooting. The cuttings were first soaked in sterile water for 2 hrs, then immersed for 5 min in 8.2% sodium hypochlorite, and then rinsed three times with sterile water. Leaves were removed from the lower 5 cm portion of the cutting. A fresh oblique cut was made at the bottom of the cutting and the cutting was rolled in hormone powder ('TakeRoot', Garden Safe Brand supplied by Schultz Company, Bridgeton, MO). The hormone-treated portions of the cuttings were gently inserted in 22 × 25 cm plastic pots filled with vermiculite that was soaked with 0.3% hydrogen peroxide made in sterile water. The cuttings were immediately sprayed with sterile water, and then covered with a large beaker to maintain high humidity. The cuttings were sprayed with water 4 times daily and 500 ml of 0.3% hydrogen peroxide were added (hydrogen peroxide provides increased oxygen and suppress microbial growth and hence promote rooting) to each pot twice a week. The pots were kept on heating pads to maintain 25 °C-27 °C, and were given 10 hrs. photoperiod using grow lights. After 7 weeks, the rooted cuttings were divided into two groups; one group was dedicated to be inoculated and other was our uninoculated control

The rooted cuttings were planted in Deepot Cells, 656 ml (Stuewe and Sons, Inc. Corvallis, OR) filled with a mixture of Premier Pro-mix BX and vermiculite (1:1). One group of rooted cuttings was inoculated with 750 spores of *Rhizophagus intraradices* in 1 ml of distilled water. The inoculated and control plants were grown for 3 months in a Conviron growth chamber, set at 10 hrs. photoperiod (daytime temp 25 °C, and night temp 18 °C). After 3 months, the plants were transferred to the greenhouse (in the month of Febraury), where temperature ranged between as low as 10 °C during the night and as high as 27 °C during the day.

R. intraradices fungus were grown in petri dishes on Ri.tDNA carrot roots (St-Arnand et al., 1996).

Nine months after the start of the experiment total number of leaves in each replicate of mycorrhizal and non-mycorrhizal plant were counted and the height of each plant was measured to document differences in plant growth as result of mycorrhizal inoculation.

2.2 Polyphenol Extraction from the Leaves

The leaves sampled from inoculated and control olive saplings were immediately frozen and stored at -80 °C and were ground in liquid nitrogen for extraction. The polyophenols were extracted from powdered leaf material in 80% methanol as described previously (Malik et al., 2015).

2.3 Separation, Quantitation and Identification of Polyphenols in Extracts

The chromatographic separation of the methanol extract was performed as reported before(Malik et al., 2015) with a Nano-Acquity (Waters, Milford, MA) ultrahigh performance liquid chromatographer (UHPLC) equipped with an Acquity UPLC BEH C18, 1.7 μm (1 × 100 mm) column (Waters) maintained at 40 °C and running at 60μl/minute. The UHPLC-UV chromatogram was obtained by attaching to the UHPLC instrument an Acquity TUV detector (Waters) set to scan at 280 nm. The solvent gradient was modified, starting with water-acetonitrile 90:10 (0.1% formic acid) for 2 minutes and ramped linearly to water-acetonitrile 70:30 (0.1% formic acid) at a final time of 18 minutes, maintained at that solvent composition for 2 minutes and followed with a columns wash of water-acetonitrile 15:85 (0.1%) formic acid) and returning to the initial condition at 25 minutes. A 10 minutes stabilization time was allowed between injections. Samples for the mycorrhizal and non-mycorrhizal plants were separately combined and mixed with 10 µl of a kaempferol solution (internal standard, 5 µg/ml). Three injections of 4 µl were made for each sample for determination of the concentration change according to the peak height determined by MassLynx v.4.1 software (Waters). The same chromatographic conditions were used for the mass spectrometry analysis.

The mass spectrometry analysis was accomplished by connecting the effluent of the UHPLC instrument to a Synapt G1 quadrupole-time of flight mass spectrometer (Waters) operating in the V mode (resolving power of 8,000 FWHM) and with an electrospray ionization (ESI) probe operated in the positive or negative mode and controlled by MassLynx v.4.1 software (Waters). The instrument parameters were 2.7 kV capillary voltage, 48 V extractor voltage, 300 L/h desolvation gas (N_2) flow, and 120 °C and 150 °C source and desolvation temperatures, respectively. The MS/MS of the deprotonated precursor ions [M-H]⁻ were obtained by collision-induced dissociation with argon gas at 0.9 ml/min with the collision energy ramped between 6 to 30 eV.

3. Results and Discussion

3.1 Effects of Mycorrhizal Fungi Inoculation on Plant Growth

Olive saplings inoculated with *R. intraradices* at early rooting stage showed increased plant growth relative to the uninoculated controls (Figure 1).

Figure 1. Growth differences in mycorrhizal (inoculated with Rhizophagus intraradices; on the right) and uninoculated (on the left) 'Arbequina' olive saplings nine months after rooting and inoculation

The picture was taken nine months after inoculation. The increase in plant height was 146% and the number of leaves were 117% more in mycorrhizal plants compared to uninoculated controls (Table 1). This increased plant growth in mycorrhizal plants is consistent with previous reports that showed increase in growth in different plants, including olives (Artursson et al., 2006; Auge, 2004; Dag et al., 2009; Farzaneh et al., 2011). The new and important findings of this study, however, are changes in polyphenol levels (Table 2).

Table 1. Differences in plant growth between mycorrhizal and uninoculated plants

Treatment	Plant Height (centimeters)	Number of Leaves	Root Colonization
Non Myc	19.6 ± 2.3	50.6 ± 10.8	0%
Myc	48.3 ± 7.6	109 ± 9.0	80.04% ± 2.3%
Percent increase over Non Myc	146%	117%	

Note. All values are Average ± SEM.

All Data Significant with a P value of less than 0.006 using t test for the significance.

Myc = Mycorrhizal plant.

Table 2. Percent change in the levels of polyphenol in the leaves of olive saplings in mycorrhizal plants compared to uninoculated control

Identification of Polyphenols	Peak number	Retention time	Percent increase over unioculated controls	Significantly different at P value using t test
Luteolin-7'-O-Glucoside	1	10.02	107.25%	0.0279
Verbascoside	2	10.27	-100%*	-
Luteolin-4'-O-Glucoside	3	11.47	42.65%	0.0327
Oleuropein	4	13.74	42.44%	< 0.0001
Oleuroside	5	14.46	67.89%	< 0.0001
Oleuropien derivative	6	14.97	32.02%	< 0.0001
Ligstroside	7	16.05	48.21%	0.0166

Note. * Not detected in mycorrhizal plants.

3.2 Effects of Mycorrhizal Fungi Inoculation on Changes in Levels of Polyphenols

Figure 2 shows an HPLC profile identifying various polyphenols in the olive leaf extract, which are similar to our previous reports (Malik & Bradford, 2006; Selin et al., 2012), but with the addition of a few more compounds due to the use of mass spectrometry.

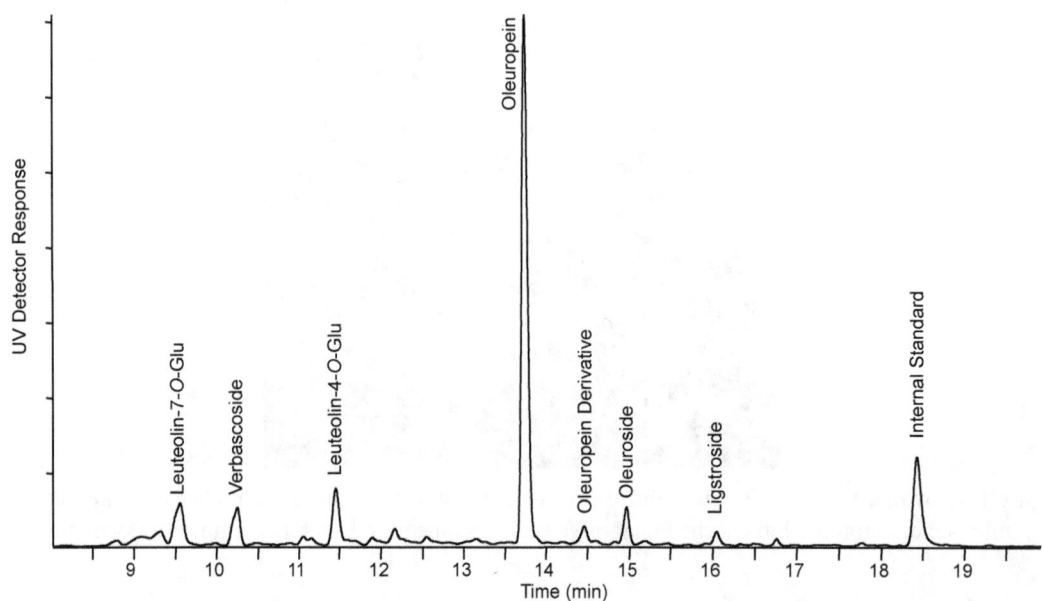

Figure 2. A representative profile of different polyphenols from olive leaf extract separated by HPLC

The identity of the compounds eluting at the corresponding retention time, in Table 2, were determined by using commercially available compounds, comparing retention time and the mass obtained with the mass spectrometer for the fallowing entrees: leuteolin-7-O-glucose ([M-H]- = 447.04; calculated 447.09, MS/MS fragments at m/z 285), verbascoside ([M-H]- = 621.14, calculated 621.18, MS/MS fragments at m/z 461, and 161), leuteolin-4'-O-glucose ([M-H]- = 447.04; calculated 447.09, MS/MS fragments at m/z 285), and oleuropein ([M-H]- = 539.15; calculated 539.18, MS/MS fragments at m/z 377, 307, 275, and 225) (Savarese et al., 2007). The peaks eluting at 14.46 and 14.97 min have the same mass spectra as oleuropein, with the same fragments at m/z 377; 307; 275; and 225 suggesting two isomers of oleuropein. Savarese et al. (2007) reported oleuroside as an isomer eluting after oleuropein, with the same mass spectrum and accordingly, we are assigning the peak at 14.46 as oleuroside based on its longer retention time in the reversed phase column as was reported before (Savarese et al., 2007; Savournin et al., 2001). No further identification was possible for the peak closer to oleuropein at 14.97 min and is reported as a derivative of oleuropein. The peak eluting at 16.05 was identified as ligstroside, ([M-H]- = 523.15; calculated 523.18, MS/MS fragments at m/z 361, 291, 259 and 101) as reported before (Savarese et al., 2007).

Although the majority of polyphenols identified in the leaves had increased levels in mycorrhizal plants compared to the uninoculated controls, the level of verbascoside decreased to a point where they were below detection (Table 2). This was the only polyphenol species whose levels decreased in mycorrhizal olive saplings, oleuropein which is a major component of olive leaf polyphenols increased by 42%, and its derivative oleuroside and ligstroside increased by 68% and 48%, respectively (Table 2). The greatest increase was found in the levels of luteolin-7'-O-glucoside (107% increase), while its sister compound luteolin-4'-O-glucoside increased by 43% (Table 2). This pattern of increases in certain polyphenol species with decreases in the levels of other polyphenol species, in mycorrhizal plants, compared to controls, has been observed in other plant species (Malik et al., 2015, 2016).

In general, our results show that inoculation of olive sapling's freshly formed roots with mycorrhizal fungi (*Rhizophagus intraradices*) produces beneficial effects on the growth of the plants and improves the levels of its major polyphenols. Olive polyphenols are well known for several health benefits, described before, and therefore, the current study is important as it provides information regarding improving the quality of commercial olive products.

References

Allen, M. F., & Boosalis, M. G. (1983). Effect of two species VA mycorrhizal fungi on drought tolerance of winter wheat. *New Phytologist, 93*, 67-76. https://doi.org/10.1111/j.1469-8137.1983.tb02693.x

Artursson, V., Finlay, R. D., & Jansson, J. K. (2006). Interactions between arbuscular mycorrhizal fungi and bacteria and their potential for stimulating plant growth. *Environ. Microbiol., 8*, 1-10 https://doi.org/10.1111/j.1462-2920.2005.00942.x

Auge, M. R. (2004). Arbuscular mycorrhizae and soil/plant water relations. *Can. J. Soil Sci., 84,* 373-381. https://doi.org/10.4141/S04-002

Baldioli, M., Servili, M., Perretti, G., & Montedoro, G. F. (1996). Antioxidant activity of tocopherols and phenolic compounds of virgin olive oil. *J. Ameri. Oil Chemist Soc., 73*, 1589-1593. https://doi.org/10.1007/BF02523530

Baslam, M., Garmendia, I., & Goicoechea, N. (2011). Arbascular mycorrhizal fungi (AMF) improved growth and greenhouse-grown lettuce. *J. Agric. Food Chem., 59*, 10067-10080. https://doi.org/10.1021/jf200501c

Castillo, P., Nico, A. I., Azcón-Aguilar, C., Del Río Rincón, C., Calvet, C., & Jim Enez Diáz, R. M. (2006). Protection of olive planting stocks against parasitism of root-knot nematodes by arbuscular mycorrhizal fungi. *Plant. Path., 55*, 705-713. https://doi.org/10.1111/j.1365-3059.2006.01400.x

Connor, D. J. (2005). Adaptation of olive (*Olea europaea* L.) to water-limited environment. *Aust. J. Agric. Res., 56*, 1181-1189. https://doi.org/10.1071/AR05169

Corcuera, L. J. (1993). Biochemical basis for resistance of barley to aphids. *Phytochem., 33*, 741-747. https://doi.org/10.1016/0031-9422(93)85267-U

Covas, M. I. (2007). Olive oil and cardiovascular system. *Phar. Res., 55*, 175-186. https://doi.org/10.1016/j.phrs.2007.01.010

Dag, A., Yermiyahu, U., Ben-Gal, A., Zipori, I., & Kapulnik, Y. (2009). Nursery and post-transplant field response of olive trees to arbuscular mycorrhizal fungi in an arid region. *Crop Pasture Sci., 60*, 427-433. https://doi.org/10.1071/CP08143

Espinosa, F., Garrido, I., Ortega, A., Casimiro, I., & Alverez-Tinaut, M. C. (2014). Redox activities and ROS, NO and phenylpropanoids production by axenically cultured olive seedling rootsafter interaction with a mycorrhizal or a pathogenic fungus. *PLoS ONE, 996*, e100132. https://doi.org/10.1371/journal.pone. 0100132

Estaun, V., Camprubi, A., Calvet, C., & Pinochet, J. (2003). Nursery and field response of olive trees inoculated with two arbuscular mycorrhizal fungi, glomus intraradices and glomus mosseae. *J. Amer. Soc. Hort. Sci., 128*, 767-775.

Farzaneh, M., Vierheilig, H., Lossl, A., & Kaul, H. P. (2011). Arbuscular mycorrhizae enhances nutrient uptake in chickpea. *Pl. Soil Environ., 57*, 465-470.

Feeny, P. P. (1976). Plant apparency and chemical defense. *Recent Adv. Phytochem., 10*, 1-40. https://doi.org/ 10.1007/978-1-4684-2646-5_1

Gerdemann, J. W. (1968). Vesicular-arbuscular mycorrhiza and plant growth. *Annu. Rev. Phytopathol., 6*, 397-418. https://doi.org/10.1146/annurev.py.06.090168.002145

Jones, K. C., & Klocke, J. A. (1987). Aphid feeding deterrency of ellagitannins, their phenolic hydrolysis products and related phenolic derivatives. *Entomol. Exp. Appl., 44*, 229-234. https://doi.org/10.1111/ j.1570-7458.1987.tb00549.x

Koide, R. T. (1991). Nutrient supply, nutrient demand and plant responses to mycorrhizal infection. *New Phytologist, 114*, 365-386. https://doi.org/10.1111/j.1469-8137.1991.tb00001.x

Lattanzio, V., Lattanzio, V. L. M., & Cardinali, A. (2006). Role of phenolics in the resistance mechanism of plants against fungal pathogens and insects. *Photochem. Advan. Res., 661*, 23-67.

Lee-Huang, S., Zhang, L., Huang, P. L., Chang, Y.-T., & Huang, P. L. (2003). Anti-HIV activity of olive leaf extract (OLE) and modulation of host cell gene expression by HIV-1infection and OLE treatment. *Bio. Bioch. Res. Comm., 307*, 1029. https://doi.org/10.1016/S0006-291X(03)01292-0

Liu, J., Maldonado-Mendoza, I., Lopez-Meyer, M., Cheung, F., Town, C. D., & Harrioson, M. J. (2007). Arbuscular mycorrhizal symbiosis is accompanied by local and systemic alterations in gene expression and an increase in disease resistance in shoot. *Plant J., 50*, 529-544. https://doi.org/10.1111/j.1365-313X.2007. 03069.x

Malik, N. S. A. (2011). Feasibility of growing olives at selected sites on coastal Texas. *J. Agric. Sci. Tech., 5*, 139-146.

Malik, N. S. A. (2014). *Olive a holy plant with several benefits to human health.* Retrieved from http://www.themuslimtimes.org/2014/04/americas/olive-a-holy-plant-with-several-benefits-to-human-health #ixzz2ysEynTxn

Malik, N. S. A., & Bradford, J. M. (2004). Genetic diversity and clonal variation among olive cultivars offer hope for selecting cultivars for Texas. *J. Am. Pom. Soc., 58*, 203-209.

Malik, N. S. A., & Bradford, J. M. (2006). Changes in oleuropein levels during differentiation and development of floral buds in 'Arbequina' olives. *Sci. Hort., 110*, 274-278. https://doi.org/10.1016/j.scienta.2006.07.016

Manna, C., D'Angelo, S., Migliardi, V., Loffredi, E., Mazzoni, O., Morrica, P., ... Zappia V. (2002). Protective effect of phenolic fraction from virgin olive oils against oxidative stress in human cells. *J. Agric. Food Chem., 50*, 6521-6526. https://doi.org/10.1021/jf020565+

Marschner, H., & Dell, B. (1994). Nutrient uptake in mycorrhiza symbiosis. *Pl. Soil, 159*, 89-102.

Menge, J. A. (1983). Utilization of vescular-arbuscuar mycorrhizal fungi in agriculture. *Can. J. Bot., 61*, 1015-1024. https://doi.org/10.1139/b83-109

Mosse, B., Powell, L. I., & Hayman, D. S. (1976). Plant growth responses to vescular-arbuscuar mycorrhiza. IX. Interactions between VA mycorrhiza, rock phosphate and symbiotic nitrogen fixation. *New Phytol., 76*, 331-342. https://doi.org/10.1111/j.1469-8137.1976.tb01468.x

Nelsen, C. E., & Safir, G. R. (1982). Increased drought tolerance of mycorrhizal onion plants caused by improved phosphorus nutrition. *Planta, 154*, 407-413. https://doi.org/10.1007/BF01267807

Nicholson, R. L., & Hammerschmidt, R. (1992). Phenolic compounds and their role in disease resistance. *Ann. Rev. Phytopath., 30*, 369-389. https://doi.org/10.1146/annurev.py.30.090192.002101

Porras-Soriano, A., Soriano-Martin, M. L., Porras-Piedra, A., & Azcon, R. (2009). Arbuscular mycorrhiza fungi increased growth, nutrient uptake and tolerance to salinity in olive trees under nursery conditions. *J. Pl. Physiol., 166*, 1350-1359. https://doi.org/10.1016/j.jplph.2009.02.010

Pozo, M. J., & Azcon-Aguilar, C. (2007). Unravelling mycorrhiza-induced resistance. *Curr. Opin. Pl. Biol., 10*, 393-398. https://doi.org/10.1016/j.pbi.2007.05.004

Rabie, G. H. (1998). Induction of fungal disease resistance in vicia faba by dual inoculation with Rhizobium leguminosarum and vesicular-arbascular mycorrhizal fungi. *Mycopathology, 141*, 159-166. https://doi.org/10.1023/A:1006937821777

Ruiz-Lozano, J. M., Azcon, R., & Gomez, M. (1995). Effects of arbuscular-mycorrhizal glomus species on drought tolerance: Physiological and nutritional responses. *Appl. Envirn. Microb., 61*, 456-460.

Savarese, M., De Marco, E., & Sacchi R. (2007). Characterization of phenolic extracts from olive (*Olea europea* cv. Pisciottana) by electrospray ionization mass spectrometry. *Food Chem., 105*, 761-770. https://doi.org/10.1016/j.foodchem.2007.01.037

Savournin, C., Baghdikian, B., Elias, R., Dargouth-Kesraoui, F., Boukef, K., & Balansard, G. (2001). Rapid high-performance liquid chromatography analysis for the quantitative determination of oleuropein in Olea europaea leaves. *J. Agric. Food Chem., 49*, 618-621. https://doi.org/10.1021/jf000596+

Sebastiani, L., d'Andria R, Motisi, A., & Caruso, T. (2006). The olive industry outside the mediterranean. In T. Caruso, A. Motisi, & L. Sebastiani (Eds.), *Proceedings of the 2nd International Seminar, Olivebioteq–Recent Advances in Olive Industry, Marsala, Italy* (pp. 183-195). Dipartimento Colture Arboree, Universita di Palermo: Palermo, Italy.

Selin, S., Malik, N. S. A., Perez, J. L., & Brockington, J. E. (2012). Seasonal changes of individual phenolic compounds in leaves of twenty olive cultivars grown in Texas. *J. Agric. Sci. Tech. B, 2*, 242-247.

Simon, L., Bousquet, J., Levesque, R. C., & Lalonde, M. (1993). Origin and diversification of endomycorrhizal fungi and coincidence with vascular plants. *Nature, 363*, 67-69. https://doi.org/10.1038/363067a0

Smith, S. E., & Read, D. J. (1997). *Mycorrhizal Symbiosis* (2nd ed.). Academic Press, San Diego, Calif.

Soler-Rivas, C., Espin, J. C., & Wichers, H. J. (2000). Oleuropein and related compounds. *J. Sci. Food. Agric., 80*, 1013-1023. https://doi.org/10.1002/(SICI)1097-0010(20000515)80:7%3C1013::AID-JSFA571%3E3.0. CO;2-C

St-Arnaud, M. C., Hamel, C., Vimard, B., Caron, M., & Fortin, J. A. (1996). Enhanced hyphal growth and spore production of the arbuscular mycorrhizal fungus Glomus intraradices in an in vitro system in the absence of host roots. *Myco. Res., 100*, 328-332. https://doi.org/10.1016/S0953-7562(96)80164-X

Tripoli, E., Giammanco, M., Tabacchi, G., Di Majo, D., Giammanco, S., & La Guardia, M. (2005). The phenolic compounds of olive oil: Structure, biological activity and beneficial effects on human health. *Nutri. Res. Rev., 18*, 98-112. https://doi.org/10.1079/NRR200495

Uccella, N. (2001). Olive biophenols: Biomolecular characterization, distribution and phytolexin histochemical localization in the drupes. *Trends Food Sci. Tech., 11*, 315-327. https://doi.org/10.1016/S0924-2244(01) 00029-2

Visioli, F., Bellosta, S., & Galli, C. (1998). Oleuropein, the bitter principle of olives, enhances nitric oxide production by mouse macrophages. *Life Sci., 62*, 541-546. https://doi.org/10.1016/S0024-3205(97)01150-8

Watanarojanaporn, N., Boonkerd, N., Wongkaew, S., Prommanop, P., & Teaumroong, N. (2011). Selection of arbuscular mycorrhizal fungi for citrus growth promotion and Phytophthora suppression. *Sci Hort, 128*, 423-433. https://doi.org/10.1016/j.scienta.2011.02.007

Zohary, D., & Spiegle-Roy, P. (1975). Beinning of fruit growing in old world. *Science, 187*, 319-327. https://doi.org/10.1126/science.187.4174.319

11

Characterization and Antibacterial Mode of a Novel Bacteriocin with Seven Amino Acids from *Lactobacillus plantarum* in Guizhou Salted Radish

Meizhong Hu[1,2], Lijuan Dang[1], Haizhen Zhao[1], Chong Zhang[1], Yingjian Lu[3], Jiansheng Yu[2] & Zhaoxin Lu[1]

[1] College of Food Science and Technology, Nanjing Agricultural University, Nanjing, Jiangsu, P.R. China

[2] Tongren Polytechnic College, Tongren Guizhou, P.R. China

[3] Department of Nutrition and Food Science, University of Maryland, College Park, Maryland, USA

Correspondence: Zhaoxin Lu, College of Food Science and Technology, Nanjing Agricultural University, No.1 Weigang Nanjing, Jiangsu 210095, P.R. China. E-mail: fmb@njau.edu.cn

Abstract

Traditional Chinese fermented vegetables are excellent probiotic food with probiotic lactic acid bacteria that are benefical to the health. A novel bacteriocin with molecular weight, 825 Da was found successfully from Lactobacillus plantarum 163, which was isolated from Guizhou salted radish. The complete amino acid sequence was speculated as YVCASPW based on the mass spectrometry, and was named as bacteriocin 163-1. The bacteriocin 163-1 was highly thermostable and stability over a broad pH range (pH 3-6), sensitive to protease K and pepsin, and exhibited a wide range of antimicrobial activity not only against lactic acid bacteria (LAB) but also against other foodborne pathogens including Gram-positive and Gram-negative bacteria. Bacteriocin 163-1 could disrupt the cell membrane of bacteria. The observations of the transmission electron microscopy and laser confocal microscopy on the cell membrane of Escherichia coli and Staphylococcus aureus showed that bacteriocin 163-1 could result in forming pores on the cell membrane and then cytolysis of the bacteria. The new bacteriocin with broad-spectrum antibacterial activity will be useful in preservation of vegetable, fruit and food as well agricultural bio-controlling.

Keywords: *Lactobacillus plantarum*, bacteriocin, action mode of antibacteria, bio-preservative

1. Introduction

Traditional Chinese fermented vegetables are excellent probiotic food with probiotic lactic acid bacteria (LAB), which has various health benefits including anticonstipation, anticancer, antioxidative and immune-boosting. The functional ingredients from fermentation by LAB such as bacteriocins could be used as bio-preservative. LAB strains are generally recognized as safe (GRAS) microorganisms (Burdock & Carabin, 2004), and bacteriocins have also achieved GRAS status. Some bacteriocins with remarkable thermostability and pH stability show a significant inhibitory activity against spoilage and pathogenic bacteria, indicating that bacteriocin could be potentially used as an effective bio-preservative in the food industry.

Some bacteriocins are already commercially available, such as nisin (produced by *Lactococcus lactis*), used to keep and extend the shelf life of food products in many countries in the world. However, nisin has some drawbacks when applied in the food industry. For instance, nisin does not perform the good antibacterial activity against Gram-negative bacteria, and it is only effective under an acidic environment. These limits are not positive for nisin to be applied in process and conservation for foods. Thus, an alternative novel bacteriocin with good antibacterial activities is essential for the food industry. Recently, the researcher is focusing on the mining of novel bacteriocin from LAB for such as *Lb. plantarum* from salted vegetable, yoghurt, fermented meat (Biscola et al., 2013; Ahmad et al., 2014; Alvarez-Sieiro et al., 2016).

Generally, most bacteriocins generated from LAB appear to share a common mode of action which form a pore in the sensitive bacterial membrane, and therefore lead to sensitive cell death (Castellano et al., 2003; Oppegård et al., 2007). In addition, other type of bacteriocins such as lantibiotic has exhibited different mode of action by binding themselves to lipid II, and therefore lead to cell death through false cell wall synthesis (Cotter et al., 2005). Furthermore, some of bacteriocins such as Pep 5 cause lysis of treated cells, which is another mode of action (Bierbaum & Sahl, 1987). The effective use of bacteriocins in food preservation requires a better understanding of their mode of action and their inhibitory action under different biochemical conditions naturally occurring in food.

In this study, the purification, identification, antimicrobial spectrum, biochemical and genetic characteristics of new bacteriocin from *Lb. plantarum* 163 in Guizhou salted radish were comprehensively investigated, and the action mode of the bacteriocin against to *S. aureus* and *E. coli* was discussed.

2. Materials and Methods

2.1 Strains, Media and Growth Conditions

Lb. plantarum 163 was screened from Chinese fermented salted radish (Hu et al., 2013) and was stored in China General Microbiological Culture Collection Center (No. 8224). The lactic acid bacteria and fungi used in this study and their respective growth conditions are listed in Table 1.

Table 1. Antimicrobial activity profile of bacteriocin 163-1 produced from the isolated *Lactobacillus plantarum* 163

Strains	Source [a]	Growth conditions	Antimicrobial activity [b]
Staphylococcus ureus	ATCC 25923	nutrient broth/37 °C	19.38±0.68
Listeria monocytogenes	ATCC 19114	nutrient broth/37 °C	16.25±0.58
Bacillus pumilus	CMCC 63202	nutrient broth/37 °C	16.28±0.47
Bacillus cereus	AS 1.1846	nutrient broth/37 °C	13.07±0.27
Micrococcus luteus	CMCC 28001	nutrient broth/37 °C	13.07±0.27
Lactobacillus thermophilus	Our Lab	MRS medium/37 °C	9.55±0.52
Lactobacillus rhamnosus	Our Lab	MRS medium/37 °C	9.60±0.54
E. coli	ATCC 25922	nutrient broth/37 °C	15.35±0.67
Pseudomonas aeruginosa	AS 1.2620	nutrient broth/37 °C	9.91±1.30
Pseudomonas fluorescens	AS 3.6452	nutrient broth/37 °C	12.92±1.46
Penicillium notatum	AS 3.4356	Potato Dextrose/30 °C	0
Aspergillus niger	AS 3.6459	Potato Dextrose/30 °C	0
Rhizopus stolonifer	AS 3.822	Potato Dextrose/30 °C	0

Note. [a] ATCC, American type culture collection; CMCC, China center of medicine culture collection; AS, China General Microbiological Culture Collection Center. [b] Well: 5 mm, Means of three replicate values.

2.2 Purification of Bacteriocin 163-1

Lb. plantarum 163 was activated in MRS broth at 37 °C for 14 h without agitation. 10 mL pre-culture was inoculated into 1,000 mL MRS broth, and incubated at 37 °C for 24 h without agitation. The bacterial cells then were harvested by centrifugation (6,000 g, 15 min) at 4 °C, and the cell-free supernatant collected was filtrated by 0.45 μm membrane. Then, the cell-free supernatant was separated by preparative chromatography (Waters, 600) with Waters SunFire OBD-C18 columns (19×150 mm) in a 40 min isocratic elution of 95% water-acetonitrile (5%) and containing 0.1% trifluoroacetic acid (TFA). The elution was monitored continuously at 267 nm and all individual peaks were collected. Then, all individual peaks were further purified by reversed-phase high-performance liquid chromatography (RP-HPLC, UltiMate 3000) using the following conditions: Agilent Eclipse XDB-C18 columns, (4.6 × 250 mm), followed by an isocratic elution using 95% water-acetonitrile (5%) and containing 0.1% TFA for 40 min. Each individual peak was collected and further used for antimicrobial test using *Bacillus pumilus* CMCC 63202 as indicator.

2.3 Determination of the Primary Structure of Bacteriocin 163-1

The molecular mass and amino acid sequence of purified compound (bacteriocin 163-1) were analyzed by a matrix-assisted laser desorption ionization-time of flight mass spectrometry (MALDI-TOF-MS) (Bruker Daltonics, Bremen, Germany) using α-cyano-4-hydroxycinnamic acid as a reference (Hu et al., 2013; Chen et al., 2014).

2.4 Antimicrobial Spectrum of Bacteriocin 163-1

The bacteriocin 163-1 from preparative chromatography were used to evaluate the antimicrobial activity against indicator organisms (seven of positive bacteria, three of gram-negative bacteria and three of fungi). Antibacterial activity of bacteriocin 163-1 was measured using agar well diffusion assay method (Ruixiang et al., 2015).

2.5 Effects of Enzyme, Temperature and pH on the Activity of Bacteriocin 163-1

The effect of Protease K (30 U/mg), Trypsin (2.5 KU/mg), α-Chymotrypsin (1 KU/mg) and Pepsin (3 KU/mg) on the activity of bacteriocin 163-1 was determined. The purified bacteriocin 163-1 from preparative chromatography was incubated at the optimum pH at 37 °C for 3 h) with the enzymes of the final concentration of 5 mg/mL. The influence of temperature on the activity of bacteriocin 163-1 was examined by treating at 60 °C, 80 °C and 100 °C for 10, 20 and 30 min, respectively in a thermostatic water bath and at 121 °C for 20 min in an autoclave. Lastly, the pH value of the purified bacteriocin 163-1 was adjusted to 2-10 and then kept at 37 °C for 3 h in a thermostatic water bath. The residual antimicrobial activities were measured after enzymatic, temperature and pH treatments using agar diffusion assay method using *Bacillus pumilus* CMCC 63202 as the indicator (Gao et al., 2010; Zhang et al., 2013).

2.6 Effect of Bacteriocin 163-1 on the Sensitive Cell Growth and Time-Kill Kinetics

The activity unit of bacteriocin 163-1 was defined as the reciprocal of the highest dilution with antimicrobial activity and was expressed in activity units (AU) per milliliter (Barefoot & Klaenhammer, 1983; Pucci et al. 1988).

In order to determine the effect of bacteriocin 163-1 on the sensitive cell growth, *S. aureus* (ATCC 25923) or *E. coli* (ATCC 25922) were activated in LB medium, then bacteriocin 163-1 were added at a concentration of 6.4 Au/mL and 12.8 Au/mL. The sterile distilled water treatment was used as a control while the sterile LB medium was used as a baseline. The OD_{600} of all treatments were measured at 2, 3, 4, 5, 6, 7, 8, 10, 12 and 16 h. In order to determine the Time-kill kinetics, *S. aureus* (ATCC 25923) or *E. coli* (ATCC 25922) was activated in LB medium and incubated at 37 °C for 6 hours with agitation. Then bacteriocin 163-1 was added at a concentration of 12.8 AU/mL and the bacteria were counted at 0.5, 1, 1.5, 2, 2.5 and 3 h.

2.7 Effect of Bacteriocin 163-1 on Release of Proteins and Nuclear Acid in Cells

To determine the effect of bacteriocin 163-1 on release of proteins and nuclear acid in bacteria cells, *S. aureus* (ATCC 25923) or *E. coli* (ATCC 25922) were activated in LB medium, and incubated at 37 °C for 6 hours with agitation, then bacteriocin 163-1 were added at a concentration of 6.4 AU/mL. The bacterial cells then were harvested by centrifugation (6,000 g, 15 min) at 4 °C. The OD_{280} and OD_{260} of the cell-free supernatants were measured by 0.5, 1, 1.5, 2, 2.5 and 3 h. The sterile distilled water was used as a control while the sterile LB medium was used as a baseline.

S. aureus (ATCC 25923) or *E. coli* (ATCC 25922) were activated and incubated at 37 °C for 5 hours with agitation, then bacteriocin 163-1 were added at a concentration of 6.4 Au/mL for the determination of the lactic dehydrogenase (LDH) release. After bacterial cells were harvested by centrifugation at 6000 g for 15 min at 4 °C, the LDH contents of the cell-free supernatants were measured at 0.5, 1, 1.5, 2, 2.5 and 3 h using LDH kit (Nanjing Jiancheng Bioengineering Institute, China). The sterile distilled water treatment was used as a control while the sterile LB medium was used as a baseline.

2.8 Transmission Electron Microscopy (TEM) Observations on the Sensitive Cell Membrane Treated with Bacteriocin 163-1

S. aureus (ATCC 25923) or *E. coli* (ATCC 25922) were activated in LB medium, and incubated at 37 °C for 8 hours (*E. coli* 4 h) with agitation, then bacteriocin 163-1 were added at a concentration of 6.4 AU/mL, incubated at 37 °C for 2 hours, bacterial cells were harvested by centrifugation at 6,000 g for 15 min at 4 °C. The bacterial cells were washed by 2.5% Gluteraldehyde twice, finally bacterial cells were fixed by 2.5% Gluteraldehyde and the cell membrane observation was done by using Transmission Electron Microscopy (HITACHI, H-600, Japan). The sterile distilled water treatment was used as a control.

2.9 Bacteriocin 163-1 Localization on Celluar

S. aureus (ATCC 25923) or *E. coli* (ATCC 25922) were activated in LB medium, and incubated at 37 °C for 8 hours (*E. coli* 6 h) with agitation, then fluorescein isothiocyanate (FITC)- Bacteriocin 163-1 were added at a concentration of 1 mg/L, and incubated at 37 °C for 0.5 hours with agitation (in dark), bacterial cells were harvested by centrifugation at 6,000 g for 15 min at 4 °C, then the bacterial cells were washed by 2.5% Gluteraldehyde twice. Then, the bacterial cells were suspended using 2.5% Gluteraldehyde (in dark) and finally observed by Laser confocal microscopy (LSM-710, Zeiss, Germany) using Fluorescence wavelength at 488 nm and 40X object lens.

3. Results and Discussion

3.1 Purification and Identification of the Bacteriocin 163-1

After the purification by preparative chromatography and RP-HPLC, one substance with antibactreia activity, where the retention time was 10.6 min (Figure 1a) was obtained. Then, the molecular mass and amino acid sequence of the bacteriocin 163-1 was identified by means of MALDI-TOF-MS, It was found that molecular mass of bacteriocin 163-1 was 825 Da (Figure 1b). MALDI-TOF-MS/MS fragment of bacteriocin 163-1 are shown in Figure 1b. Based on the previous report about peptide fragmentation nomenclature (Roepstorff & Fohlman, 1984), the primary sequence of bacteriocin 163-1 was determined to be: Y-V-C-A-S-P-W.

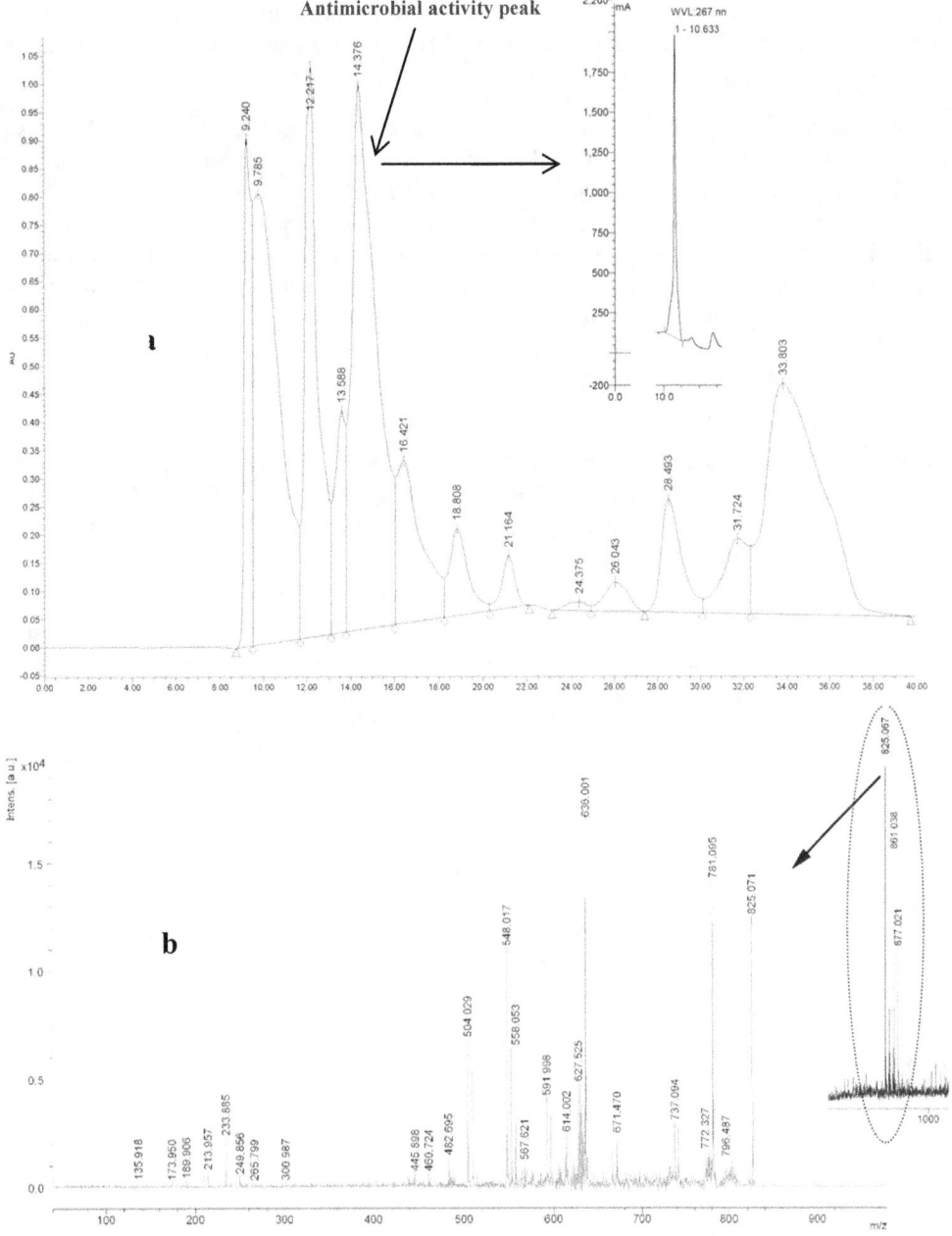

Figure 1. a) The purified chromatograms of bacteriocin 163-1 with retention time of 10.633 s as identified; b) MALDI-TOF-MS of bacteriocin 163-1, the molecular ion peak is represented by a major peak at 825 Da, 861: [4*CHCA-3H+1K+3Na]+, 877: [4*CHCA-3H+2K+2Na]+, CHCA: α-cyano-4-hydroxycinnamic acid

Bacteriocin is an active peptide produced by bacteria. From previous studies, many bacteriocins produced by *Lb. plantarum* have been found and identified, such as plantaricin Y produced by *Lb. plantarum* 510 (Chen et al., 2014), plantaricin LD1 Produced by *Lb. plantarum* LD1 (Gupta & Tiwari, 2014). Some of *Lb. plantarum* could produce more than one type of plantaricin, for instance *Lb. plantarum* C11 could produce plantaricin A, plantaricin F, plantaricin E (Hauge et al., 1999). In our previous work, *Lb. plantarum* 163 could produce plantaricin 163 which consist of 32 amino acid with the molcecular weight of 3553 Da (Hu et al., 2013).

Bacteriocin 163-1 was another antibacterial substance from *Lb. plantarum* 163. The molecular weight of bacteriocin produced by *Lb. plantarum* are between 2 kDa-5 kDa, for example, plantaricin C19 (3.8 KDa) (Atrih et al., 2001), plantaricin UG1 (3-10 KDa) (Enan et al., 1996), plantaricin MG (2180 Da) (Gong et al., 2010), plantaricin-149 (2.2 KDa) (Kato et al., 1994), plantaricin ASM1 (5045.7 Da) (Hata et al., 2010), and plantaricin 163 (3553 Da) (Hu et al., 2013). In this study, we discovered a new bacteriocin 163-1with 825 KDa of molecular weight, which was similar to those found in molecular level, such as Acidocin NX2 (824 Da) (Zhang et al., 2014). And it was confirmed that the sequence of bacteriocin 163-1 was no homology by search of protein BLAST (BLAST) against the GenBank database (http://www.ncbi.nlm.nih.gov/BLAST) and Bactibase (http://bactibase.pfba-lab-tun.org), thus suggesting that bacteriocin 163-1 may be a novel bacteriocin.

3.2 Antimicrobial Spectrum of Bacteriocin 163-1

The antimicrobial activities of bacteriocin 163-1 was shown in Table 1. The bacteriocin 163-1 significantly inhibited the growth of Gram-positive bacteria (*S. aureus, Listeria monocytogenes, Bacillus pumilus, Bacillus cereus, Micrococcus luteus, Lactobacillus thermophilus, Lactobacillus rhamnosus*) and the Gram-negative bacteria (*E. coli, Pseudomonas aeruginosa,* and *Pseudomonas fluorescens*), but had no antimicrobial activity against fungi such as *Penicillium notatum, Aspergillus niger,* and *Rhizopus nigricans,* indicated that bacteriocin 163-1 had a broad antimicrobial spectrum. As shown in previous literature, most of bacteriocin, such as plantaricin F, plantaricin E, plantaricin J could only showed antibacterial activity against homologous species (Anderssen et al., 1998), whereas only a few plantaricin appeared to be a broad spectrum of antimicrobial activity. A bacteriocin with broad antibacterial activity may be more useful and valuable in agro-food industry.

3.3 Effects of Enzyme, Temperature and pH on the Activity Bacteriocin 163-1

Effects of enzyme, temperature and pH on the activity bacteriocin 163-1 were investigated and the results were shown in Table 2. The antimicrobial activity of bacteriocin 163-1 was found to be fully lost after enzymatic treatments with protease K, pepsin, and partially lost after enzymatic treatments with trypsin, α-Chymotrypsin, thus highlighting the typical property of a peptide.

Further, the stability of bacteriocin 163-1 was tested at 60 °C, 80 °C, 100 °C and 121 °C and under a different pH (2-10). Surprisingly, it was observed that the activity recovery of bacteriocin 163-1was more than 90% by treatments of 60-121 °C in pH 4, appeared to be good thermostability. On other hand, its recovery was more than 70% when tested at pH 3-10, of which showed remarkable pH stability in pH 3-8. The result was consistent with previous findings about bacteriocin peptides such as plantaricin MG (Gong et al., 2010), plantaricin 163 (Hu et al., 2013). The good physicochemical properties of bacteriocin 163-1 could offer an essential promise for its application in the processing and preservation of foods as well bio-controlling of plant diseases.

3.4 Effect of Bacteriocin 163-1 on Cell Growth and Time-Kill Kinetics S. aureus and E. coli

The effect of bacteriocin 163-1 on cells growth and Time-kill kinetics of *S. aureus* and *E. coli* was shown in Figure 2. The growths of *S. aureus* and *E. coli* were partly inhibited when they were treated with bacteriocin 163-1 at a concentration of 6.4 AU/mL (Figures 2a and 2b). However, *S. aureus* growth was fully inhibited, but *E. coli* growth was done before 8 h of culture when bacteriocin 163-1 was added at a concentration of 12.8 AU/mL (Figures 2a and 2b), suggested *S. aureus* was more sensitive to bacteriocin 163-1 than *E. coli* and it needed higher concentration of the bacteriocin to fully inhibit growth of *E. coli*. The growth of *S. aureus* was highly inhibited after 1 h with the treatment of 12.8 AU/mL (Figure 2c). The growth of *E. coli* was significantly reduced from 0.5 h to 3 h, and a 5-log reduction was observed after 3 h when treated with bacteriocin 163-1 at a concentration of 12.8 AU/mL (Figure 2d).

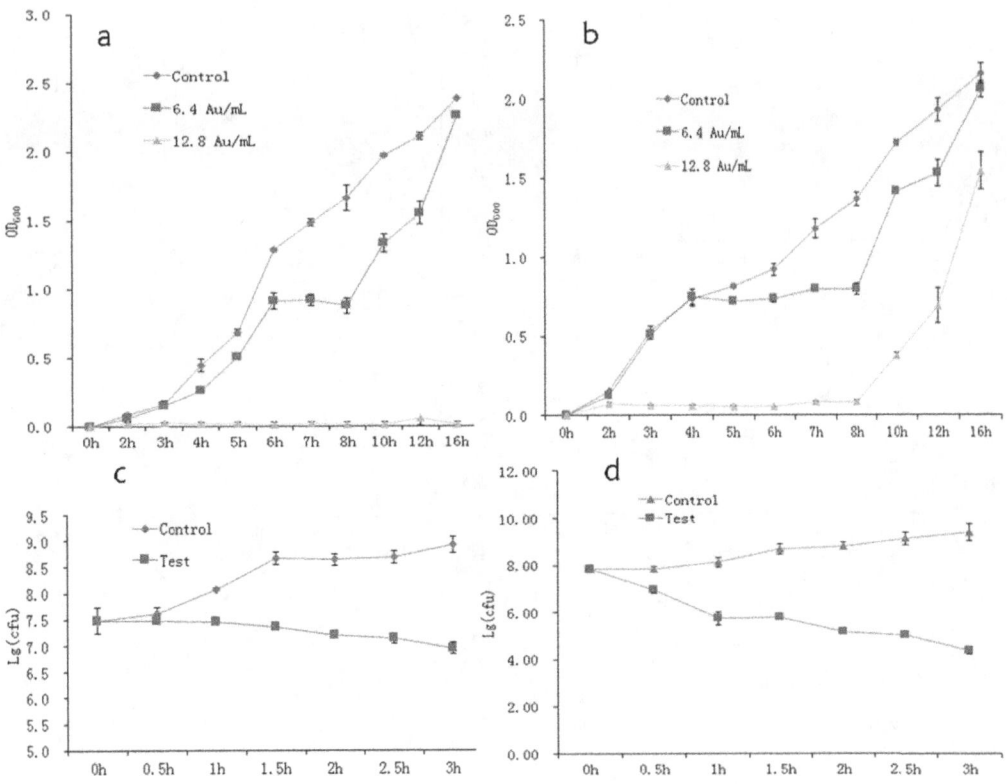

Figure 2. a) The effect of *S. aureus* biological growth by Bacteriocin 163-1; b) The effect of E. coli biological growth by Bacteriocin 163-1; c) Time-kill kinetics of Bacteriocin 163-1 against *S. aureus*; d) Time-kill kinetics of Bacteriocin 163-1 against *E. coli*

3.5 Effect of Bacteriocin 163-1 on Cell Conponents

To determine whether the bacteriocin 163-1 had an impact on the proteins, nuclear acid and LDH of *S. aureus* and *E. coli*, the bacteriocin 163-1 were added to the medium at a concentration of 6.4 Au/mL. It was found OD_{280} (extracellular proteins), OD_{260} (extracellular nuclear acid) and extracellular LDH activity were significantly increased after adding the bacteriocin (Figure 3), suggested that the the addition of bacteriocin 163-1 resulted in massive leakage of proteins, nuclear acid and LDH from the cells. The result implied that the bacteriocin may be disrupted the cell structure of the bacteria as cell membrane (Castellano et al., 2003; Oppegård et al., 2007).

In order to confirm above speculation, the cell structure was observed by Transmission Electron Microscopy (TEM). The non-treated *E. coli* cell grow normally its cell wall was smooth and not broken (Figure 4a), whereas the structures of cell membrane with the bacteriocin-treated were damaged (Figure 4b), and the holes were observed in the cell membrane (Figure 4b, arrow tip) and even the cells were deformed and distorted. In the case of *S. aureus* cells, similar effects were observed (Figure 4c), moreover, a larger number of cavity cells due to leakage out of cell component were found by treatment of the bacteriocin (Figure 4d). The results of TEM suggested that bacteriocin 163-1 could damage the cell wall and cross the cell membrane of the bacteria. Then, cytoplasms and components inside the cells were leaked to form empty hole, finally resulted in cell death.

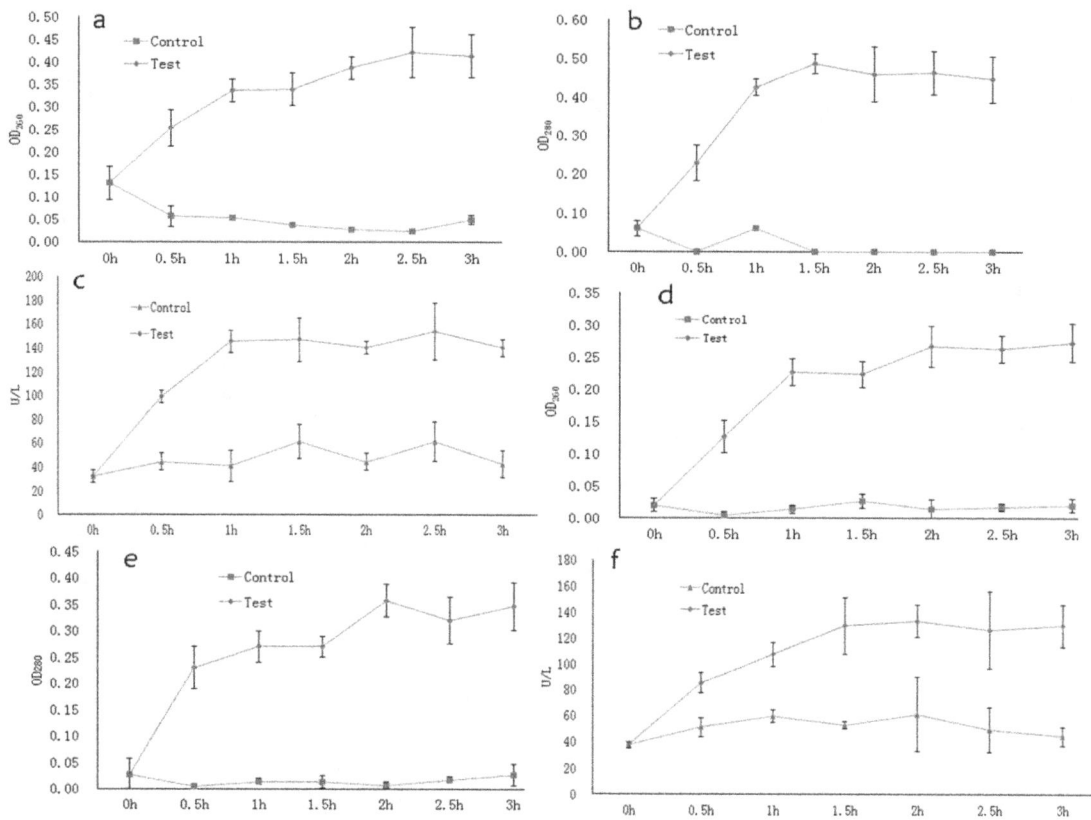

Figure 3. a) Change of the cell-free supernatant OD$_{260}$ of *S. aureus* by Bacteriocin 163-1-treated ; b) T Change of the cell-free supernatant OD$_{280}$ of *S. aureus* by Bacteriocin 163-1-treated; c) Change of the cell-free supernatant LDH of *S. aureus* by Bacteriocin 163-1-treated; d) Change of the cell-free supernatant OD$_{260}$ of *E. coli* by Bacteriocin 163-1-treated; e) Change of the cell-free supernatant OD$_{280}$ of *E. coli* by Bacteriocin 163-1-treated; f) Change of the cell-free supernatant LDH of *E. coli* by Bacteriocin 163-1-treated

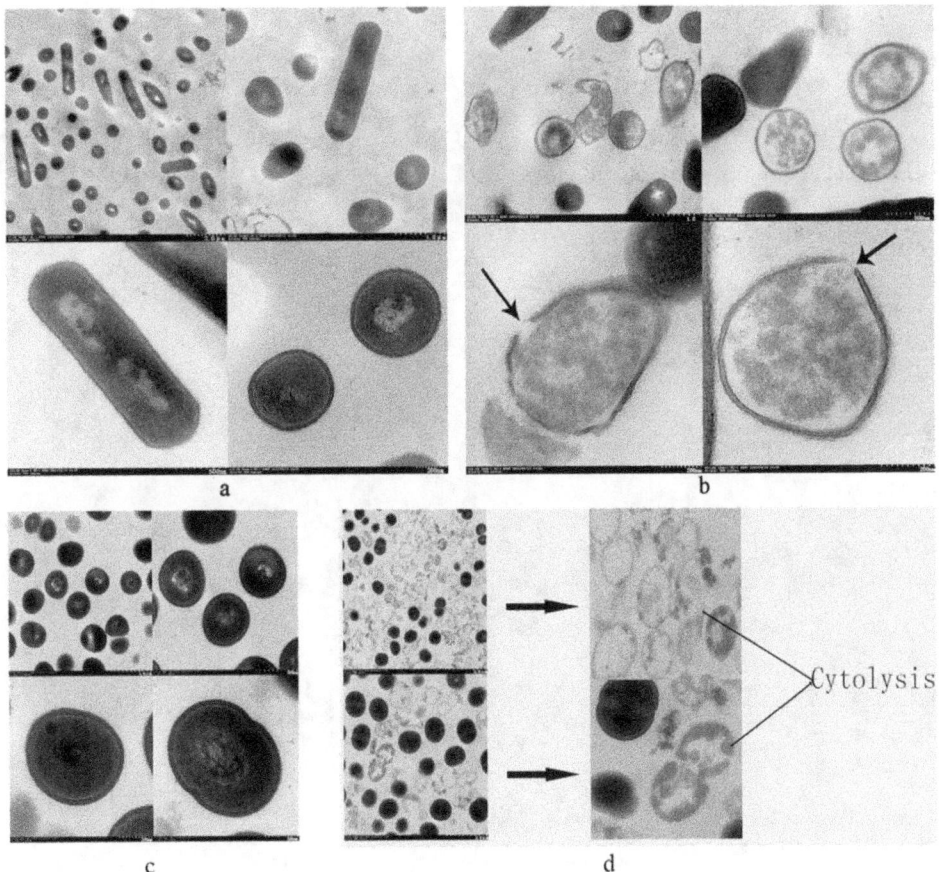

Figure 4. a) E. Coli treated without bacteriocin 163-1; b) E. Coli treated with bacteriocin 163-1(Circular for slitting, rod for crosscutting); c) S. aureus treated without bacteriocin 163-1; d) S. aureus treated with bacteriocin 163-1

3.6 Bacteriocin 163-1 Localization in Cell

FITC-Bacteriocin 163-1was subjected to observe action part of bacteriocin 163-1 in the bacteria cell by laser confocal microscopy (LCM). The observation results were shown in Figure 5. The fluorescence with FITC-Bacteriocin 163-1 was observed in whole cells of *E. coli* (Figures 5a and 6b) it was only done in cell wall or membrane of *S. aureus* due to leakage up of cell component (Figures 5c and 5d), indicating that the bacteriocin entered into *E. coli* cells and *S. aureus* cells by across the membrane. Some bacteriocins could form a pore in the sensitive bacterial membrane that leads to sensitive cell death, such as nisin, subtilin, lacticin 3147 (Bierbaum & Sahl, 2009; Chatterjee, Paul, Xie, & van der Donk, 2005; Wiedemann, Benz, & Sahl, 2004). Therfore, the results also suggested that the action mode of bacteriocin 163-1 for *E. coli* and *S. aureus* might be pore mode, then cytolysis.

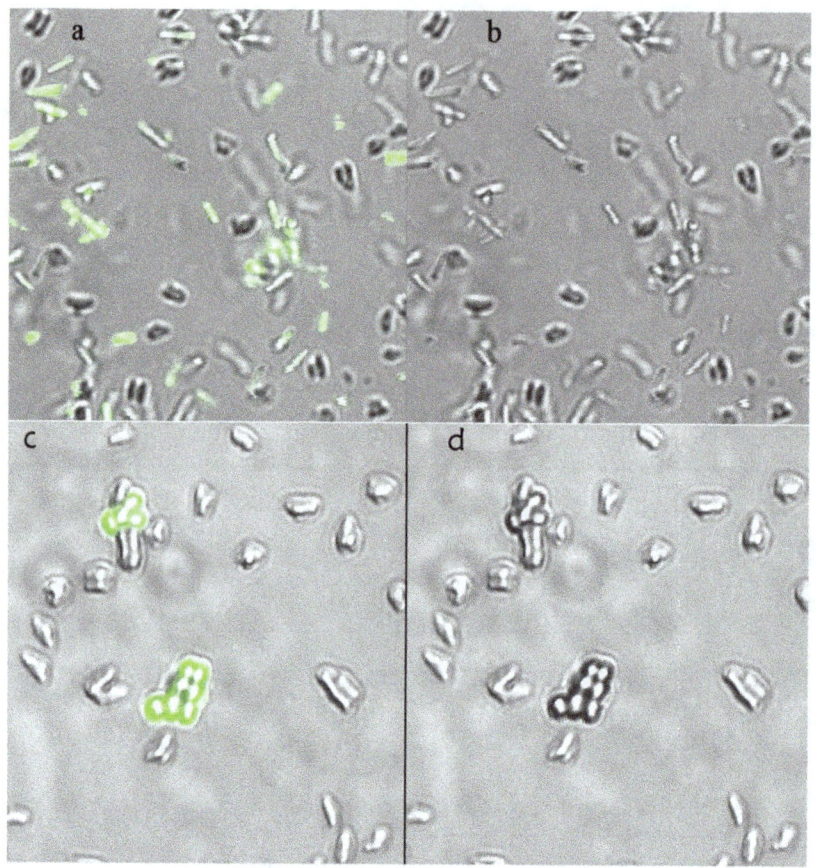

Figure 5. Fluorescence observations of bacteriocin 163-1-treated E.coli and S.aureus cells. a) *E. coli* treated with bacteriocin 163-1 were observed with fluorescence mode by Laser confocal microscopy; b) *E. coli* treated with bacteriocin 163-1 were observed with no fluorescence mode; c) *S. aureus* treated with bacteriocin 163-1 were observed with fluorescence mode by Laser confocal microscopy; d) *S. aureus* treated with bacteriocin 163-1 were observed with no fluorescence mode

4. Conclusion

In this study, bacteriocin 163-1 was purified through the preparative chromatography and reversed-phase chromatography, and its molecular mass (825 Da) was determined by MALDI-TOF-MS. The primary sequence of bacteriocin 163-1 was deduced as Y-V-C-A-S-P-W. Bacteriocin 163-1 was sensitive to the proteases and exhibited a broad-spectrum antimicrobial activity not only against lactic acid bacteria, but also against some foodborne pathogens. In addition, bacteriocin 163-1 appeared as a remarkable thermostability and pH stability. The fluorescence test with FITC-Bacteriocin 163-1 showed bacteriocin 163-1 acted on the cell membrane of bacteria and resulted in pore formation on the cell membrane and then cytolysis. The results indicated that the novel bacteriocin might be a potential candidate as a bio-preservative in agro-food industry in the future.

Acknowledgements

The authors would like to thank the Priority Academic Program Development of the Jiangsu Higher Education Institutions (PAPD) and Jiangsu Collaborative Innovation Center of Meat Production and Processing, Quality and Safety Control for financial support.

References

Ahmad, V., Muhammad Zafar Iqbal, A. N., Haseeb, M., & Khan, M. S. (2014). Antimicrobial potential of bacteriocin producing Lysinibacillus jx416856 against foodborne bacterial and fungal pathogens, isolated from fruits and vegetable waste. *Anaerobe, 27*, 87-95. http://dx.doi.org/10.1016/j.anaerobe.2014.04.001

Alvarez-Sieiro, P., Montalbán-López, M., Mu, D., & Kuipers, O. (2016). Bacteriocins of lactic acid bacteria: Extending the family. *Applied Microbiology & Biotechnology, 100*(7), 2939-2951. http://dx.doi.org/ 10.1007/s00253-016-7343-9

Anderssen, E. L., Diep, D. B., Nes, I. F., Eijsink, V. G., & Nissen-Meyer, J. (1998). Antagonistic activity of *Lactobacillus plantarum* C11: two new two-peptide bacteriocins, plantaricins EF and JK, and the induction factor plantaricin A. *Applied and Environmental microbiology, 64*(6), 2269-2272.

Atrih, A., Rekhif, N., Moir, A., Lebrihi, A., & Lefebvre, G. (2001). Mode of action, purification and amino acid sequence of plantaricin C19, an anti- Listeria bacteriocin produced by *Lactobacillus plantarum* C19. *International Journal of Food Microbiology, 68*(1), 93-104. http://dx.doi.org/10.1016/S0168-1605 (01)00482-2

Barefoot, S. F., & Klaenhammer, T. R. (1983). Detection and activity of lactacin B, a bacteriocin produced by Lactobacillus acidophilus. *Applied and Environmental microbiology, 45*(6), 1808-1815.

Bierbaum, G., & Sahl, H. (1987). Autolytic system of Staphylococcus simulans 22: Influence of cationic peptides on activity of N-acetylmuramoyl-L-alanine amidase. *Journal of Bacteriology, 169*(12), 5452-5458.

Bierbaum, G., & Sahl, H.-G. (2009). Lantibiotics: Mode of action, biosynthesis and bioengineering. *Current Pharmaceutical Biotechnology, 10*(1), 2-18. http://dx.doi.org/10.2174/138920109787048616

Biscola, V., Todorov, S. D., Capuano, V. S. C., Abriouel, H., Gálvez, A., & Franco, B. D. G. M. (2013). Isolation and characterization of a nisin-like bacteriocin produced by a Lactococcus lactis strain isolated from charqui, a Brazilian fermented, salted and dried meat product. *Meat Science, 93*(3), 607-613. http://dx.doi.org/10.1016/j.meatsci.2012.11.021

Burdock, G. A., & Carabin, I. G. (2004). Generally recognized as safe (GRAS): History and description. *Toxicology Letters, 150*(1), 3. http://dx.doi.org/10.1016/j.toxlet.2003.07.004

Castellano, P., Raya, R., & Vignolo, G. (2003). Mode of action of lactocin 705, a two-component bacteriocin from Lactobacillus casei CRL705. *International Journal of Food Microbiology, 85*(1), 35-43. http://dx.doi.org/10.1016/S0168-1605(02)00479-8

Chatterjee, C., Paul, M., Xie, L., & van der Donk, W. A. (2005). Biosynthesis and mode of action of lantibiotics. *Chemical Reviews, 105*(2), 633-684. http://dx.doi.org/10.1021/cr030105v

Chen, Y.-S., Wang, Y.-C., Chow, Y.-S., Yanagida, F., Liao, C.-C., & Chiu, C.-M. (2014). Purification and characterization of plantaricin Y, a novel bacteriocin produced by *Lactobacillus plantarum* 510. *Archives of Microbiology, 196*(3), 193-199. http://dx.doi.org/10.1007/s00203-014-0958-2

Cotter, P. D., Hill, C., & Ross, R. P. (2005). Bacteriocins: Developing innate immunity for food. *Nature Reviews Microbiology, 3*(10), 777-788. http://dx.doi.org/10.1038/nrmicro1273

Enan, G., El-Essawy, A. A., Uyttendaele, M., & Debevere, J. (1996). Antibacterial activity of *Lactobacillus plantarum* UG1 isolated from dry sausage: Characterization, production and bactericidal action of plantaricin UG1. *International Journal of Food Microbiology, 30*(3), 189-215.

Gao, Y., Jia, S., Gao, Q., & Tan, Z. (2010). A novel bacteriocin with a broad inhibitory spectrum produced by Lactobacillus sake C2, isolated from traditional Chinese fermented cabbage. *Food Control, 21*(1), 76-81. http://dx.doi.org/10.1016/j.foodcont.2009.04.003

Gong, H. S., Meng, X. C., & Wang, H. (2010). Plantaricin MG active against Gram-negative bacteria produced by *Lactobacillus plantarum* KLDS1.0391 isolated from "Jiaoke", a traditional fermented cream from China. *Food Control, 21*(1), 89-96. http://dx.doi.org/10.1016/j.foodcont.2009.04.005

Gupta, A., & Tiwari, S. K. (2014). Plantaricin LD1: A Bacteriocin Produced by Food Isolate of *Lactobacillus plantarum* LD1. *Applied Biochemistry and Biotechnology, 172*(7), 3354-3362. http://dx.doi.org/1007/s12010-014-0775-8

Hata, T., Tanaka, R., & Ohmomo, S. (2010). Isolation and characterization of plantaricin ASM1: A new bacteriocin produced by *Lactobacillus plantarum* A-1. *International Journal Of Food Microbiology, 137*(1), 94-99. http://dx.doi.org/10.1016/j.ijfoodmicro.2009.10.021

Hauge, H. H., Mantzilas, D., Eijsink, V. G., & Nissen-Meyer, J. (1999). Membrane-mimicking entities induce structuring of the two-peptide bacteriocins plantaricin E/F and plantaricin J/K. *Journal of Bacteriology, 181*(3), 740-747.

Hu, M., Zhao, H., Zhang, C., Yu, J., & Lu, Z. (2013). Purification and Characterization of Plantaricin 163, a Novel Bacteriocin Produced by *Lactobacillus plantarum* 163 Isolated from Traditional Chinese Fermented Vegetables. *Journal of Agricultural and Food Chemistry, 61*(47), 11676-11682. http://dx.doi.org/10.1021/jf403370y

Kato, T., Matsuda, T., Ogawa, E., Ogawa, H., Kato, H., Doi, U., & Nakamura, R. (1994). Plantaricin-149, a bacteriocin produced by *Lactobacillus plantarum* NRIC 149. *Journal of Fermentation and Bioengineering, 77*(3), 277-282.

Oppegård, C., Rogne, P., Emanuelsen, L., Kristiansen, P. E., Fimland, G., & Nissen-Meyer, J. (2007). The two-peptide class II bacteriocins: Structure, production, and mode of action. *Journal of Molecular Microbiology and Biotechnology, 13*(4), 210-219. http://dx.doi.org/10.1159/000104750

Pucci, M. J., Vedamuthu, E. R., Kunka, B. S., & Vandenbergh, P. A. (1988). Inhibition of Listeria monocytogenes by using bacteriocin PA-1 produced by Pediococcus acidilactici PAC 1.0. *Applied and Environmental Microbiology, 54*(10), 2349-2353.

Roepstorff, P., & Fohlman, J. (1984). *Biomed. Mass Spectrometry* (Vol. 11, p. 601).

Ruixiang, Z., Gaili, D., Tianyou, Y., Shengyang, N., & Ying, W. (2015). Purification, Characterization and Antibacterial Mechanism of Bacteriocin from Lactobacillus Acidophilus XH1. *Tropical Journal of Pharmaceutical Research, 14*(6), 989-995. http://dx.doi.org/10.4314/tjpr.v14i6.8

Wiedemann, I., Benz, R., & Sahl, H.-G. (2004). Lipid II-mediated pore formation by the peptide antibiotic nisin: A black lipid membrane study. *Journal of Bacteriology, 186*(10), 3259-3261. http://dx.doi.org/10.1128/JB.186.10.3259-3261.2004

Zhang, H., Liu, L., Hao, Y., Zhong, S., Liu, H., Han, T., & Xie, Y. (2013). Isolation and partial characterization of a bacteriocin produced by *Lactobacillus plantarum* BM-1 isolated from a traditionally fermented Chinese meat product. *Microbiology and Immunology, 57*(11), 746-755. http://dx.doi.org/10.1111/1348-0421.12091

Zhang, Q., Lu, Y., Liu, X., Bie, X., Lv, F., & Lu, Z. (2014). Preservative Effect of Food-Based Fermentate from Lactobacillus acidophilus NX2-6 on Chilled Pork Patties. *Journal of Food Protection, 77*(3), 459-465. http://dx.doi.org/10.4315/0362-028X.JFP-13-359

12

Identification of Resistance Sources to Wheat Stem Rust from Introduced Genotypes in Kenya

Molly O. Akello[1], Felister Nzuve[1], Florence Olubayo[1], Godwin Macharia[2] & James Muthomi[1]

[1] Department of Plant Science and Crop Protection, University of Nairobi, Nairobi, Kenya

[2] Kenya Agricultural and Livestock Research Organization, Njoro, Kenya

Correspondence: Molly O. Akello, Department of Plant Science and Crop Protection, University of Nairobi, P.O. Box 29053-0625, Nairobi, Kenya. E-mail: mollyakello2@gmail.com

The research is financed by Alliance for a Green Revolution in Africa (AGRA).

Abstract

Stem rust *Puccinia graminis* Pers. f. sp. *tritici* of wheat is the most important disease in Kenya. Emergence of race Ug99 and other variants virulent to host resistance genes including *Sr31* has rendered 95% of Kenyan cultivars susceptible. This study aimed to identify new sources of resistance to stem rust in a collection of exotic genotypes. Three hundred and sixteen wheat genotypes were screened at the Kenya Agricultural and Livestock Research Organization (KALRO) in Njoro for two seasons in 2015. The host reaction to disease was evaluated based on the modified Cobb scale. The relative Final Rust Severity (rFRS), Average Coefficient of Infection (ACI) and relative Area Under Disease Progress Curve (rAUDPC) were used to characterize the genotypes for stem rust resistance. Agronomic traits were also recorded. Six genotypes namely ALBW-100, ALBW- 204, EPCBW-261, EPCBW-295, PCHP-309 and PCHPBW-310 with significantly low ACI, rAUDPC and rFRS were identified. Thirty five genotypes showed Pseudo-Black Chaff (PBC) phenotype associated with resistant gene *Sr2*, a source of partial resistance in wheat. The genotypes also showed low disease severity (20-25%) and Moderately Susceptible (MS) – Susceptible (S) infection types in both seasons. Genotypes had significant differences ($p \leq 0.05$) on plant height, 1000-kernel weight and number of tillers indicating genetic variation which could be exploited in breeding for resistance to stem rust. The negative relationship between agronomic variables involving plant height, spikelet length and 1000-kernel weight showed harmful effects of stem rust on plant characteristics including yield. The stem rust resistant genotypes with good agronomic traits could be introgressed into adapted Kenyan backgrounds while the genotypes showing presence of PBC could be utilized to develop durable stem rust resistant wheat. Inheritance studies to elucidate the exact genes conferring resistance to stem rust could be conducted for breeders to exploit their genetic variability.

Keywords: adult plant resistance (APR), introduced genotypes, *Sr2* genes, stem rust

1. Introduction

Wheat (*Triticum aestivum* spp. *aestivum* L.) is an important food grain source worldwide. Its demand in developing countries is rising in recent decades and is expected to reach 60% by 2050 (FAO, 2016). However, climate change associated with erratic temperatures, droughts, floods, pests and disease epidemics will reduce wheat production by 29% (Rosegrant et al., 1995). Ever since wheat was introduced in Kenya-nearly a century ago, stem rust has remained a key production challenge. Strategic introgression and deployment of resistance genes in commercial varieties throughout the 1950s greatly circumvented major stem rust epidemics. However, in more recent years, the evolution and selection for new races with increased virulence has become undesirably frequent (Velu & Singh, 2013). In the last decade, the threats due to stem rust disease have re-emerged owing to a new set of races, called the "Ug99 family". The first recognizable variant in this family is the initially characterized race Ug99, also designated as TTKSK, based on its effects on select host resistance genes differentials (Jin et al., 2008). First reported in Uganda in 1998, the race TTKSK was found in Kenya in 2001. This race has unique virulence to *Sr31* and *Sr38* resistance genes widely utilized in wheat worldwide and for which virulence had not been reported previously in the world (Pretorius et al., 2000).

Globally, since the discovery of Ug99, about 80% to 90% of the wheat grown is susceptible to stem rust often leading to up to 70% yield losses (Singh et al., 2006). In Kenya, stem rust associated yield losses of 100%, have been reported (Njau et al., 2010). Over 95% of local commercial varieties are susceptible or highly susceptible, while only a few older varieties showing some level of 'adult plant resistance' (APR) (Wanyera et al., 2006; Njau et al., 2010) have been identified. Currently, more than 15 confirmed races in the Ug99 lineage have been reported in Africa and beyond (Singh et al., 2015). Through airborne transmission, Ug99 and its variants have reached Asia one of the main global wheat bread baskets; having been reported in Yemen in 2006, Iran in 2007 and Pakistan in 2009 (Hodson et al., 2009; Nazari et al., 2009; Admassu et al., 2009).

While chemicals can be used to manage stem rust, the main challenge is the high costs and the detrimental effects posed on the environment (Beard et al., 2006). Genetic resistance is the most economically viable method of controlling stem rust. To date over 70 genes have been designated for resistance to stem rust (McIntosh et al., 2014). Of those, 34 are ineffective against race Ug99 (Singh et al., 2015). Among the adult plant resistance genes, *Sr2* gene is the only well studied. A combination of *Sr2* gene and other unknown slow rusting resistance genes forms the "*Sr2* complex" which provides durable resistance to stem rust (McIntosh, 1988). The concept-"breeding for durable resistance in wheat" championed by Dr. Norman Borlaug has led to a global search for new genes or gene combinations to combat the new stem rust race Ug99 that could be released as new varieties.

A shuttle breeding scheme with the goal of phenotyping breeding populations for adult plant resistance to Ug99 family of races was initiated in 2005 between KALRO-Njoro and the International Maize and Wheat Improvement Centre (CIMMYT) (http://www.globalrust.org/). Through the initiative, several genotypes, with potentially good sources of APR genes have been identified. A study conducted at KALRO-Njoro on elite advanced CIMMYT bread wheat indicated that 30% of the materials were susceptible at the seedling stage while various levels of APR were identified in the field tests (Njau et al., 2010). Invariably, all APR in the studied material was associated with the *Sr2* complex. Bhavani et al. (2011) reported detection of *Sr2* gene based on mapping studies on six CIMMYT parental lines. These reports indicate that exotic wheat genotypes could be good sources of resistance to stem rust that could be deployed in the national wheat improvement program at KALRO-Njoro. A study on 30 vintage Kenyan varieties for resistance to stem rust both at the seedling and adult plant growth stage revealed that none of them except variety *Bonny* was resistant at the seedling stage, while a few had APR attributed to *Sr2* gene in their backgrounds (Njau et al., 2009).

Given the devastating nature of Ug99 family of stem rust races to wheat productivity regionally, effort to explore for resistance sources and incorporation of effective genes into new high yield commercial varieties is paramount. Accordingly, the objective of this study was to identify suitable sources of resistance to stem rust among exotic wheat genotypes.

2. Materials and Methods

2.1 Experimental Site

The research was carried at KALRO-Njoro which is about 2185 meters above sea level, approximately 0°20'S; 35°56'E. Average temperatures ranges between 9.7 °C and 23.5 °C. Mean annual precipitation is 900 mm (Ooro et al., 2009). KALRO-Njoro hosts the global phenotyping facility for characterizing and selection of wheat genotypes resistant to stem rust (Singh et al., 2006), under the auspices of the BGRI project (http://www.globalrust.org/).

2.2 Genotypes

Three hundred and sixteen exotic genotypes namely: 250 "Aluminum Bread Wheat (AL BW)", 47 "Elite Bread Wheat (EPC BW)" and 19 "PC-Harvest plus Bread Wheat (PCHP BW)". These are genetically fixed lines, bred targeting high yields and superior grain quality including specific nutritional needs (Velu & Singh, 2013). The universally stem rust susceptible genotype CACUKE was included to monitor proliferation of the disease epidemic. A mixture of seven rust susceptible genotypes including Morocco, Robin, and PBW343 were used.

2.3 Experimental Procedures

The study was undertaken twice; off-season (January to June 2015) and main-season nursery (July to November 2015). Experiments in both seasons were established based on augmented alpha lattice square design with no replication. Experimental plots were 6 rows by 2 m length and 20 cm inter-row spacing. Diammonium phosphate (DAP) was applied at the rate of 125 kg/ha while planting, followed by an application of Urea (75 kg/ha) as a source of Nitrogenn at late tillering and booting stages of the plants.

2.4 Data Collection

2.4.1 Agronomic Traits

These included: plant height (cm), spike length (cm) and 1000-kernel weight (g) evaluated at harvest maturity. Presence of *Sr2* gene was recorded by observing occurrence of pseudo-black chaff (PBC) on the stems and heads of the plant. A plus (+) was used to indicated presence of PBC while a minus (-) indicated absence of PBC.

2.4.2 Disease Scoring

This was done in season one and two when the susceptible check CACUKE expressed 50% rust severity, and 70% severity respectively. The stem rust severity was recorded based on the modified Cobb's scale on a 0-100% scale (Peterson et al., 1948). The host response which included the trace (TR), resistant (R), moderately resistant (MR), moderately susceptible (MS), and susceptible (S) were recorded based on Roelfs et al. (1992) scale. The disease severity scores were converted into area under the disease progress curve (AUDPC) values following Wilcoxson et al. (1975) Equation 1.

$$AUDPC = \sum_{i=1}^{n} \frac{(y_i + y_{i+1})}{2}(t_{i+1} - t_i) \tag{1}$$

Where, y_i = average coefficient of infection of the ith reading; y_{i+1} = average coefficient of infection of $i + 1^{th}$ reading; $(t_{i+1} - t_i)$ = number of days between the ith and the $i + i^{th}$ reading, and n = number of observations.

The susceptible check CACUKE was used as a reference to obtain the relative AUDPC (rAUDPC) in Equation 2, as well as relative Final Rust Severity (rFRS), in Equation 3.

$$rAUDPC = \frac{AUDPC\ of\ Genotypes}{AUDPC\ of\ Susceptible\ Check} \times 100 \tag{2}$$

$$rFinal\ Rust\ Severity = \frac{Final\ Rust\ Severity\ of\ Genotype}{Final\ Rust\ Severity\ of\ Susceptible\ Check} \times 100 \tag{3}$$

The coefficient of infection (CI) was obtained by multiplying the final disease severity for each individual score by the numerical value where TR = 0.1; R = 0.2; MR = 0.4; M = 0.6; MS= 0.8; and averaged to give average coefficient of infection (ACI) (Roelfs et al., 1992).

2.5 Data Analysis

The analysis of variance (ANOVA) was used to discern differences among genotypes. Genotypes were considered fixed while the seasons were considered as random effects. Genotypic means were separated based on Fisher's protected least significance difference (LSD) at $p \leq 0.05$. A Pearson correlation coefficient test was done to determine the relationship between the different disease parameters and the agronomic traits.

3. Results

3.1 Reactions of Genotypes to Weather Conditions across the Seasons

There was high disease pressure among the genotypes during both seasons. The susceptible check, CACUKE showed 90% and 100% final rust severity in season one and two respectively.

3.2 Reaction of Genotypes to Stem Rust and Other Agronomic Traits

ANOVA revealed significant differences at $p \leq 0.05$ among the genotypes for all the agronomic traits, apart from TKW (Table 1). Seasonal variations were significant at ($p \leq 0.05$) for AUDPC, rAUDPC, TKW and PH. Significant genotypes × season interactions were revealed among the disease parameters but not the agronomic traits. Dwarf genotypes with heights ranging from 62.5 to 68.2 cm relative to the tallest genotype (ALBW-135) - 94 cm tall were identified in both seasons. These were: ALBW-38, ALBW-65, ALBW-72, ALBW-81, ALBW-82, ALBW-99, ALBW-108, ALBW-207, ALBW-208, ALBW-210, EPCBW-292, PCHP-300, PCHP-312 and PCHP-316. They also recorded high TKW ranging between 32.7 to 39.2 g in both seasons (Table 2). The longest spikes averaged from 12.4 cm among the ALBW genotypes. ALBW-98 had the highest average TKW (47.4 g). Notably, rust susceptible genotypes ALBW-16 depicting high disease severities of 40% and 50% in season one and two respectively also had a relatively low average TKW of 24.4 g, only slightly higher than that of the susceptible check CACUKE 20.2 g in season 2 (Table 2).

Genetic variations for resistance to stem rust within each season were observed. Six genotypes: ALBW-100, ALBW-204, EPCBW-261, EPCBW-295, PCHP-309 and PCHPBW-310, showed R-MR response in season one and trace responses to stem rust in season two. Among the ALBW, ACI ranged from 1-60, while the AUDPC ranged from 15-455 in season one and 0-795 in season two. Over half (54%) of the two hundred and fifty ALBW recorded moderately susceptible-susceptible (MSS) response with AUDPC, ranging from 90-650, in season two, lower than the susceptible check CACUKE, whose AUDPC was 700 in season two. Only 6 genotypes from the ALBW set namely ALBW-4, ALBW-100, ALBW-106, ALBW-173, ALBW-174 and ALBW-204 showed R or MR infection types in both seasons (Table 2).

Among the 47 genotypes in the EPCBW set, the AUDPC ranged from 20-230 in season one and 0-605 in season two. EPCBW genotypes depicted lower rAUDPC, than ALBW genotypes ranging between 0-46 in season one and 0-86 in season two. The EPCBW genotypes had the highest number of intermediate infection type (M) in (twenty two genotypes) (Table 2). Among the PCHP accessions, AUDPC ranged from 78-303 in season one and 0-415 in season two lower than those in the ALBW set. The rFRS ranged from 14-35 in season one and 0-42 in season two. Fourteen of the 19 PCHPBW genotypes evaluated (74%) showed moderately susceptible to susceptible infection types.

Across all test material thirty five wheat genotypes depicted the pseudo black chaff (PBC) on the spikes and the necks of the plants in season two.

3.3. Correlation Coefficients among Disease Parameters and Agronomic Traits

The Pearson Correlation coefficients considered between pairs of the respective disease parameters were highly significant ($p \leq 0.05$) (Table 3). Among the agronomic traits, negative values for Pearson correlation coefficient were observed with respect to stem rust severity. Specifically, Pearson's correlation coefficients with rust severity were -0.0475, -0.0253 and -0.3401 respectively for plant height, spikelet length and TKW (Table 4). This indicated that as the stem rust severities increased, this had negative effects on the agronomic traits. There were significant differences in the relationships between spikelet length and plant height; thousand kernel weight and spikelet length and stem rust and thousand kernel weights ($p \leq 0.05$). The p-values of all the other agronomic traits were not significantly different from each other (Table 4).

Table 1. ANOVA table highlighting levels of significance observed among various disease and agronomic parameters

SOURCE	Df	AUDPC	rAUDPC	ACI	rFRS	TKW (g)	PH (cm)	SL (cm)
Genotype	316	20167*	411.6*	126.7*	320.2*	34.46	67.79*	1.441*
Season	1	37712*	769.6*	83.27	360.2	4037.23*	687.7*	0.3222
Genotype by Season	316	13136*	268.1*	69.09*	208.4*	33.08	18.02	0.5211
Residual	316	8943	182.5	52.38	134.6	27.72	33.22	0.8384
Total	633							

Note. * Significance at $p \leq 0.05$. df = degree of freedom; AUDPC = area under the disease progress curve; rAUDPC = relative area under the disease progress curve; ACI = average coefficient of infection; rFRS = relative final rust severity; TKW = thousand kernel weight in grams; PH = plant height; SL = spikelet length in centimeters.

Table 2. Means among different disease and agronomic traits parameters among 316 tested genotypes

Genotype	ACI	AUDPC	rAUDPC	rFRS	Stem rust season 1		Stem rust season 2		TKW (g)	PH (cm)	SL (cm)	PBC
					Severity	FR	Severity	FR				
AL BW-1	6	114	32.6	11.8	10	MSS	10	MR	32.2	77.4	8.5	-
AL BW-2	10	134	38.3	14.7	5	MSS	20	MSS	37.6	73.7	9.2	-
AL BW-3	44	627.5	179.3	44.1	60	MSS	15	MSS	37.3	73.5	8.5	-
AL BW-4	4	47.5	13.6	5.9	10	MSS	0	TR	39.3	61.5	7.2	-
AL BW-5	9	105	30	14.7	20	MSS	5	MR	36.8	71.5	6.5	-
AL BW-6	8	141.5	40.4	11.8	10	MSS	10	MS	44.9	72.3	6.8	-
AL BW-7	11	170	48.6	17.6	20	MSS	10	M	33.8	76.9	7.7	-
AL BW-8	18	256.5	73.3	26.5	15	MSS	30	MSS	35.1	81.7	8.9	-
AL BW-9	18	229	65.4	26.5	15	MSS	30	MSS	39.2	76.5	7.6	-
AL BW-10	20	285	81.4	29.4	30	MSS	20	MSS	35.4	73.5	9.2	-
AL BW-11	18	247.5	70.7	26.5	5	MSS	40	MSS	29.8	80.9	8.8	-
AL BW-12	19.5	274	78.3	26.5	15	S	30	MSS	31.6	86.5	8.6	-

Genotype	ACI	AUDPC	rAUDPC	rFRS	Stem rust season 1		Stem rust season 2		TKW (g)	PH (cm)	SL (cm)	PBC
					Severity	FR	Severity	FR				
AL BW-13	24	360	102.9	35.3	30	MSS	30	MSS	33.1	72.4	8.6	-
AL BW-14	28	397.5	113.6	41.2	20	MSS	50	MSS	35	64.5	7.8	-
AL BW-15	30	419	119.7	44.1	25	MSS	50	MSS	33	79.5	7.6	-
AL BW-16	40	521.5	149	52.9	40	S	50	MSS	24.4	75.2	8.1	-
AL BW-17	20	275	78.6	29.4	10	MSS	40	MSS	35.5	78.4	8.4	-
AL BW-18	22	286.5	81.9	32.4	15	MSS	40	MSS	28.2	81	7.3	-
AL BW-19	20	275.5	78.7	29.4	20	MSS	30	MSS	27.3	83.9	8.5	-
AL BW-20	20	294	84	29.4	10	MSS	40	MSS	29.6	73.3	8.7	-
AL BW-21	18	320	91.4	26.5	5	MSS	40	MSS	33.5	89.5	8.4	-
AL BW-22	15	191.5	54.7	23.5	10	M	30	MSS	29.6	68.3	8.8	-
AL BW-23	9.5	151.5	43.3	14.7	5	M	20	MS	29.7	74.4	8.1	-
AL BW-24	16	191.5	54.7	23.5	10	MSS	30	MSS	33.6	71.5	8.1	-
AL BW-25	14	190	54.3	20.6	5	MSS	30	MSS	29.7	79	8.5	-
AL BW-26	13.5	192	54.9	20.6	5	M	30	MSS	32.2	78.4	9.3	-
AL BW-27	28	406.5	116.1	41.2	20	MSS	50	MSS	33	80.2	7.9	-
AL BW-28	16	272.5	77.9	23.5	20	MSS	20	MSS	28.3	76.7	7.9	-
AL BW-29	10	170.5	48.7	14.7	5	MSS	20	MSS	31.3	69.7	9.5	-
AL BW-30	19	294	84	29.4	10	M	40	MSS	36.9	74	9	-
AL BW-31	16	200	57.1	23.5	10	MSS	30	MSS	29.4	73.5	8.1	-
AL BW-32	16	227.5	65	23.5	20	MSS	20	MSS	34.5	81.2	8.8	-
AL BW-33	12	216.5	61.9	17.6	20	MSS	10	MSS	28.3	77.9	9.2	-
AL BW-34	15	219	62.6	23.5	30	MSS	10	M	32.6	84.3	9.3	-
AL BW-35	16	245	70	23.5	25	MSS	15	MS	35.5	83.8	7.7	-
AL BW-36	28	305.5	87.3	35.3	40	S	20	MS	37	69.7	9	-
AL BW-37	11	153	43.7	17.6	20	MSS	10	M	30.1	67.5	8.7	-
AL BW-38	14	207.5	59.3	20.6	20	MSS	15	MS	34.2	66	10.2	-
AL BW-39	8	124	35.4	14.7	15	MSS	10	MR	36.7	75.7	8.5	+
AL BW-40	22	331.5	94.7	32.4	15	MSS	40	MSS	28.1	74.5	8.1	-
AL BW-41	18	229	65.4	26.5	15	MSS	30	MSS	33.6	77.5	8.8	-
AL BW-42	16	227.5	65	23.5	10	MSS	30	MSS	32.8	81.7	9.8	-
AL BW-43	12.5	226.5	64.7	20.6	20	MSS	15	M	31.8	71.3	8.6	-
AL BW-44	18	229	65.4	26.5	15	MSS	30	MSS	32.3	76.2	8.9	-
AL BW-45	4.3	55	15.7	6.5	10	M	10	MS	38	80.7	8.8	-
AL BW-46	6	104	29.7	8.8	5	MS	10	MS	29.8	81.5	7.9	-
AL BW-47	22	312.5	89.3	32.4	15	MSS	40	MSS	29.6	80.3	8.6	-
AL BW-48	5.5	104	29.7	8.8	5	M	10	MS	32.2	77	9.6	-
AL BW-49	22	276.5	79	29.4	40	S	10	MR	35.7	71.4	9.3	+
AL BW-50	14	255.5	73	23.5	20	M	20	MSS	34.2	81.2	8.7	-
AL BW-51	6	131.5	37.6	8.8	10	MSS	5	MS	37.2	76.9	8.7	-
AL BW-52	9	114	32.6	11.8	10	S	10	MSS	34.4	72.5	9.2	-
AL BW-53	5	104	29.7	8.8	5	MSS	10	M	35.3	80	8.1	-
AL BW-54	12	161.5	46.1	17.6	10	MSS	20	MSS	33.4	76	8.6	-
AL BW-55	5.5	112.5	32.1	8.8	5	M	10	MS	37.7	84.5	8.2	-
AL BW-56	38	490	140	55.9	25	MSS	70	MSS	25.8	73.5	8.5	-
AL BW-57	3.5	131.5	37.6	8.8	10	MR	5	M	37.7	78.2	9.9	+
AL BW-58	14	171.5	49	20.6	15	MSS	20	MSS	26.9	73.2	8.5	-
AL BW-59	8	95	27.1	11.8	10	MS	10	MS	39.6	79.5	7.8	-
AL BW-60	13	134	38.3	17.6	10	S	20	MSS	35.6	80.7	9.1	-
AL BW-61	4.5	114	32.6	11.8	5	M	15	MR	34.2	85.5	8.8	+
AL BW-62	8	122.5	35	11.8	10	MSS	10	MSS	28.4	77.4	8.6	-

Genotype	ACI	AUDPC	rAUDPC	rFRS	Stem rust season 1		Stem rust season 2		TKW (g)	PH (cm)	SL (cm)	PBC
					Severity	FR	Severity	FR				
AL BW-63	12	180	51.4	17.6	10	MSS	20	MSS	36.9	79.3	8	-
AL BW-64	13.5	190	54.3	20.6	5	M	30	MSS	30.2	80.4	8.4	-
AL BW-65	22	295.5	84.4	32.4	15	MSS	40	MSS	33	66.9	8.7	-
AL BW-66	7	141.5	40.4	11.8	10	MSS	10	M	29.5	75.7	9.8	-
AL BW-67	8	114	32.6	11.8	15	MSS	5	MS	31	71.9	9	-
AL BW-68	24	314	89.7	35.3	30	MSS	30	MSS	32.1	77	8.8	-
AL BW-69	18	275	78.6	26.5	15	MSS	30	MSS	36.3	81.4	7.3	-
AL BW-70	17	218	62.3	20.6	30	S	5	MSS	34	77.5	8.8	-
AL BW-71	7	141.5	40.4	11.8	10	M	10	MSS	29.8	73.4	7.1	-
AL BW-72	4.5	112.5	32.1	8.8	10	M	5	M	35.7	62.5	8.1	-
AL BW-73	5	86.5	24.7	11.8	10	M	10	MR	37.5	84.5	8.3	+
AL BW-74	8	105.5	30.1	11.8	15	MSS	5	MS	37.2	71.9	8.1	-
AL BW-75	20	294	84	29.4	35	MSS	15	MSS	26.8	76.5	8.3	-
AL BW-76	12	161.5	46.1	17.6	20	MSS	10	MSS	26.7	86.5	7.8	-
AL BW-77	13	209	59.7	20.6	25	MSS	10	M	28.9	74.2	9.2	-
AL BW-78	14	190	54.3	20.6	15	MSS	20	MSS	25.4	79.9	10.2	-
AL BW-79	8.5	163	46.6	20.6	15	M	20	MR	33.3	80.9	7.9	+
AL BW-80	12.5	190.5	54.4	20.6	20	MSS	15	M	32	71.9	10.2	-
AL BW-81	16	209	59.7	23.5	20	MSS	20	MSS	26.6	66.7	9	-
AL BW-82	9	151.5	43.3	14.7	15	MSS	10	M	38.2	68.8	8.6	-
AL BW-83	14	217.5	62.1	20.6	15	MSS	20	MSS	32.1	77	8.7	-
AL BW-84	14	218	62.3	20.6	15	MSS	20	MSS	28.4	76.3	8.1	-
AL BW-85	23	266.5	76.1	29.4	40	S	10	M	35	81.7	8.5	-
AL BW-86	28	341.5	97.6	35.3	40	S	20	MSS	39.7	85.4	7.7	-
AL BW-87	18	303	86.6	29.4	20	M	30	MSS	28.4	82.5	8.9	-
AL BW-88	10	153	43.7	17.6	20	M	10	MS	26	83.3	9.7	-
AL BW-89	12	255	72.9	23.5	20	M	20	M	31.4	78.9	8	-
AL BW-90	12	216.5	61.9	17.6	20	MSS	10	MSS	35.6	73.4	8.3	-
AL BW-91	10	179	51.1	14.7	20	MS	5	MS	30	87.2	9.3	-
AL BW-92	5	76.5	21.9	8.8	10	M	5	MS	31.7	71.3	9.6	-
AL BW-93	3.5	58	16.6	5.9	5	M	5	M	29.3	77.9	9.1	-
AL BW-94	5	78	22.3	11.8	10	M	10	MR	31.8	81.3	8.2	+
AL BW-95	3.5	95.5	27.3	8.8	5	M	10	MR	34.1	73	8.7	+
AL BW-96	5.5	54.5	15.6	8.8	5	M	10	MS	36.9	73	8.1	-
AL BW-97	3.5	68	19.4	8.8	10	MR	5	M	34.2	71.7	8.5	+
AL BW-98	3.5	104	29.7	8.8	10	MR	5	M	47.4	72	8.1	+
AL BW-99	2.5	47.5	13.6	5.9	5	MR	5	M	30.7	64.7	8.4	+
AL BW-100	3	40	11.4	5.9	10	M	0	TR	30.2	72.3	9.1	-
AL BW-101	8	105.5	30.1	11.8	10	MSS	10	MS	32.6	79.8	8.9	-
AL BW-102	12	227.5	65	23.5	20	M	20	M	32.6	86.5	10.1	-
AL BW-103	10	124	35.4	14.7	5	MSS	20	MSS	28.7	71.3	8.7	-
AL BW-104	12	171.5	49	17.6	15	MSS	15	MS	29.1	81.7	9.8	-
AL BW-105	8	141.5	40.4	11.8	10	MSS	10	MS	28.6	76.3	8	-
AL BW-106	8	123	35.1	11.8	20	MSS	0	TR	34.5	62.8	7.5	-
AL BW-107	5.5	85	24.3	8.8	10	MSS	5	M	36.6	71.2	8.5	-
AL BW-108	11	197.5	56.4	17.6	10	M	20	MSS	30.9	61.5	9.1	-
AL BW-109	20	320	91.4	29.4	20	MSS	30	MSS	32.2	82	8.6	-
AL BW-110	14	218	62.3	20.6	30	MSS	5	MS	37.8	83.8	8.9	-
AL BW-111	7	141.5	40.4	11.8	10	M	10	MS	33.4	81.5	9.2	-
AL BW-112	12	189	54	17.6	25	MSS	5	MS	30	84.7	8.3	-

Genotype	ACI	AUDPC	rAUDPC	rFRS	Stem rust season 1		Stem rust season 2		TKW (g)	PH (cm)	SL (cm)	PBC
					Severity	FR	Severity	FR				
AL BW-113	13.5	181.5	51.9	20.6	30	MSS	5	M	36.3	79.8	9.2	-
AL BW-114	20	230.5	65.9	29.4	30	MSS	20	MSS	32.6	71.4	9.7	-
AL BW-115	23	257.5	73.6	29.4	30	S	20	MSS	30.3	76.7	10	-
AL BW-116	8.5	134	38.3	14.7	15	M	10	MS	33.4	71.5	9.8	-
AL BW-117	12.5	209	59.7	20.6	15	M	20	MSS	31.9	80	8.2	-
AL BW-118	16	229	65.4	23.5	20	MSS	20	MSS	26.7	76	10.6	-
AL BW-119	14	190.5	54.4	20.6	20	MSS	15	MSS	29.2	72.3	8.4	-
AL BW-120	10	170.5	48.7	14.7	20	MSS	5	MS	33.6	79.9	9.8	-
AL BW-121	16	200.5	57.3	23.5	20	MSS	20	MSS	26.3	72.7	9.9	-
AL BW-122	8	151.5	43.3	14.7	10	MR	15	MSS	30.2	84.5	8.6	-
AL BW-123	16	219	62.6	23.5	10	MSS	30	MSS	32.9	75	9.1	-
AL BW-124	4	104	29.7	8.8	5	MS	10	MR	27.8	74.2	9	+
AL BW-125	10	132.5	37.9	14.7	15	MSS	10	MSS	36.3	76.2	8.4	-
AL BW-126	14	226.5	64.7	20.6	25	MSS	10	MSS	31.9	83.7	9.5	-
AL BW-127	16	228	65.1	23.5	25	MSS	15	MSS	28.3	82.7	9.2	-
AL BW-128	11	180.5	51.6	17.6	20	MSS	10	M	36.2	81.2	7.9	-
AL BW-129	20	247.5	70.7	29.4	20	MSS	30	MSS	37.8	88.3	8.1	-
AL BW-130	6.5	160.5	45.9	11.8	15	M	5	MS	31.1	85.5	9.1	-
AL BW-131	7	141.5	40.4	11.8	10	M	10	MSS	34.1	79.8	9.2	-
AL BW-132	12	216.5	61.9	17.6	20	MSS	10	MSS	32.5	76.9	8.6	-
AL BW-133	9	199	56.9	17.6	20	M	10	M	36.5	77.7	8.9	+
AL BW-134	11	207.5	59.3	17.6	10	M	20	MSS	28.7	79.5	7.7	-
AL BW-135	9	245.5	70.1	17.6	20	M	10	M	34.1	94	9.8	-
AL BW-136	9.5	218	62.3	20.6	25	M	10	MR	29.8	88.5	10.5	+
AL BW-137	24.5	304	86.9	32.4	25	S	30	MSS	30.8	77.2	9.1	-
AL BW-138	14.5	180	51.4	17.6	25	S	5	MS	29.5	82.2	8.7	-
AL BW-139	10	143	40.9	14.7	15	MSS	10	MS	32.9	78.5	8.2	-
AL BW-140	16	200.5	57.3	23.5	30	MSS	10	MS	29.4	73.7	9.3	-
AL BW-141	14	173	49.4	20.6	25	MSS	10	MS	30.4	67.2	8	-
AL BW-142	5.5	76.5	21.9	8.8	5	M	10	MS	37.5	72.5	9.6	-
AL BW-143	7	114	32.6	11.8	10	M	10	MSS	33.8	71.7	8.4	-
AL BW-144	32	435	124.3	41.2	40	S	30	MSS	31.2	83.5	9.5	-
AL BW-145	9.5	153	43.7	14.7	5	M	20	MSS	30.4	84.7	10.4	-
AL BW-146	10	151.5	43.3	14.7	15	MSS	10	MS	34.1	85	8.7	-
AL BW-147	24.5	321.5	91.9	32.4	25	S	30	MSS	35.6	72.9	7.5	-
AL BW-148	26	360.5	103	35.3	20	S	40	MSS	32.3	76.9	9.8	-
AL BW-149	12	180.5	51.6	17.6	10	MS	20	MSS	35	75.7	8.5	-
AL BW-150	16	237.5	67.9	23.5	20	MSS	20	MSS	31.1	81.5	10	-
AL BW-151	6	133	38	11.8	10	M	10	M	30	73.7	9.7	-
AL BW-152	3.5	85.5	24.4	5.9	5	MSS	5	M	36	77.9	9.6	-
AL BW-153	4	75	21.4	5.9	5	MSS	5	MS	38.4	83.5	10.3	-
AL BW-154	6	105.5	30.1	11.8	10	M	10	M	39.4	83.7	9.7	-
AL BW-155	4	95.5	27.3	8.8	10	M	5	MR	34.8	79.4	8.2	+
AL BW-156	7	124	35.4	11.8	15	MSS	5	MR	30	70.5	8.6	-
AL BW-157	15.5	218	62.3	20.6	25	S	10	M	30.4	76.7	8.4	-
AL BW-158	6	133	38	11.8	15	M	5	M	29.1	76.4	9.4	+
AL BW-159	10	114	32.6	14.7	15	MSS	10	MS	35.5	76.2	10.6	-
AL BW-160	5	112.5	32.1	8.8	10	M	5	MSS	36.6	82	8.3	-
AL BW-161	8.5	143	40.9	14.7	15	M	10	MSS	36.4	80.4	7.9	-
AL BW-162	7.5	151.5	43.3	14.7	15	M	10	M	40.1	89.5	9.5	+

Genotype	ACI	AUDPC	rAUDPC	rFRS	Stem rust season 1		Stem rust season 2		TKW (g)	PH (cm)	SL (cm)	PBC
					Severity	FR	Severity	FR				
AL BW-163	5.5	85	24.3	8.8	5	M	10	MSS	37.2	63.4	7.4	-
AL BW-164	6.5	105.5	30.1	11.8	15	M	5	MS	38.4	88.2	8.5	-
AL BW-165	13.5	200.5	57.3	20.6	30	MSS	5	M	32.7	80.2	8.4	-
AL BW-166	11.5	197.5	56.4	17.6	25	MSS	5	M	33.3	79.8	8.5	-
AL BW-167	11.5	199	56.9	20.6	25	M	10	MS	27.5	74.2	8.3	-
AL BW-168	31	371.5	106.1	41.2	30	S	40	MSS	30.7	81.2	9	-
AL BW-169	18	116	33.1	23.5	20	S	20	MSS	36.4	87.5	9	-
AL BW-170	7	133	38	11.8	10	M	10	MS	30.7	91.7	9.7	-
AL BW-171	5.5	76.5	21.9	8.8	10	MSS	5	M	37.7	88.9	9.9	-
AL BW-172	5	104	29.7	8.8	10	M	5	MSS	37.2	84.2	8.9	-
AL BW-173	1.5	29	8.3	2.9	5	M	0	TR	38.9	74.7	8.4	-
AL BW-174	3	66.5	19	5.9	10	M	0	TR	35.1	71.8	8.7	-
AL BW-175	6	76.5	21.9	8.8	10	MSS	5	MSS	39.2	64.4	9.2	-
AL BW-176	15.5	163	46.6	20.6	15	S	20	MSS	35.7	72.9	10.2	-
AL BW-177	3	85.5	24.4	5.9	5	M	5	M	36.4	74	9.2	-
AL BW-178	18	275	78.6	26.5	15	MSS	30	MSS	30.5	81.7	9.7	-
AL BW-179	8	114	32.6	11.8	10	MSS	10	MS	31.9	85.3	8.8	-
AL BW-180	7	114	32.6	11.8	10	MS	10	M	34.4	81	8.7	-
AL BW-181	9	190.5	54.4	17.6	10	M	20	M	33.7	73.5	8.2	+
AL BW-182	6.5	187.5	53.6	11.8	15	M	5	MS	30.9	69.9	9	-
AL BW-183	14	217.5	62.1	20.6	15	MSS	20	MSS	26.7	79.5	9.4	-
AL BW-184	9	161.5	46.1	17.6	10	M	20	M	32.2	86.7	8.5	-
AL BW-185	12	180.5	51.6	17.6	20	MSS	10	MSS	30.5	85.2	9.8	-
AL BW-186	7.5	105.5	30.1	11.8	15	MSS	5	M	32.7	74.8	9.6	-
AL BW-187	16	246.5	70.4	23.5	20	MSS	20	MSS	41.4	86.3	9.3	-
AL BW-188	16	219	62.6	23.5	20	MSS	20	MSS	32.5	77.5	8.9	-
AL BW-189	24	331.5	94.7	35.3	30	MSS	30	MSS	31.3	71.5	9	-
AL BW-190	16	227.5	65	23.5	20	MSS	20	MSS	26.2	79.7	10.8	-
AL BW-191	35	455	130	47.1	30	S	50	MSS	25	79	10.2	-
AL BW-192	37.5	511.5	146.1	50	35	S	50	MSS	26.9	72.9	8.4	-
AL BW-193	7	133	38	11.8	10	MS	10	M	34.4	79	8.4	-
AL BW-194	5	95.5	27.3	8.8	5	MS	10	M	34.6	75.2	8.6	-
AL BW-195	8	114	32.6	11.8	10	MSS	10	MSS	34.8	78	8.4	-
AL BW-196	24	367.5	105	35.3	10	MS	50	MSS	27.1	81.2	9.2	-
AL BW-197	5.5	104	29.7	8.8	5	M	10	MSS	31.9	78.9	9.3	-
AL BW-198	11	134	38.3	17.6	10	M	20	MSS	28.8	72.5	8.5	-
AL BW-199	11	125.5	35.9	17.6	10	M	20	MSS	29.1	87.8	8	-
AL BW-200	14	164	46.9	23.5	20	M	20	MSS	32.2	71.7	7.4	-
AL BW-201	16.5	229	65.4	26.5	15	M	30	MSS	36.2	77.4	7.7	-
AL BW-202	12.5	209	59.7	20.6	15	15M	20	MSS	34.4	77.5	9.9	-
AL BW-203	5	114	32.6	11.8	10	M	10	MR	30.9	80	10.2	+
AL BW-204	4.5	76.5	21.9	8.8	15	M	0	TR	40	76.7	8.4	-
AL BW-205	11	153	43.7	17.6	10	M	20	MSS	32.6	75	8.4	-
AL BW-206	17	265	75.7	26.5	5	MR	40	MSS	36.8	73.8	7.3	-
AL BW-207	14	245	70	23.5	20	M	20	MSS	32.7	67.5	8.3	-
AL BW-208	8	122.5	35	11.8	10	MSS	10	MSS	33.8	68.2	9	-
AL BW-209	9	151.5	43.3	14.7	15	MSS	10	M	33	74	8.3	-
AL BW-210	7	114	32.6	11.8	15	MSS	5	MR	35.7	65	8.4	-
AL BW-211	10	160	45.7	14.7	15	MSS	10	MSS	31.7	78.2	10.2	-
AL BW-212	16	161.5	46.1	23.5	20	MSS	20	MSS	32.9	72.4	9.7	-

Genotype	ACI	AUDPC	rAUDPC	rFRS	Stem rust season 1		Stem rust season 2		TKW (g)	PH (cm)	SL (cm)	PBC
					Severity	FR	Severity	FR				
AL BW-213	20	285	81.4	29.4	20	MSS	30	MSS	33.7	76.3	9.3	-
AL BW-214	28	389	111.1	41.2	30	MSS	40	MSS	35.8	77.2	8.8	-
AL BW-215	32	329	94	47.1	30	MSS	50	MSS	27.3	75.9	8.6	-
AL BW-216	26	452.5	129.3	41.2	20	M	50	MSS	29.5	77.2	8.1	-
AL BW-217	12	209	59.7	23.5	20	M	20	M	32.6	73.9	8.1	+
AL BW-218	4	131.5	37.6	8.8	10	M	5	MR	37.5	76.5	7.2	+
AL BW-219	6.5	143	40.9	14.7	15	M	10	MSS	40.2	80.9	9.7	+
AL BW-220	6	114	32.6	11.8	10	M	10	M	37.7	83.5	9.4	-
AL BW-221	4.5	104	29.7	8.8	5	M	10	M	36.2	74.2	9	-
AL BW-222	3.5	76.5	21.9	8.8	5	M	10	MR	35.9	74.2	8.2	+
AL BW-223	6.5	143	40.9	14.7	15	M	10	MR	29.8	78	8.6	+
AL BW-224	12	180.5	51.6	17.6	15	MS	15	MSS	36.5	70	8.7	-
AL BW-225	14	209	59.7	20.6	15	MSS	20	MSS	30.6	67.7	8.2	-
AL BW-226	18	246.5	70.4	26.5	15	MSS	30	MSS	26.6	76	8.8	-
AL BW-227	16	190	54.3	23.5	20	MS	20	MSS	26.5	78.5	10.3	-
AL BW-228	15	220.5	63	23.5	10	M	30	MSS	34.5	74.4	10	-
AL BW-229	20	275.5	78.7	29.4	20	MSS	30	MSS	24.9	80	8.9	-
AL BW-230	7.5	95	27.1	11.8	5	M	15	MSS	40.9	79.7	8.1	-
AL BW-231	21	246.5	70.4	26.5	30	S	15	MSS	35.3	80.2	8.1	-
AL BW-232	14	209	59.7	20.6	5	MSS	30	MSS	28.3	84	8.7	-
AL BW-233	16	191.5	54.7	23.5	10	MSS	30	MSS	34.1	78.2	10	-
AL BW-234	6	133	38	11.8	10	M	10	M	33.6	81.7	9.2	-
AL BW-235	3	20	5.7	5.9	5	M	5	M	33.7	80.8	8.3	-
AL BW-236	2.5	66.5	19	5.9	5	M	5	MR	30.2	75.5	7.3	+
AL BW-237	2.5	66.5	19	5.9	5	M	5	MR	32.6	78	8.3	+
AL BW-238	12	170	48.6	17.6	20	MSS	10	MS	34	81.4	8	-
AL BW-239	16	246.5	70.4	23.5	20	MSS	20	MSS	38.7	76.2	10.4	-
AL BW-240	15	237.5	67.9	23.5	10	M	30	MSS	32.2	74	9.8	-
AL BW-241	6	133	38	11.8	10	M	10	M	38	73.7	8.1	-
AL BW-242	5.5	85	24.3	8.8	10	MSS	5	M	40.8	74.7	8.9	-
AL BW-243	5	85	24.3	8.8	10	M	5	MS	32.1	76.2	8.6	-
AL BW-244	10	151.5	43.3	14.7	15	MSS	10	MSS	29.1	77.8	9.4	-
AL BW-245	16	246.5	70.4	23.5	20	MSS	20	MSS	30.4	68	9.7	-
AL BW-246	4.5	104	29.7	8.8	5	M	10	M	30.9	78.9	9.2	-
AL BW-247	8	114	32.6	11.8	10	MSS	10	MS	33.7	71.2	9.7	-
AL BW-248	6	85	24.3	8.8	10	MSS	5	MSS	31.2	79.4	12.1	-
AL BW-249	2.5	66.5	19	5.9	5	M	5	MR	30.4	77.7	8.4	+
AL BW-250	5	95.5	27.3	8.8	10	M	5	MS	33.2	72.5	8.8	-
EPC BW-251	4	68	19.4	8.8	10	M	5	MR	33.5	70.3	8.4	+
EPC BW-252	11	180	51.4	17.6	10	M	20	MSS	25.9	77.9	10.1	-
EPC BW-253	15	274	78.3	23.5	10	M	30	MSS	29.6	77.9	9.8	-
EPC BW-254	13.5	162.5	46.4	20.6	5	M	30	MSS	26.6	79.3	9.7	-
EPC BW-255	15	191.5	54.7	23.5	10	M	30	MSS	26.7	78.2	9.7	-
EPC BW-256	12	285	81.4	29.4	10	M	40	M	30	76.4	10.3	+
EPC BW-257	8	141.5	40.4	11.8	10	MSS	10	MS	25.8	70.8	8.7	-
EPC BW-258	4.5	104	29.7	8.8	10	M	5	M	25.4	77.9	9.2	+
EPC BW-259	6	141.5	40.4	11.8	10	M	10	M	26	79.7	10.6	+
EPC BW-260	10	151.5	43.3	14.7	20	MSS	5	MS	33	72.9	8.4	-
EPC BW-261	8	114	32.6	11.8	20	MSS	0	TR	30.1	83.5	9.5	-
EPC BW-262	4	76.5	21.9	8.8	10	M	5	MR	37.4	69.5	10.5	+

Genotype	ACI	AUDPC	rAUDPC	rFRS	Stem rust season 1		Stem rust season 2		TKW (g)	PH (cm)	SL (cm)	PBC
					Severity	FR	Severity	FR				
EPC BW-263	4	76.5	21.9	8.8	10	M	5	MR	33.4	75.5	7.9	+
EPC BW-264	18	284	81.1	26.5	15	MSS	30	MSS	25.9	80.3	8.4	-
EPC BW-265	18	246.5	70.4	26.5	25	MSS	20	MSS	29.5	83.8	10.7	-
EPC BW-266	5.5	85	24.3	8.8	5	M	10	MSS	33.8	71.7	9.2	-
EPC BW-267	28	425	121.4	41.2	20	MSS	50	MSS	25	73.4	8.3	-
EPC BW-268	22	312.5	89.3	32.4	15	MSS	40	MSS	28.3	74.8	9.7	-
EPC BW-269	18	246.5	70.4	26.5	15	MSS	30	MSS	32.9	79.2	9.2	-
EPC BW-270	12	254	72.6	17.6	20	MSS	10	MSS	38.1	74.4	7.7	-
EPC BW-271	12	216.5	61.9	17.6	20	S	10	MR	32.3	75	8.8	-
EPC BW-272	12	189	54	17.6	20	MSS	10	MSS	32.2	78.7	8.3	-
EPC BW-273	6	122.5	35	11.8	10	MSS	10	MR	31.5	82	8.5	-
EPC BW-274	9.5	151.5	43.3	14.7	20	MSS	5	M	28.5	76.7	9.8	-
EPC BW-275	5.5	141.5	40.4	11.8	15	MSS	5	MR	35	69.7	8.7	+
EPC BW-276	9	170	48.6	14.7	20	MSS	5	MR	37.2	84.9	7.9	-
EPC BW-277	10	160	45.7	14.7	20	MSS	5	MS	34.6	77.9	8	-
EPC BW-278	5	104	29.7	8.8	10	M	5	MS	33	69.3	9.3	-
EPC BW-279	4	47.5	13.6	5.9	5	MSS	5	MSS	25.9	80.7	9.9	-
EPC BW-280	4	66.5	19	5.9	5	MSS	5	MSS	27	73.9	10.3	-
EPC BW-281	15	219	62.6	23.5	10	M	30	MSS	27.9	71.9	9.1	-
EPC BW-282	26	416.5	119	38.2	25	MSS	40	MSS	25.7	73	8.5	-
EPC BW-283	5	85	24.3	8.8	10	M	5	MSS	32.7	82.4	9.2	-
EPC BW-284	12.5	217.5	62.1	20.6	15	M	20	MSS	28	72.3	10.2	-
EPC BW-285	8	133	38	11.8	15	MSS	5	MS	32.5	77.7	10	-
EPC BW-286	14	217.5	62.1	20.6	15	MSS	20	MS	31.8	80.4	10.7	-
EPC BW-287	8	160	45.7	14.7	20	M	5	MS	41.6	79.8	8.8	-
EPC BW-288	8.5	151.5	43.3	14.7	15	M	10	MSS	32.2	82.4	9.2	-
EPC BW-289	20	375	107.1	29.4	20	MSS	30	MSS	29	74.2	7.8	-
EPC BW-290	15	227.5	65	23.5	10	M	30	MSS	32.1	80.2	8.2	-
EPC BW-291	12.5	190	54.3	20.6	15	M	20	MSS	28.8	71.8	8.8	-
EPC BW-292	12	216.5	61.9	17.6	20	MSS	10	MS	27.8	63.8	8	-
EPC BW-293	10.5	161.5	46.1	17.6	15	M	15	MSS	29.8	72.5	8.6	-
EPC BW-294	9	151.5	43.3	14.7	15	MSS	10	M	28.9	77.4	9.7	-
EPC BW-295	6	122.5	35	11.8	20	M	0	TR	30.5	73.4	9.4	-
EPC BW-296	8	133	38	11.8	15	MSS	5	MS	31.4	80.2	8	-
EPC BW-297	10	124	35.4	14.7	20	MSS	5	MS	31.3	74.7	9.9	-
PCHP BW-298	12	180.5	51.6	17.6	20	MSS	10	MSS	35.8	79.5	10.1	-
PCHP BW-299	18	284	81.1	26.5	25	MSS	20	MSS	31.3	77.5	8.2	-
PCHP BW-300	18	238	68	26.5	15	MSS	30	MSS	36.4	68.4	7.9	-
PCHP BW-301	14	217.5	62.1	20.6	15	MSS	20	MSS	35.7	71.9	7.8	-
PCHP BW-302	7	141.5	40.4	11.8	10	M	10	MSS	31.8	79.4	9.2	-
PCHP BW-303	16	246.5	70.4	23.5	20	MSS	20	MSS	32.3	72.7	9.3	-
PCHP BW-304	20	312.5	89.3	29.4	20	MSS	30	MSS	30.7	83.9	11	-
PCHP BW-305	14	209	59.7	20.6	15	MSS	20	MSS	34.2	74.4	10.4	-
PCHP BW-306	7	141.5	40.4	11.8	10	M	10	MS	34.6	77.4	8.9	-
PCHP BW-307	8.5	151.5	43.3	14.7	15	M	10	MSS	38.3	69.2	8.5	-
PCHP BW-308	6	133	38	11.8	10	M	10	M	32.7	74.2	8.7	+
PCHP BW-309	4	39	11.1	5.9	10	MSS	0	TR	29.3	74.7	9.6	-
PCHP BW-310	8	114	32.6	11.8	20	MSS	0	TR	32.3	71	8.2	-
PCHP BW-311	10	151.5	43.3	14.7	20	MSS	5	MS	36.5	77.2	7.4	-
PCHP BW-312	12	197.5	56.4	17.6	20	MSS	10	MSS	40.6	62.5	8.6	-

Genotype	ACI	AUDPC	rAUDPC	rFRS	Stem rust season 1		Stem rust season 2		TKW (g)	PH (cm)	SL (cm)	PBC
					Severity	FR	Severity	FR				
PCHP BW-313	10	151.5	43.3	14.7	15	MSS	10	MS	40.9	81	9.4	-
PCHP BW-314	12	197.5	56.4	17.6	20	MSS	10	MS	34.9	78.5	10.1	-
PCHP BW-315	9	179	51.1	17.6	20	M	10	M	34	78	8.7	+
PCHP BW-316	18	238	68	26.5	25	MSS	20	MSS	36.4	67.5	9.1	-
CACUKE	56	350	100	100	90	MSS	100	MSS	20.2	60.7	9.5	-
Grand Mean	12.3	187	26.7	22.9	15.7		20		32.5	76.8	8.8	
LSD	10.7	139.8	19.97	17.15	12		16		7.7	8.2	1.4	
CV (%)	8.8	5.6	5.6	5.7	5.7		7.2		16.2	7.5	10.4	

Note. ACI = average coefficient of infection; AUDPC = area under the disease progress curve; rAUDPC = relative area under the disease progress curve; rFRS = relative final rust severity; Disease Severity based on Modified Cobb's (0-100%) scale (Peterson et al., 1948); IT = Infection Type based on (Roelfs et al., 1992); TR = trace , R = resistant, MR = moderately resistant, RMR = resistant to moderately resistant, MRMS (M) = moderately resistant to moderately susceptible, MSS = moderately susceptible to susceptible, MS = moderately susceptible and S = susceptible; TKW = thousand kernel weight in grams; PH = Plant Height in centimeters; SL = Spike length in centimeters TKW = thousand kernel weight; PBC = Pseudo Black Chaff, (+) predictive of Sr2 gene, (-) absence of Sr 2 gene; LSD = Least Significant Difference CV (%) = Percentage Coefficient of Variation.

Table 3. Pearson's correlation coefficient for the disease parameters among the wheat genotypes across seasons

	ACI	AUDPC	rAUDPC	rFRS
ACI	-			
AUDPC	0.9242***	-		
rAUDPC	0.9242***	1.0000***	-	
rFRS	0.9745***	0.8982***	0.8982***	-

Note. ***: large positive relationship between the variables at $p \leq 0.05$; ACI = average coefficient of infection; AUDPC = area under the disease progress curve; rAUDPC = relative area under the disease progress curve; rFRS = relative final rust severity.

Table 4. Pearson's correlation coefficient for the different agronomic traits among the wheat genotypes across seasons

	Number of tillers	Plant height	Spikelet length	Thousand kernel weight	Stem rust
Number of tillers	-				
Plant height	0.0475	-			
Spikelet length	-0.0924	0.1472	-		
Thousand kernel weight	0.0848	0.0269	-0.2167	-	
Stem rust severity	0.0559	-0.0475	-0.0253	-0.3401	-

4. Discussion

Seasonal effects of temperature and precipitation on disease development were discernible. In season one when the temperatures were higher and the precipitation was lower, the infection rate was apparently lower. Abiotic stresses such embodied in high temperatures and drought in general induce the plant defense pathway leading to increased plant resistance to stem rust (Mittler, 2006). During the second season in which relatively lower temperatures and high precipitation were recorded, higher disease pressure was noted. Besides supporting vigorous plant growth, providing increased surface area for spore landing and infection, lower temperatures and "free water" through precipitation favors rust infection process per se. Similar findings have been extensively reported in rust epidemiology literature *e.g.* as highlighted in GRDC (2011). The present results underscores that KALRO-Njoro sits in an environment conducive to stem rust proliferation. In the larger East African region stem rust epidemics are driven by a combination of factors including favorable weather conditions for the disease, volunteer host plants, and presence of a green crop of wheat at any one time, also referred to as a "green bridge" that enhance survival build-up and spread of the stem rust pathogen through the seasons. In combination with growing of stem rust susceptible varieties especially among the small scale resource poor farmers, these factors have aggravated the Ug99 threat in the region leading to seasonal stem rust outbreaks (GRDC, 2011).

Different reactions to stem rust observed between the genotypes suggested that the material had diverse genetic backgrounds. It can be inferred that the six genotypes namely ALBW-106, ALBW-204, EPCBW-261, EPCBW-295, PCHPBW-309 and PCHPBW-310 that only showed trace response to the diseases (TR) with no visible stem rust infections could either be carrying single effective major genes or a combination of those. Singh et al. (2005) reported that a combination of 4-5 minor effect genes with race non-specific responses provided near immunity reaction to leaf rust. Accordingly, the 6 highly resistant lines could also be harboring a combination of minor effect genes. Leonard and Szabo (2005) suggested that the presence of effective major genes in a variety limit infection process by triggering necrosis of the host cells in the neighborhood of the infective structures.

Among the 250 ALBW accessions, 22 and 26 genotypes were susceptible during season one and two respectively, while 135 genotypes showed moderately susceptible (MS) response. Their severities however were low, compared to the susceptible check variety CACUKE whose final severity was 90 S and 100 S during season one and two respectively. The ALBW are among old varieties bred between the 70s and the 80s by CIMMYT Mexico (Kohli & Rajaram, 1988). The high frequency of MS to S genotypes among the ALBW genotypes suggested presence of ineffective stem rust resistance genes in their backgrounds, probably $SrTmp$, $Sr24$, $Sr31$, to which the current family of Ug99 races are highly virulent. Results on previous rust resistance studies on genotypes within the ALBW set are consistent with the present study. For instance, Chaves et al. (2011) reported high frequency of moderately susceptible to susceptible infection types within Brazilian genotypes with aluminium tolerance backgrounds. Notably, 19 genotypes among the ALBW set showed resistant to moderately resistant (RMR) infection types. Among those, 6 genotypes (ALBW-4, ALBW-100, ALBW-106, ALBW-173, ALBW-174 and ALBW-204) showed resistant (R) infection type with no visible or compatible interaction between the host genotypes and the stem rust fungus despite the heavy disease pressure in season two. This suggests that these lines could be carrying stem rust resistance genes which are still effective against the Ug99 and its variants.

Among the 47 EPCBW genotypes, nearly half showed intermediate infection type (M), and relatively low severities ranging from 5% to 20% during both seasons implying that these genotypes could be containing effective stem rust resistance genes in their backgrounds. Despite, some level of disease, which in fact suggests incomplete resistance, these lines produced plump grain with no apparent stem rust associated yield loss.

While over 70% of the 19 PCHPBW genotypes evaluated during both seasons, showed MSS infection types, those nonetheless depicted lower severities. Moreover, those genotypes expressed the pseudo black chaff phenotype, implying the presence of the adult plant resistance gene-$Sr2$. These lines could serve as useful genetic resources in breeding for durable resistance to the prevailing stem rust races, especially when combined with effective major genes. Such a strategy will not only counteract the wheat yield losses currently common in many wheat growing zones of Kenya but equally important, it will counter the rapid evolution of new stem rust races due to delayed step-wise mutations triggered by over cultivation of single resistance varieties over extensive acreages and across seasons (Tsilo et al., 2010).

Across the two seasons of evaluation, 14 genotypes exhibiting dwarfing traits also had relatively high TKW values. The TKW is considered a "yield component" and a good proxy for yield potential of a genotype (Dill-Macky et al., 1990). The semi-dwarf stature of these genotypes suggested presence of dwarfing genes (Rht), which reduce height. Semi dwarf genotypes unlike the traditionally taller varieties are tolerant to lodging, and hence are more responsive to high nitrogen-fertilization and irrigated cropping systems especially under intensive management. Development of semi-dwarf types of wheats was initiated at CIMMYT by Norman Borlaug in the early 1960s through crosses made between the double-dwarf Japanese cultivar Norin and taller breeding lines (Gale et al., 1981; Kihara, 1984). A report by Sayre et al. (1997) indicated that there has been an annual increase in yield by 1% among semi-dwarf wheat varieties due to incorporation of dwarfing genes in their backgrounds.

The large negative correlation between TKW and stem rust can be attributed to the fact that the fungus damages the vascular system of the susceptible host plant extensively limiting transportation of water and nutrients from the soil to the developing kernel and other organs as well as interfering with translocation of photosynthates, which leads to shrivelled grains (Singh et al., 2006; Everts et al., 2001). Similar results have been reported by numerous previous research groups (Tadesse et al., 2010; Taye et al., 2015). Among the highly susceptible varieties, the endosperm barely forms and resultant grains are invariably completely shrivelled.

5. Conclusion

The six exotic wheat genotypes with high resistance to stem rust could be used as donors for introgression of resistance to the adapted Kenyan wheat backgrounds. This will also help improve Kenyan germplasm with regard to aluminum tolerance and micronutrient fortification (Velu & Singh, 2013). The 35 genotypes with low MSS response and which also showed presence of PBC could be integrated in the Durable Rust Resistance Wheat (DRRW) pipeline to develop durable sources of resistance to stem rust. Further greenhouse studies involving seedlings coupled with marker assisted selection needs be carried out to identify the exact genes conferring the resistance to stem rust among the exotic varieties. Inheritance studies could also be done among the elite wheat genotypes to elucidate the exact genes and their effects especially in conditioning the stem rust resistance. This will ensure the effective utilization of the resistance sources in the wheat breeding program through their deployment into adapted but susceptible wheat varieties.

Acknowledgements

This publication was made possible through support provided by Alliance for a Green Revolution in Africa (AGRA). The opinions expressed herein are those of the author(s) and do not necessarily reflect the views of AGRA. My heartfelt gratitude also goes to the Borlaug Global Rust Initiative (BGRI) under the Durable Rust Resistant Wheat (DRRW) project, KALRO-Njoro, my supervisors and other technicians for assisting me in setting up my experiments.

References

Admassu, B., Lind, V., Friedt, W., & Ordon, F. (2009). Virulence Analysis of *Puccinia graminis* f. sp. *Tritici* Populations in Ethiopia with Special Consideration of Ug99. *Plant Pathology, 58*(2), 362-369. http://dx.doi.org/10.1111/j.1365-3059.2008.01976.x

Beard, C., Jayasena, K., Thomas, G., & Loughman, R. (2006). Managing stem rust of wheat. Plant Pathology, Department of Agriculture, Western Australia. *Farmnote, 73*. Retrieved from https://www.agric.wa.gov.au

Chaves, M. S., Martinelli, J., da Silva, P. R., Scagliusi, S., & Brammer, S. (2011). *Surveillance of Race Ug99 in Brazil and the Search for Effective Resistance* (pp. 2-23). Paper presented at the Embrapa Trigo CPACT.

Dill-Macky, R., Rees, R. G., & Platz, G. J. (1990). Stem Rust Epidemics and Their Effects on Grain Yield and Quality in Australian Barley Cultivars. *Crop and Pasture Science, 41*(6), 1057-1063. http://dx.doi.org/10.1071/AR9901057

Everts, K. L., Leath, S., & Finney, P. L. (2001). Impact of Powdery Mildew and Leaf Rust on Milling and Baking Quality of Soft Red Winter Wheat. *Plant Disease, 85*(4), 4. http://dx.doi.org/10.1094/PDIS.2001.85.4.423

FAO. (2016). *Agricultural commodities profiles and relevant WTO negotiations isssues.* Economic and Social Development Department. Retrievd September 1, 2016, from http://www.fao.org/economic/ess/ess-home/en

Gale, M. D., Marshall, G. A., & Rao, M. V. (1981). A Classification of the Norin 10 and Tom Thumb Dwarfing Genes in British, Mexican, Indian and Other Hexaploid Bread Wheat Varieties. *Euphytica, 30*(2), 355-361. http://dx.doi.org/10.1007/BF00033997

Grain Research and Development Cooperation (GRDC). (2011). Stem rust of wheat fact sheet. *Seasonal conditions drive outbreaks.* Retrieved from http:// www.grdc.com.au/rustlinks

Hodson, D. P. (2015). *Field Survey Protocols and Tools.* Paper presented at the 2015 SAARC Wheat Rust Training Course. CIMMYT-Ethiopia.

Jin, Y., Szabo, L. J., Pretorius, Z. A., Singh, R. P., Ward, R., & Fetch, J. T. (2008). Detection of virulence to resistance gene *Sr24* within race TTKS of *Puccinia graminis* f. sp. *tritici. Plant Disease, 92*(6), 923-926. http://dx.doi.org/10.1094/PDIS-92-6-0923

Kihara, H. (1984). Origin and History of 'Daruma', A Parental Variety of Norin 10. In S. Sakamoto (Ed.), *Proceedings of the 6th International Wheat Genetics Symposium.* Plant Germplasm Institute, University of Kyoto, Kyoto, Japan.

Kohli, M. M., & Rajaram, S. (1988). Review of Brasilian/CIMMYT collaboration 1974-1986. In D. G. Tanner, M. van Ginkel, & W. Mwangi (Eds.), *Wheat Breeding for Acid Soils.* Sixth Regional Wheat Workshop for Eastern, Central and Southern Africa. Mexico. D.F.: CIMMYT.

Leonard, K. J., & Szabo, L. J. (2005). Stem Rust of Small Grains and Grasses Caused by *Puccinia graminis. Molecular Plant Pathology, 6*(2), 99-111. http://dx.doi.org/10.1111/J.1364-3703.2004.00273.x

McIntosh, R. A. (1988). The role of specific genes in breeding for durable stem rust resistance in wheat and triticale. In N. W. Simmonds & R. Sanjaya (Eds.). *Breeding strategies for resistance to the rusts of wheat* (pp. 1-9). CIMMYT, Mexico.

McIntosh, R. A., Dubcovsky, J., Rogers, W. J., Morris, C., Appels, R., Xia, X. C., & Azul, B. (2014). *Catalogue of Gene Symbols for Wheat: 2013-2014.*

Mittler, R. (2006). Abiotic Stress, the Field Environment and Stress Combination. *Trends in Plant Science, 11*(1), 15-19. http://dx.doi.org/10.1016/j.tplants.2005.11.002

Nazari, K., Mafi, M., Yahyaoui, A., Singh, R. P., & Park, R. F. (2009). Detection of Wheat Stem Rust (*Puccinia graminis* f. sp. *tritici*) Race TTKSK (Ug99) in Iran. *Plant Disease, 93*(3), 317-317. http://dx.doi.org/10.1094/PDIS-93-3-0317B

Njau, P. N., Jin, Y., Huerta-Espino, J., Keller, B., & Singh, R. P. (2010). Identification and Evaluation of Sources of Resistance to Stem Rust race Ug99 in Wheat. *Plant Disease, 94*(4), 413-419. http://dx.doi.org/10.1094/PDIS-94-4-0413

Njau, P. N., Wanyera, R., Macharia, G. K., Singh, J. M. R., & Keller, B. (2009). Resistance in Kenyan bread wheat to recent eastern African isolate of stem rust, *Puccinia graminis* f. sp. *tritici*, *Ug99*. *Journal of Plant Breeding and Crop Science, 1*(2), 022-027. Retrieved from http://www.academicjournals.org/jpbcs

Ooro, P. A., Bor, P. K., Amadi, D. O. K., Tenywa, J. S., Joubert, G. D., Marais, D., … Nampala, M. P. (2009). Evaluation of wheat genotypes for improved drought tolerance through increased seedling vigour. *9th African Crop Science, Conference Proceedings*, Cape Town, South Africa, 28 September-2 October 2009. Retrieved from https://www.cabdirect.org/cabdirect/abstract/20133232444

Peterson, R. F., Campbell, A. B., & Hannah, A. E. (1948). A Diagrammatic Scale for Estimating Rust Intensity on Leaves and Stems of Cereals. *Canadian Journal of Research, 26*(5), 496-500. http://dx.doi.org/10.1139/cjr48c-033

Pretorius, Z. A., Singh, R. P., Wagoire, W. W., & Payne, T. S. (2000). Detection of Virulence to Wheat Stem Rust Resistance Gene *Sr*31 in *Puccinia graminis* .f. sp. *Tritici* in Uganda. *Plant Dis., 84*, 203. http://dx.doi.org/10.1094/PDIS.2000.84.2.203B

Roelfs, A. P., Singh, R. P., & Saari, E. E. (1992). *Rust Diseases of Wheat: Concepts and Methods of Disease Management* (2nd ed.). CIMMYT, Mexico, D.F.

Rosegrant, M. W., Agcaoili-Sombilla, M., & Perez, N. D. (1995). *Global Food Projections to 2020: Implications for Investment*. International Food Policy Research Institute, Washington, D.C.

Sayre, K. D., Rajaram, S., & Fischer, R. A. (1997). Yield Potential Progress in Short Bread Wheats in Northwest Mexico. *Crop science, 37*(1), 36-42. http://dx.doi.org/10.2135/cropsci1997.0011183X003700010006x

Singh, R. P., Hodson, D. P., Jin, Y., Lagudah, E. S., Ayliffe, M. A., Bhavani, S., … Basnet, B. R. (2015). Emergence and spread of new races of wheat stem rust fungus: Continued threat to food security and prospects of genetic control. *Phytopathology, 105*(7), 872-884. http://dx.doi.org/10.1094/PHYTO-01-15-0030-FI

Singh, R. P., Huerta-Espino, J., & William, H. M. (2005). Genetics and breeding for durable resistance to leaf and stripe rusts in wheat. *Turkish Journal of Agriculture and Forestry, 29*(2), 121-127. Retrieved from http://journals.tubitak.gov.tr/agriculture/abstract.htm?id=7487

Singh, R. P., Kinyua, M. G., Wanyera, R., Njau, P., Jin, Y., & Huerta-Espino, J. (2006). Spread of a Highly Virulent Race of *Puccinia graminis Tritici* in Eastern Africa: Challenges and Opportunities CAB Reviews. *Perspectives in Agriculture, Veterinary Science, Nutrition and Natural Resources, 1*, 054. Retrieved from http://dx.doi.org/10.1079/PAVSNNR20061054

Tadesse, K., Ayalew, A., & Badebo, A. (2010). Effect of Fungicide on the Development of Wheat Stem Rust and Yield of Wheat Varieties in Highlands of Ethiopia. *African Crop Science Journal, 18*(1). http://dx.doi.org/10.4314/acsj.v18i1.54194

Taye, T., Fininsa, C., & Woldeab, G. (2015). Yield Variability of Bread Wheat under Wheat Stem Rust Pressure at Bore Field Condition of Southern Oromia. *Journal of Agricultural Science and Food Technology, 1*(2), 11-15. Retrieved from http://pearlresearchjournals.org/journals/jmbsr/index.html

Tsilo, T. J., Jin, Y., & Anderson, J. A. (2010). Identification of Flanking Markers for the Stem Rust Resistance Gene *Sr6* in Wheat. *Crop Science, 50*, 1967-1970. http://dx.doi.org/10.2135/cropsci2009.11.0648

Velu, G., & Singh, R. P. (2013). Phenotyping for Plant Breeding. In S. K. Pangaluri & A. A. Kumar (Eds.), *Applications of Phenotyping Methods for Crop Improvement*. Springer Science & Business Media.

Wanyera, R., Kinyua, M. G., Jin, Y., & Singh, R. P. (2006). The Spread of Stem Rust Caused by *Puccinia graminis* f. sp. *tritici*, with Virulence on *Sr31* in Wheat in Eastern Africa. *Plant Disease, 90*(1), 113-113. http://dx.doi.org/10.1094/PD-90-0113A

Wilcoxson, R. D., Skovmand, B., & Atif, A. H. (1975). Evaluation of Wheat Cultivars for Ability to Retard Development of Stem Rust. *Annals of Applied Biology, 80*(3), 275-281. http://dx.doi.org/10.1111/j.1744-7348.1975.tb01633.x

13

Combining Ability and Heterosis of Selected Grain and Forage Dual Purpose Sorghum Genotypes

Sally Chikuta[1,2], Thomas Odong[1], Fred Kabi[1] & Patrick Rubaihayo[1]

[1] Department of Agricultural Production, School of Agricultural Sciences, College of Agriculture and Environmental Sciences, Makerere University, Kampala, Uganda

[2] Department of Agriculture, Ministry of Agriculture, Chibombo District, Zambia

Correspondence: Sally Chikuta, Department of Agricultural Production, School of Agricultural Sciences, College of Agriculture and Environmental Sciences, Makerere University, P.O. Box 7062, Kampala, Uganda. E-mail sallychikuta@yahoo.com

Support for this research was made possible through a capacity building competitive grant Training the next generation of scientists provided by Carnegie Cooperation of New York through the Regional Universities Forum for Capacity Building in Agriculture (RUFORUM).

Abstract

Sorghum is an important food and feed source in mixed crop-livestock production systems where its dual usage is a preferred option, especially among the resource poor small-scale farmers. Attempts to improve fodder quality traits in maize have been at the expense of grain traits and vice versa, but other studies demonstrated that it was possible to select for high stem biomass without compromising the improvement of grain yields in sorghum. As a follow up to this effort, this study was undertaken to estimate the combining ability of grain and forage sorghum genotypes and determine heterosis for several traits as a criteria for improving dual purpose sorghum cultivars. Four grain and four forage sorghum cultivars were crossed to generate 23 crosses following the half diallel mating design scheme at Makerere University Agricultural Research institute Kabanyolo (MUARIK) in 2013. The crosses were evaluated at three locations in Uganda during two rainy seasons of 2014. Data were taken and analysed on leaf area, leaf-stem ratio, plant height, seed weight, grain yield, and biomass. Results indicated that the gene action for the traits under observation was controlled by both additive and non additive genetic effects. Majority of the parental lines had significant GCA estimates for all traits except line 20 for grain yield, lines 22 and 34 for plant height, line 35 for leaf-stem ratio, and line 22 for days to flowering. Significant ($P \leq 0.05$) SCA estimates were prominent in most of the individual parental combinations for all traits except leaf area and leaf-stem ratio indicating the role of dominance gene action. Bakers ratio and heritability coefficients were > 52% for biomass, flowering duration and plant height indicating that genetic gains can be achieved by conventional breeding for the three traits. Heterosis in grain yield and biomass over both the mid and better parents was shown by more than half of the crosses studied. This study suggested that both inter and intra allelic interactions were involved in the expression of the traits.

Keywords: biomass, gene action, grain yield, heterosis, heritability coefficients

1. Introduction

Sorghum (*Sorghum bicolor* (L) Moench) is an important food and feed crop of dry land agriculture because of its wide range of adaptability to various agro-ecological conditions. It is a self pollinating, diploid ($2n = 2x = 20$) with a genome 25% of the size of maize or sugar cane (Rai et al., 1999). Although sorghum is the fifth most important cereal crop in the world after wheat, rice, maize and barley, it is ranked second following maize in Africa (Kenga et al., 2005). In developing countries, sorghum is primarily used as a food crop (Bawazir, 2009), and has been improved to a great extent for grain (Williams et al., 1997). However, in the developed countries, it is used primarily as a feed crop (Chakauya et al., 2006). Given that crop- livestock production systems are the most common form of land use in semi-arid areas of Africa (Mativavarira et al., 2011) among the resource poor small scale farmers who rely on crop residues as livestock feed (Sibanda et al., 2011), genetic improvement of this crop for dual usage as grain and fodder is cardinal.

Traits like grain and fodder yield are governed by polygenes with complex gene action (Jain & Patel, 2014), hence understanding the gene action would help plant breeders in selecting appropriate breeding methods. In addition, efficient transmission of desirable genes from selected parents to their progeny needs firm knowledge about gene action (Falconer & Mackay, 1996). Combining ability studies provide useful information regarding the selection of suitable parents for effective hybridization programmes and indicate the nature and magnitude of various types of gene action involved in the expression of quantitative characters (Bernardo, 2014). The process also helps in ensuring accumulation of desirable unfixable or fixable gene effects (Nadarajan & Gunasegaram, 2005). General combining ability (GCA) was described by Falconer (1989) as the mean performance of a genotype when crossed with a series of other genotypes. The performance of a cross can deviate from the average general combining ability of two parental lines due to genetic effects that are specific to that cross and this deviation is referred to as specific combining ability (SCA) (Bernardo, 2014). The differences in GCA are mainly due to additive effects and higher order additive interactions while differences in SCA may be attributed to non-additive gene effects. The analysis of combining ability, therefore, allows broad inferences on the nature of gene effects for a trait under selection. Analysis of diallel data partitions variation into GCA of the parents and SCA of the crosses (Yan & Hunt, 2002). The estimation of GCA effects helps to identify good combiners which may be hybridised to exploit heterosis and select better crosses for further breeding (Singh & Chaudhary, 1985). The grain and fodder yields are primary traits targeted for improvement of dual purpose sorghum productivity through exploitation of heterosis. The desirable tendency is to have progeny that perform better than the parental lines for traits of interest.

This study was undertaken to estimate the general and specific combining ability and heterosis of different grain and forage sorghum genotypes in F_1 combinations for grain yield, biomass and related traits as a criteria for developing superior dual purpose sorghum cultivars.

2. Materials and Method

2.1 Planting Materials

Eight sorghum genotypes were selected based on their performance for high grain and fodder yield in a prior diversity study and used as parental lines in this study. The genotypes comprised four grain and four forage sorghum cultivars which were crossed to generate 23 crosses following the half diallel mating design scheme (Griffing, 1956) at Makerere University Agricultural Research institute Kabanyolo (MUARIK) in 2013 season B.

2.2 The Experiment and Site Descriptions

The 23 crosses and parents were sown in randomized complete block design with three replications at MUARIK in 2014 (Season A and B), Mbarara Zonal Agricultural Research Station in 2014 season B and National Semiarid Agricultural Research institute (NaSARRI) in 2014 season B. MUARIK is located at 0°28′N; 32°37′E and is 1200 m asl with mean daily temperatures of 20 °C. NaSARRI is located at 1°39′N; 33°27′E, and is 1038 masl with mean daily temperatures of 24 °C andMbarara is located at 0.6°13′S; 30°65′E and is 1445 masl. Each genotype was planted in four 3 m rows, 0.6 m apart with an intra row spacing of 0.3 m. A distance of 1 m was left between plots and 2 m between replications.Data was collected on days from planting to 50% flowering, grain yield, 1000 seed weight, plant height, above ground biomass, Leaf-stem ratio and Leaf area (Leaf number × Leaf length × Leaf width × 0.75) following recommended sorghum descriptors (IBPGR/ICRISAT, 1993). Number of days from planting to flowering for each genotype were recorded when half the number of plants in the plots had flowered. To estimate plant height, the height of ten randomly selected plants was measured at the 50% plant flowering stage from the ground to the panicle tip. Leaf-stem ratio was obtained by destructive sampling at the soft dough stage and stripping leaves off the stems of five randomly selected plants. Each sample was oven dried at 65 °C for 72 hours and weighed to compute the ratio.

2.3 Data Analysis

The data were first analysed separately for each location and a combined analysis over locations was computed. Analysis of variance was done prior to computing combining ability estimates according to Griffing's model I (fixed model for parental effects), method 4 (exclusion of parents and reciprocal F1's) diallel analysis procedures (Griffing, 1956). Important combining ability effects were revealed through F-tests, the restrictions imposed on combining ability estimates were: $Sg_i = 0$ and $Ss_{ij} = 0$, for all GCA and SCA effects respectively (Bernardo, 2014).These combining ability estimates were tested for deviation by using two tailed t-tests as described by

Singh and Chaudhary (1985). The combining ability ratio (Baker's ratio) $2\sigma^2_{gca}/(2\sigma^2_{gca+}\sigma^2_{sca})$, was derived following Baker (1978) while broad and narrow sense coefficients of genetic determination were calculated following Abney et al. (2000). Mid and better parent heterosis were estimated for days to flowering, leaf stem ratio, grain yield and biomass following method of Singh and Narayanan (1993).

Combining ability Model:

$Y_{ij} = \mu + g_i + g_j + s_{ij} + e_{ij}$

Where, Y_{ij} = mean of the F1 resulting from crossing i^{th} parent and j^{th} parent, μ = population mean, g_i = GCA effect of ith parent, g_j = GCA effect of jth parent, s_{ij} = SCA effect of the cross between i and j parent. The GCA of the ith line in the diallel can be defined as the mean performance of the crosses having i^{th} line as one of its parents. The SCA of the ith and j^{th} cross in array can be defined as the deviation in mean of this cross from the mean of that array.

Narrow Sense Coefficient of Genetic Determination (NSCGD) ~ $h^2 = 2\sigma^2_{gca}/(2\sigma^2_{gca} + \sigma^2_{sca} + \sigma^2_e)$

Broad Sense Coefficient of Genetic Determination (BSCGD) ~ $H^2 = 2\sigma^2_{gca} + \sigma^2_{sca}/(2\sigma^2_{gca} + \sigma^2_{sca} + \sigma^2_e)$

Where, σ is the variance of the respective subscript.

Mid-parent heterosis (%) = $(F_1\text{-MP})/MP$

Better-parent Heterosis (%) = $(F_1\text{-BP})/BP$

Where, F_1 is the performance of the cross, MP is the average performance of the parents and BP is the performance of the better parent.

All data analysis was done using GenStat statistical package (VSN International, 2011).

3. Results

3.1 Performance of Parental Lines and Crosses across Three Locations

The results from the analysis of variance and heritability estimates of parental lines and crosses across the three locations are presented in Table 1. The effects of genotypes were significant ($P \leq 0.001$) for all traits. These variations in all traits except leaf area were mainly due to the additive gene effects of the parents as indicated by the significant ($P \leq 0.05$) GCA mean squares. However, significant ($P \leq 0.05$) SCA effects were also observed for leaf-stem ratio, grain yield and biomass implying that these traits were controlled by either or both additive and non additive gene actions. The effects of locations were significant ($P \leq 0.05$) for leaf area, leaf-stem ratio, plant height, grain yield and biomass. All the interaction effects were significant ($P \leq 0.05$) for all traits except GCA × Location effect for grain yield.

The baker's ratio and narrow sense coefficient of genetic determination (NSCGD) for all traits ranged between 22 to 87% while the broad sense coefficient of genetic determination for heritability (BSCGD) was high for all traits ranging between 66 to 99%.

Table 1. Analysis of variance and heritability coefficients for seven traits measured across three locations in Uganda

Sources of Variation	DF	Days to 50% Flowering	Leaf area (m^2)	Leaf-stem Ratio	Plant height (m)	1000 Seed Wt (g)	Grain Yld (ton ha^{-1})	Biomass (ton ha^{-1})
Replications	2	5.11 ns	0.15 ns	0.00 ns	306ns	2.85	0.02 ns	3.27 ns
Genotypes (G)	30	414.08 ***	0.26***	0.06***	17966.1***	109.07***	5.47***	1072.92***
Crosses (C)	22	108.03 ***	0.10 ns	0.02***	4926***	24.91***	1.49***	344.71***
GCA	7	237.71***	0.13 ns	0.01***	10876.42*	31.33**	1.42***	572.95***
SCA	15	47.51 ns	0.08 ns	0.02***	2150 ns	21.9 ns	1.53***	238.2***
Location (Loc)	2	20.4 ns	0.41**	0.00***	82455*	7.49 ns	0.15*	174.98***
G × Loc	60	57.81***	0.16**	0.00***	7884***	14.97***	0.23***	66.82***
GCA × Loc	14	8.67***	0.07*	0.00*	3693***	7.89***	0.00 ns	18.05***
SCA × Loc	30	24.73***	0.05*	0.00***	1715***	4.29*	0.06***	24.47***
C × Loc	44	19.62***	0.06**	0.00***	2344***	5.44***	0.52***	22.43***
Residual	60	1.38	0.03	0.00	48.77	2.60	0.01	0.87
Bakers ratio		0.80	0.47	0.22	0.87	0.35	0.28	0.52
NSCGD		0.79	0.31	0.22	0.86	0.32	0.28	0.52
BSCGD		0.98	0.66	0.99	0.98	0.91	0.99	0.99

Note. ***= significant at 0.001, ** = significant at 0.01, * = significant at 0.05.

3.2. Estimates of General Combining Ability (GCA) Effects

Table 2. Estimates of general combining ability effects for eight parental lines

	Days to 50% Flowering	Leaf area (m^2)	Leaf-stem Ratio	Plant height (m)	1000 Seed Wt (g)	Grain Yld (ton ha^{-1})	Biomass (ton ha^{-1})
22 (F)	-0.35	0.01	-0.03***	0.71	0.69	-0.10**	1.98***
24 (F)	2.72***	0.04	-0.06***	43.55***	0.18	0.53***	11.37***
29 (F)	1.35***	0.16*	-0.01*	13.01***	1.97 **	-0.31***	5.94***
34 (F)	4.08***	-0.04	0.04***	1.03	0.59	-0.17***	3.27***
35 (G)	5.13***	0.05	0.00	22.77***	0.69	0.04*	0.99*
41 (G)	-2.12***	0.01	-0.01**	10.09**	1.23	0.38***	4.76***
20 (G)	-7.50***	-0.18*	0.04***	-38.63***	2.71***	0.07*	-9.53***
42 (G)	-2.77***	-0.07	0.02***	-45.61***	1.26	0.26***	6.63***

Note. (F) = Forage sorghum, (G) = Grain sorghum, ***, **, *= significant at 0.001, 0.01 and 0.05 respectively.

The estimates of general combining ability effects of parents for seven traits are presented in Table 2. The significant (P ≤ 0.05) GCA effects for parents 20 and 29 indicated that these lines were good combiners for 1000 seed weight. Significant (P ≤ 0.05) estimates of GCA effects were observed for all parental lines except 22 for days to 50% flowering although only lines 20, 41, and 42 significantly (P ≤ 0.001) reduced the flowering dates. Lines 20 and 29 had significant (P ≤ 0.05) GCA for Leaf area although only line 29 contributed significantly (P ≤ 0.05) to higher Leaf area in the crosses. Significant (P ≤ 0.05) GCA effects for leaf-stem ratio were observed in all the lines except line 35. All the parental lines had significant GCA effects for Plant height, grain yield and biomass except lines 20 and 34 for grain yield and plant height respectively.

3.3 Estimates of Specific Combining Ability Effects

The estimates for specific combining ability are presented in Table 3. Significant non-additive effects were observed in some of the crosses. Of the five crosses that showed significant (P ≤ 0.05) estimates of SCA effects only cross 41×42 was positive for 1000 seed weight. SCA effects were significant (P ≤ 0.05) in 15 crosses for days to 50% flowering although reduction in flowering duration was only seen in 8 crosses. Only 42×29 cross showed significant (P ≤ 0.001) non-additive effects for leaf area while all crosses but 3 showed significance (P ≤ 0.05) for leaf- stem ratio with 9 showing positive (P ≤ 0.05) significant effects. 10 crosses had significant (P ≤

0.05) estimates of SCA effects for plant height. All but five crosses showed significant (P ≤ 0.05) estimates of SCA effects for grain yield and biomass.Eight crosses had positive significant (P ≤ 0.05) SCA effects.

Table 3. Estimates of specific combining ability effects

Cross	Days to 50% Flowering	Leaf area (m²)	Leaf-stem Ratio	Plant height (m)	1000 Seed Wt (g)	Grain Yld (ton ha⁻¹)	Biomass (ton ha⁻¹)
22×20	-3.02**	-0.04	-0.10***	2.32	-1.4	0.14	-2.73***
24×20	1.66	0.02	0.07***	7.18	1.33	0.18*	7.04***
29×20	-2.96**	-0.12	0.03***	-8.28	-4.31**	-0.07	-9.03***
29×22	0.67	-0.09	-0.07***	-10.22	-0.36	-0.38***	1.06
29×24	1.93	-0.08	0.00	4.54	1.84	-1.03***	5.69***
34×20	-0.02	0.10	0.03***	-16.9**	0.51	-0.64***	1.14
34×22	-3.83***	0.11	-0.01*	31.47***	1.07	-0.52***	0.87
34×29	4.46***	-0.15	-0.05***	0.46	1.62	1.28***	1.45
35×20	6.71***	0.10	-0.12***	56.65***	1.83	-0.43***	2.37**
35×22	-0.55	0.01	0.08***	-41.78***	-0.03	0.81***	0.38
35×24	-1.4	-0.01	-0.08***	-28.53***	-3.09*	0.17*	1.73*
35×29	-1.11	-0.01	0.10***	4.11	2.32	0.69***	-4.82***
35×34	1.57	0.05	0.03***	5.6	-2.05	-0.67***	4.50***
41×20	-2.37*	-0.06	0.09***	-40.97***	2.02	0.82***	1.21
41×22	6.04***	0.10	-0.02**	26.6***	4.01**	0.13	8.54***
41×24	-3.14**	0.13	-0.03***	13.75*	-2.04	-0.04	-6.04***
41×29	-4.66***	-0.06	0.05***	-6.01	-3.28*	-0.43***	-13.37***
42×22	0.69	-0.09	0.12***	-8.39	-3.29*	-0.18*	-8.13***
42×24	0.95	-0.06	0.04***	3.06	1.96	0.73***	-8.42***
42×29	1.65*	0.51***	-0.07***	15.4**	2.17	-0.05	19.02***
42×34	-2.18*	-0.11	0.00	-20.62***	-1.15	0.54***	-7.97***
42×35	-5.24***	-0.13	-0.01	3.94	1.02	-0.57***	-4.16***
42×41	4.13***	-0.12	-0.09***	6.62	-0.71	-0.47***	9.66***

3.4 Performance of Crosses

The estimates of mid and better parent heterosis for flowering duration, leaf-stem ratio, grain yield and biomass are presented in Table 4. An earlier flowering date is desirable than late flowering. Mid parent and better parent heterosis for days to 50% flowering ranged between -10 to 10% and from -8 to 13% respectively. The flowering duration of forage sorghums was reduced in majority of the crosses. A negative heterosis estimate for days to flowering is desirable because it implies that the crosses flowered earlier than the parents. Sixteen crosses were better than the mid-parent while only five surpassed the better- parent.

A higher leaf stem ratio is desirable when considering a crop as a feed. The mid-parent values ranged between -29 to 34% whereas the better parent heterosis estimate was between -32 to 33%. Twelve crosses exhibited positive heterosis over the mid-parent. Some of the outstanding crosses were 29×20, 29×24, 34×20, 35×29, 41×20, 41×29, and 41×22 as they performed better than both the mid and better parent.

A positive heterosis value for grain yield is desirable because it implies that the crosses outperformed the parents. The mid-parent and better-parent heterosis ranged between -0.3 to 52% and -23 to 32% respectively. Fourteen out of the twenty three crosses performed better than the better parent.

A positive heterosis value was desirable for biomass and all the crosses exhibited heterosis over the mid parent by 4.1 to 82% while only 6 of the crosses performed below the better parent.

Table 4. Estimates of mid-parent and better-parent heterosis (%) for Days to 50% flowering, leaf-Stem ratio, grain yield and biomass

Cross	Days to 50% Flowering		Leaf-stem ratio		Grain yield (ton ha^{-1})		Biomass (ton ha^{-1})	
	MP	BP	MP	BP	MP	BP	MP	BP
22×20	-10.3	0.8	-29.7	-32.4	21.8	-11.1	25.3	5.9
24×20	-1.8	10.9	34.1	-3.2	34.3	8.8	63.1	15.0
29×20	-8.3	3.1	17.6	0.1	-5.6	-23.7	55.9	5.1
29×22	-6.2	-6.1	-18.5	-28.3	15.0	0.4	30.3	-2.0
29×24	-2.1	-1.7	28.4	4.8	-0.3	-0.6	45.6	34.7
34×20	-3.1	10.5	6.4	3.6	10.5	-6.4	13.2	-16.5
34×22	-9.2	-8.0	-12.6	-18.2	37.6	14.5	46.3	23.6
34×29	1.0	2.3	-4.5	-20.5	21.7	15.0	18.2	2.4
35×20	8.9	20.6	-23.9	-29.9	7.2	1.3	39.5	28.4
35×22	-2.2	-0.8	11.4	-1.7	36.2	-1.0	69.8	54.6
35×24	-0.3	1.6	-11.7	-32.6	30.4	1.2	67.9	24.9
35×29	-1.0	0.4	38.5	27.0	12.6	12.7	5.9	-25.1
35×34	3.4	6.2	6.2	-4.6	7.6	13.0	60.8	25.9
41×20	-3.5	-0.6	32.1	11.7	33.6	23.1	39.4	15.9
41×22	3.7	13.0	-6.6	-18.3	49.7	15.7	72.5	25.6
41×24	-3.7	5.5	15.6	-5.1	52.2	32.1	40.7	-10.6
41×29	-6.5	1.9	34.7	33.7	2.6	-11.1	11.9	-31.1
42×22	-1.5	9.3	16.8	12.6	49.5	0.1	14.4	2.1
42×24	1.8	13.6	22.5	-11.4	46.4	6.7	4.1	-23.7
42×29	1.5	12.7	-11.2	-24.2	16.1	-15.5	75.3	22.4
42×34	-1.2	11.3	-2.6	-5.5	40.8	6.4	44.4	11.3
42×35	-0.6	8.8	-2.5	-10.0	11.6	2.0	7.9	10.3
42×41	9.5	11.4	-16.5	-29.2	33.3	8.1	81.5	43.6

Note. MP = Mid-parent, BP Better-parent.

4. Discussion

4.1 Performance of Parental Lines and Crosses across Locations

Combining analysis of variance over the three locations confirmed the diversity of the genotypes and their differences in locational responses. The significant (P ≤ 0.05) interactions for majority of the traits indicated the differences of genotypes in environmental responses for the traits. Similar observations were made by Girma et al. (2010). Analysis of the crosses for seven traits indicated that most genetic variation for each trait measured was associated with significant (P ≤ 0.05) general combining ability effects except for leaf area. However, the significant (P ≤ 0.05) SCA effects for leaf stem ratio, grain yield and biomass indicated that non-additive effects were also important for the three traits. Nevertheless, for traits where the mean squares of GCA were larger than SCA, even with significant SCA such as biomass, the role of the additive genetic effects was more important.Flowering duration, 1000 seed weight, and plant height were largely governed by additive gene action.Girma et al. (2010) reported similar results for seed weight. Environmental effects could have played a major role on Leaf area as neither additive nor non additive effects were significant. The bakers ratio was moderate to high for leaf area, biomass, days to 50% flowering and plant height indicating the preponderance of additive gene effects in the variance expressed as these were also the traits that had higher GCA mean squares values. The closer the bakers ability ratio is to unity, the larger the importance of additive genetic control, and hence, the greater the capacity to predict the performance of progeny based exclusively on GCA effects (Baker, 1978).

4.2 Estimates of General Combining Ability

The primary criteria for selection of desirable parents are usually based on mean values and additive gene action (Nguyen et al., 1997). Girma et al. (2010) suggested that crossing two parents showing the highest general combining ability for a desirable trait may produce the best performing cross due to an increased frequency of favorable genes. Based on the estimates of GCA effects, it was observed that parental lines 20 and 29 would be the best combiners for 1000 seed weight. Only the grain sorghums contributed to reducing the flowering days because of the negative significant GCA estimates suggesting that the grain sorghums generally tended to flower earlier than the forage sorghums owing to their inherent genetic makeup. This makes grain sorghums useful in reducing the flowering date of forage sorghums. Neither additive nor non-additive effects were statistically significant for Leaf area in the analysis of variance possibly because it was derived from the leaf parameters which were largely influenced by the environment. However, parental line 29 showed significant estimate of GCA implying that it was a good combiner for the leaf area. Girma et al. (2010) reported significant estimates of GCA for leaf area in some of the induced sorghum mutants. The estimates of GCA for leaf-stem ratio indicated that lines 20, 34 and 42 were the best combiners for this trait because of the positive GCA. Lines 24, 35 and 41 were the best combiners for both plant height and grain yield due to the positive significant ($P \leq 0.05$) GCA estimates.All the parental lines had positive significant($P \leq 0.05$) GCA effects for biomass except lines 20 and 35 which had negative GCA estimates indicating that theses two were not the best for this trait. Similar results were reported for fodder yield and its components by Prakash et al. (2010).

Lines 29, 41 and 42 were generally good combiners for four different traits out of seven traits while lines 20 and 35 were good general combiners for at least three different traits. However, lines 24, 41 and 42 were the best general combiners for grain yield and biomass due to the positive significant GCA effects. The superior combining ability of best combiners could be exploitedin hybrid or recurrent selection programmes. Additive variance is associated with effective response to selection (Valiolla, 2012) hence small numbers of parents with desired GCAs can be used to generate crossesfor sorghum improvement.

4.3 Estimates of Specific Combining Ability

The performance of a cross can deviate from the average general combining ability of two parental lines and this deviation is referred to as specific combining ability. Only Cross 41 × 22 out of the twenty three crosses had positive significant ($P \leq 0.01$) estimate of SCA effects for 1000 seed weight implying that GCA effects were more important for this trait. Nguyen et al. (1997) reported similar findings for 100 seed weight. The estimates of SCA for leaf-stem ratio were significant ($P \leq 0.05$) in twenty out of twenty three crosses but only nine crosses were desirable as they had positive SCA estimates. Although this trait was controlled by both additive and non additive gene action, SCA effects had a slightly higher influence as observed from the slightly larger mean square value than GCA (Table 1). Of the eighteen crosses that showed significant ($P \leq 0.05$) estimates for grain yield and biomass only eight crosses had positive significant estimates for SCA suggesting that non additive effects were important for these two traits. Mwije et al. (2014) indicated that parents with the best GCA effects did not necessarily produce crosses with desirable SCA effects as was observed in this study.From Table 1, the mean sum of squares of SCA for grain yield was higher than that of GCA although both were significant ($P \leq 0.001$) because grain yield is a complex trait which results from the contribution of many grain yield components each adding varying levels of genetic effects (Umakanth et al., 2002). The heritability estimate of below 30% (Table 1) implied that yield could not be enhanced through direct effects alone.Biomass was, however, largely controlled more by additive gene action.

4.4 Performance of Crosses

Superior cross combinations could be selected based on heterosis. Quantitative genetic theory states that heterosis is a function of increasing genetic diversity among the parents (Falconer, 1989). More than half the number of crosses resulting from the grain and forage sorghums were characterized by heterosis over the mid-parent for days to 50% flowering, leaf-stem ratio, grain yield and biomass although it varied from cross to cross. The expression of such value of heterosis clearly indicated the agronomic potential of these lines for breeding to enhance grain and forage yield. The results of this study indicated that heterosis could be exploited and that different parental combinations (grain by forage, forage by grain) showed high specific combining ability thus indicating the role of dominance gene action. These results clearly showed that both inter and intra allelic interactions were involved in the expression of the traits.

References

Baker, R. J. (1978). Issues in diallel analysis. *Crop Science, 18*, 533-536. https://doi.org/10.2135/cropsci1978. 0011183X001800040001x

Bawazir, A. A. (2009). Genetic Analysis for Yield and Yield Components in Grain Sorghum (*Sorghum bicolor* L.). *Jordan Journal of Agricultural Sciences, 5*, 273-281.

Bernardo, R. (2014). *Essentials of plant breeding*. Stemma Press, Woodbury, Minnesota.

Chakauya, E., Tongoona, P., Matiburi, E. A., & Grum, M. (2006). Genetic diversity assessment of sorghum landraces in Zimbabwe using microsatellites and indigenous local names. *International Journal of Botany, 2*, 29-35. https://doi.org/10.3923/ijb.2006.29.35

Griffing, B. (1956). Concept of general and specific combining ability in relation to diallel crossing systems. *Journal of Biological Sciences, 9*, 463-493. https://doi.org/10.1071/bi9560463

Girma, M., Ayana, A., & Belete, K. (2010). Combining ability of yield and its components in Ethiopian sorghum landraces. *East African Journal of Sciences, 4*(1), 34-40.

Falconer, S. P. (1989). *Introduction to quantitative genetics*. New York: The Ronald Press Company.

Falconer, D. S., & Mackay, T. F. C. (1996). *Quantitative genetics* (4th ed., p. 464). Longman Group Limited. UK.

IBPGR/ICRISAT. (1993). *Descriptors for sorghum (Sorghum bicolor (L.) Moench)*. International Board for Plant Genetic Resources, Rome, Italy; International Crops Research Institute for Semi-Arid Tropics, Patancheru, India.

Jain, S. K., & Patel, P. R. (2014). Combining ability and heterosis for grain yield, fodder yield and other agronomic traits in sorghum (*Sorghum bicolor* (L) Moench). *Journal of Plant Breeding, 5*(2), 152-157.

Prakash, R., Ganesamurthy, K., Nirmalakumari, A., & Nagarajan, P. (2010). Combining ability for fodder yield and its components in Sorghum (*Sorghum bicolor* L. Moench). *Electron. Journal of Plant Breeding, 1*, 124-128.

Nadarajan, N., & Gunasegaram, M. (2005). *Quantitative genetics and biometrical techniques in plant breeding* (p. 258). Kalyani Publishers, New Delhi, India.

Nguyen, D. C., Nakamura, S., & Yoshida, T. (1997). Combining ability and genotype by environment interaction in early maturing grain sorghum for summer seeding. *Japan Journal of Crop Science, 66*(4), 34-40.

Mativavarira, M., Dimes, J., Masikati, P., Van Rooyen, A., Mwenje, E., Sikosana, J. L. N., & Tui, S. H. K. (2011). Evaluation of water productivity, stover feed quality and farmers' preferences on sweet sorghum cultivar types in the semi-arid regions of Zimbabwe. *Journal of SAT Agricultural Research, 9*(1), 1-9.

Mwije, A., Mukasa, S. B., Gibson, P., & Kyamanya, S. (2014). Heritability analysis in putative drought adaptation traits in sweet potato. *African Crop Science Journal, 22*(1), 79-87.

Rai, K. N., Murty, D. S., Andrews, D. J., & Bramel-Cox, P. J. (1999). Genetic enhancement of pearl millet and sorghum for the semi-arid tropics of Asia and Africa. *Genome, 42*, 617-628. https://doi.org/10.1139/g99-040

Sibanda, A., Homann-KeeTui, S., Van Rooyen, A., Dimes, J., Nkomboni, D., & Sisito, G. (2011). Understanding user communities' perception of changes in rangeland use and productivity: Evidence from Nkayi district, Zimbabwe. *Experimental Agriculture, 47*(S1), 153-168. https://doi.org/10.1017/S001447971000092X

Singh, R. K., & Chaudhary, B. D. (1985). *Biometrical Methods in Quantitative Genetic Analysis* (p. 318). Kalyani Publishers, New Delhi, India.

Umakanth, A. V., Madhusudhana, K., Madhavi Latha, P., Hema, K., & Kaul, S. (2002). Genetic architecture of yield and its contributing characters in postrainy-season sorghum. *International Sorghum and Millets Newsletter, 44*, 37-40.

Valiolla, R. (2012). Combining ability analysis of plant height and yield components in spring type of rapeseed varieties (*Brassica napus* L.) using line × tester analysis. *International Journal of Agriculture and Forestry, 2*(1), 58-62. https://doi.org/10.5923/j.ijaf.20120201.10

VSN International. (2011). *GenStat for Windows* (14th ed.). VSN International, Hemel Hempstead, UK.

Williams, T. O., Fernandez-Rivera, S., & Kelley, T. G. (1997). The influence of socio economic factors on the availability and utilization of crop residues as animal feeds. In C. Renard (Ed.), *Proceedings of the International Workshop on Crop Residues in Sustainable Mixed Crop-Livestock Farming Systems* (pp. 25-39), April 22-26, 1996, Patancheru, India. CAB International, Wallingford.

Yan, W., & Hunt, L. A. (2002). Biplot analysis of diallel data. *Crop Science Journal, 42*, 21-30. https://doi.org/10.1201/9781420040371.ch9

14

Analysis of Genotype by Environment Interaction of Improved Pearl Millet for Grain Yield and Rust Resistance

G. Lubadde[1], P. Tongoona[2], J. Derera[2] & J. Sibiya[2]

[1] National Semi-Arid Resources Research Institute, Soroti, Uganda

[2] University of KwaZulu Natal (UKZN), Pietermaritzburg Campus, Scottsville, South Africa

Correspondence: G. Lubadde, National Semi-Arid Resources Research Institute, P.O. Box 56, Soroti, Uganda. E-mail: glubadde@gmail.com

Abstract

Pearl millet is grown by inhabitants of the semi-arid zones. Due to the unpredictable climatic conditions the genotype-by-environment interaction (GEI) makes it hard to select genotypes adapted to such conditions. The study objectives therefore were to analyse the patterns of GEI and to identify superior genotypes for grain yield and rust resistance. Seventy six genotypes were planted in four environments in 4×19 alpha design with two replications. The ANOVA results showed that main effects of environments were significant ($p \leq 0.05$) for grain yield and highly significant ($p \leq 0.001$) for rust resistance while the main effects of the genotypes and their interactions with environments were also important for grain yield and rust severity at 50% physiological maturity. The GGE biplot analysis revealed that environments associated with more rains received during vegetative phase performed better than those receiving more rains during post-anthesis phase. The winner in the best environment for grain yield was ICMV3771×SDMV96053 while Shibe×CIVT9206 and Shibe×GGB8735 were the best for rust resistance.

Keywords: GGE biplot, grain yield, pearl millet, rust resistance, Uganda

1. Introduction

Pearl millet is adapted to environmentally marginalised conditions worldwide (Bashir et al., 2014) and a multipurpose (IFAD, 1999) cereal for people living in semi-arid areas in Uganda (Lubadde et al., 2014). However, on-farm productivity is low partly due to the effect of rust disease. The economical approach to control rust is through resistance breeding (Singh, 1990) and selecting genotypes adapted to low input and drought-prone environments (Vadez et al., 2012). Unfortunately, the potential performance of improved genotypes under marginal conditions is always obscured by the effect of genotype by environment interaction (GEI) (Yan & Racjan, 2002); leading to selection of genotypes not suitable for particular environments (Cooper & Delacy, 1994) and subsequently leading to low yield. It is therefore important to assess GEI effect before releasing varieties (Gupta & Ndoye, 1991; Haussmann et al., 2012). Several methods have been adopted to assess GEI in pearl millet breeding but the GGE-biplot analysis was used in this study because of the ability to graphically better explain the genotype and genotype by environment components of variation and being more efficient in discriminating genotypes and environments (Yan et al., 2007).

2. Materials and Methods

2.1 Experimental Materials and Study Environments

The experimental materials are shown in Table 1. The seventy six genotypes were evaluated in four pearl millet growing environments in Uganda. They included 60 single cross hybrids developed by crossing six male parents with ten female parents in a North Carolina2 design. The environments were defined as seasons by sites combinations. Environments; E1 was Kitgum site and 2012 second rains; E2 was Kitgum site and 2013 first rains; E3 was Serere site and 2012 second rains while E4 was Serere site and 2013 first rains.

Table 1. Genotypes evaluated

Female parents	Male parents	Crosses									
7=Exbornu	1=ICMV3771	1×7	1×8	1×9	1×10	1×11	1×12	1×13	1×14	1×15	1×16
8=CIVT9206	2=Manganara	2×7	2×8	2×9	2×10	2×11	2×12	2×13	2×14	2×15	2×16
9=GGB8735	3=Okashana2	3×7	3×8	3×9	3×10	3×11	3×12	3×13	3×14	3×15	3×16
10=ICMV221	4=ITMV8001	4×7	4×8	4×9	4×10	4×11	4×12	4×13	4×14	4×15	4×16
11=ICMV221white	5=SDMV94001	5×7	5×8	5×9	5×10	5×11	5×12	5×13	5×14	5×15	5×16
12=KatPM1	6=Shibe	6×7	6×8	6×9	6×10	6×11	6×12	6×13	6×14	6×15	6×16
13=Okoa											
14=SDMV96053											
15=Sosank											
16=Okollo											

2.2 Experimental Sites and Field Layout

The Kitgum location was 03°13′N, 032°47′E, 969 m.a.s.l. while Serere location was 01°32′N, 033°27′E, 1140 m.a.s.l. The genotypes were replicated twice in 4×19 alpha mating design. The materials were planted in four rows each of 5 m long and 60 cm × 30 cm spacing. Fertiliser application was according to Khairwal et al. (2007) and inoculation procedure done according to Thakur et al. (2011).

2.3 Data Collection and Analysis

Data was collected on 36 randomly selected plants per plot using the IBPGR and ICRISAT (1993) manual and traits were rust resistance at 50% physiological maturity (PSM_{50}) determined according to Tooley and Grau (1984) and grain yield (kg ha^{-1}). The rust resistance was determined from rust severity data collected from the third leaf from top of the plant. This PSM_{50} trait was used instead of the area under disease progress curve (AUDPC) since it had a significant effect on grain yield. The trait also seems to be more realistic since it is determined when there is no more change in grain yield. The analysis was done using the Breeding Management System 3.0 (IBP-BMS, 2014) and Genstat (Payne et al., 2012).

3. Results and Discussion

3.1 Combined Analysis of Variance for Assessing GEI and Performance of Environments

The combined analysis of variance (ANOVA) results (Table 2) indicate the main effects of environments being significant ($p \leq 0.05$) for grain yield and highly significant ($p \leq 0.001$) for rust resistance at 50% physiological maturity. The main effects of the genotypes and the interaction effects between genotypes and environments were also significantly ($p \leq 0.05$) important for the two traits. Results further show that generally the coefficients of determination (R^2) estimated from the AMMI model were low for the traits; an indicator that a greater variation was due to the environments. This is corroborated by the rainfall pattern variation observed in each environment (Figure 1). The performance of environments was influenced by the rainfall amount received where the best performing environments in terms of grain yield and rust resistance received lower rainfall amounts. Coincidentally in the best performing environments (E1 and E3), most rainfall was received during the vegetative phase while in the poor performing environments most rainfall was received during the flowering phase. When rainfall is received during flowering, there is disruption of the pollination process since the pearl millet is predominantly outcrossing, with support by wind, a probable reason why the environments performed poorly in terms of grain yield. Heavy rainfall during flowering also causes reduced seed set and poor grain quality (DPP, 2011) in addition to promoting rust and consequently low grain yield. The variation in performance highlights the importance of environments in genotype performance and consequently GEI in trait expression. Rainfall pattern is one of the factors also reported by Gebre (2014) as being a source of variable performance of improved genotypes. The environments being important in genotype performance has also been reported in several pearl millet studies (Ezeaku et al., 2014; Misra et al., 2010; Gupta et al., 2013). The ANOVA adequately identified GEI as a significant source of variation but it is not able to explore the nature (Matus-Cadiz et al., 2003) of the GEI which may mask the true performance of genotypes in certain environments (Crossa, 1990) and thus the need to explore more methods; for which case GGE biplot was adopted.

Table 2. Combined mean squares for grain yield and rust severity

Source of variation	DF	Traits	
		Grain yield (kg ha⁻¹)	Rust severity (%)
Environments	1	3235216.58*	5686.29**
Genotype	72	1269642.59*	55.07*
Environments*Genotype	75	732365.59*	61.02*
Error	447	982045.5	74.24
R^2		0.32	0.33
%CV		5.93	14.35

Note. LSD testing done at α = 0.05; ** = highly significant with p ≤ 0.001, * = significant with p ≤ 0.05.

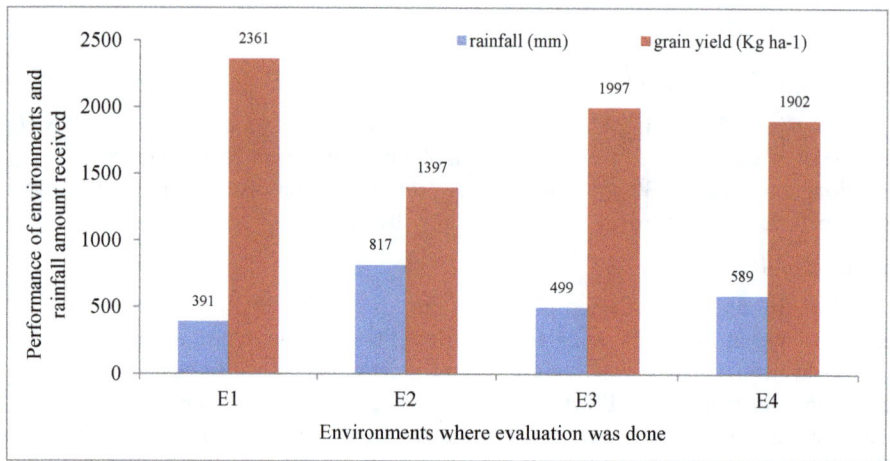

Figure 1. Total amount of rainfall received during the evaluation period and performance of environments
Source for rainfall data: Department of Meteorology, Ministry of Water and Environment, Uganda.

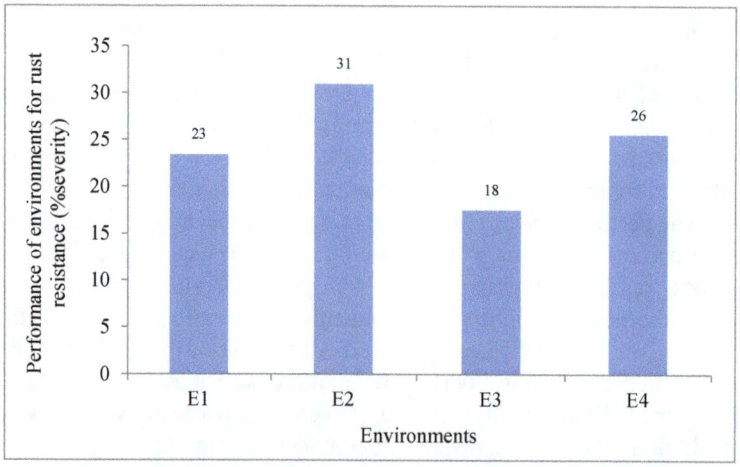

Figure 2. Performance of environments in terms of rust severity

3.2 Assessing GEI Using GGE Biplot Analysis

Figure 3 shows that E1 was the ideal environment while E3 was desirable for grain yield. Both environments were positively correlated and associated with high grain yield and grouped as one mega environment (Figure 4); implying the two environments had similar discriminating ability and so either can be used for selecting best performers with minimal loss of information (Yan & Kang, 2003). E2 was not important in discriminating genotypes due to being at the origin while E4 was the most unstable for grain yield. The genotype 6×8 (Shibe×CIVT9206; grain yield = 2387 kg ha⁻¹, rust severity = 24.36%) was the most ideal for grain yield while

1×14 (ICMV3771×SDMV96053; grain yield = 2355 kg ha^{-1}, rust severity = 24.63%) was the winner in the E1E3 mega environment. Figure 4 also indicates genotypes 6×7 (Shibe×Exbornu), 6×8 (Shibe×CIVT9206), 1×14 (ICMV3771×SDMV96053) and 5×8 (SDMV94001×CIVT9206) performing averagely the same in the E1E3 mega environment while 6×9 (Shibe×GGB8735; grain yield = 2371 kg ha^{-1}, rust severity = 23.85%) and 2×15 (Manganara×Sosank; grain yield = 2169 kg ha^{-1}, rust severity = 20.68%) won in the low yield unstable E4 mega environment and thus the source of crossover GEI effect. The high yielding genotypes were also moderately susceptible to rust. The stable and high yielding genotypes were different from those resistant to rust. The discrimination polygon (Figure 5) is a view for the environments for rust resistance. E2 and E4 were extremely discriminatory for rust resistance. The two environments were in this case the sources of crossover GEI relative to E1 and E3. The genotypes in mega environment E4 were associated with relatively high rust resistance and they included; 6×11 (Shibe×ICMV221white; grain yield = 2030 kg ha^{-1}, rust severity = 27.58%), 6×10 (Shibe×ICMV221; grain yield = 2506 kg ha^{-1}, rust severity = 24.76%), 6×7 (Shibe×Exbornu; grain yield = 2149 kg ha^{-1}, rust severity = 22.39%), 6×8 (Shibe×CIVT9206; grain yield = 2387 kg ha^{-1}, rust severity = 24.36%). Generally the genotypes associated with high rust resistance were also highly unstable in terms of grain yield and associated with the unstable environments E2 and E4 (Figure 6). These observations emphasize the importance of GEI and adopting selection for specific environments. In many pearl millet studies the GGE biplot has also been used to identify pearl millet mega environments to reduce number of test environments (Gupta et al., 2013; Ishaq et al., 2014). Mashiri et al. (2014) adopted the GGE biplot technique to estimate environmental effects for days to flowering, plant height and physiological maturity (Bashir et al., 2014). In addition, Gebre (2014) and Mustapha and Bakari (2014) used the GGE biplot analysis to identify stable genotypes with high grain yield while Bashir et al. (2014) identified best performers for grain yield. Thus, the practicality in using the GGE biplot merits its use in selecting for stability and adaptability of genotypes for grain yield and other yield-related traits.

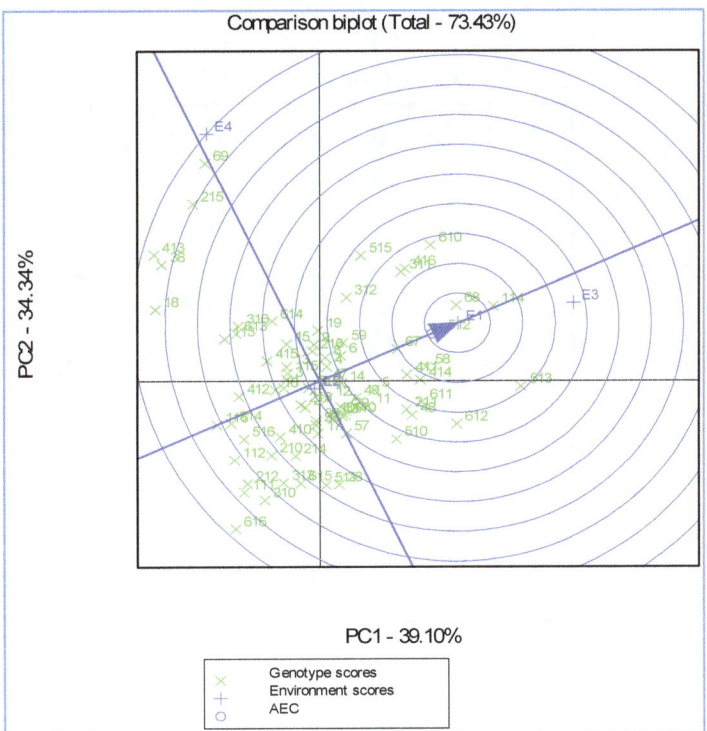

Figure 3. Ranking genotypes based on both means and stability for grain yield across environments

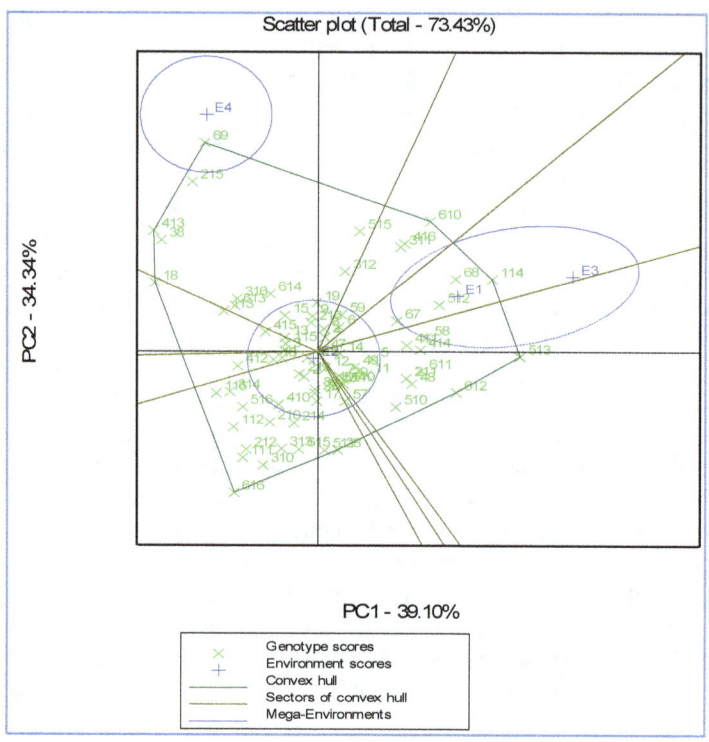

Figure 4. 'Which won where' with mega environments for grain yield

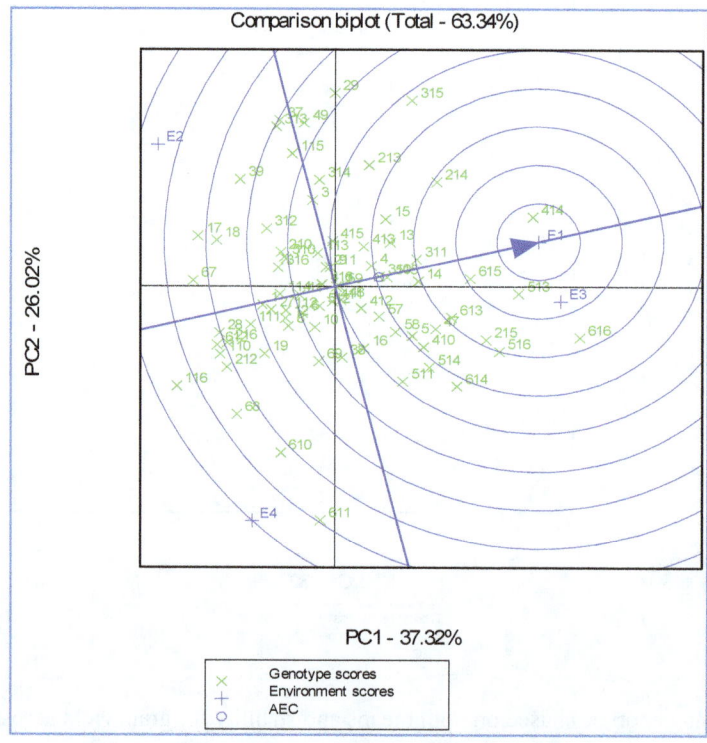

Figure 5. Association between environments for rust severity

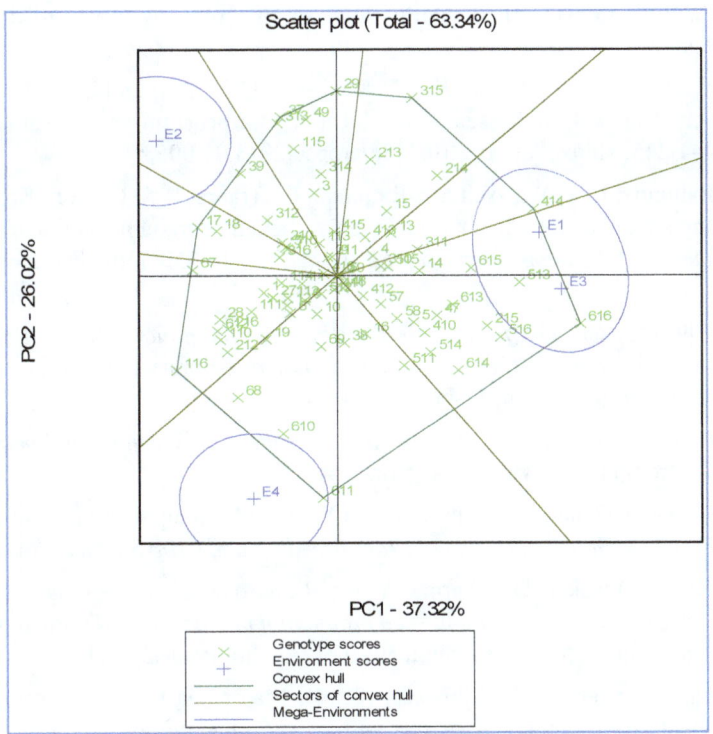

Figure 6. 'Which won where' with mega environments for rust severity

4. Conclusion

The study focused on establishing the genotype by environment interaction effect, characterising environments and genotypes. The ANOVA results showed that the effects of environments, genotypes and genotype x environment interaction (GEI) were important in trait expression and performance of genotypes. In addition, it was observed that amount of rainfall received at both vegetative and post-anthesis phases had an effect on grain yield and disease severity. Finally, the GGE biplot was useful in concisely characterising the environments and the genotypes. It characterised the environments in terms of stability and productivity. This resulted in grouping of mega environments with E2 and E4 being ideal for rust discrimination while E1E3 was the best for grain yield; implying that environment-specific selection should be adopted.

References

Bashir, E. M. A., Ali, A. M., Adam-Ali, M., Ismail, M. I., Parzies, H. K., & Haussmann, B. I. G. (2014). Patterns of pearl millet genotype-by-environment interaction for yield performance and grain iron (Fe) and zinc (Zn) concentrations in Sudan. *Field Crops Research, 166*, 82-91. https://doi.org/10.1016/j.fcr.2014.06.007

CIMMYT. (2014). *Integrated Breeding Platform's Breeding Management System version 3.0 (IBP-BMS).* Integrated Breeding Platform, CIMMYT, Veracruz, Mexico.

Cooper, M., & DeLacy, I. H. (1994). Relationships among analytical methods used to study genotypic variation and genotype-by-environment interaction in plant breeding multi-environment experiments. *Theoretical and Applied Genetics, 88*(5), 561-572. https://doi.org/10.1007/BF01240919

Crossa, J. (1990). Statistical analysis of multilocational trials. *Advances in Agronomy, 44*, 55-85. https://doi.org/10.1016/S0065-2113(08)60818-4

Directorate of Plant Protection (DPP). (2011). *Pearl millet production guide.* DPP, South Africa.

Ezeaku, I. E., Angarawai, I. I., Aladele, S. E., & Mohammed, S. G. (2014). Genotype by environment interactions and phenotypic stability analysis for yield and yield components in parental lines of pearl millet (*Pennisetum glaucum* [L.] R. Br). *African Journal of Agricultural Research, 9*(37), 2827-2833. https://doi.org/10.5897/AJAR2014.8930

Gebre, W. (2014). Evaluation of pearl millet (*Pennisetum glaucum* L.) genotypes for yield and yield stability in South Omo and West Hararghe. *Journal of Biology, Agriculture and Healthcare, 4*(8), 99-121.

Gupta, S. C., & Ndoye, A. T. (1991). Yield stability analysis of promising pearl millet genotypes in Senegal. *Maydica, 36*, 83-86.

Gupta, S. K., Rathore, A., Yadav, O. P., Rai, K. N., Khairwal, I. S., Rajpurohit, B. S., & Das, R. R. (2013). Identifying mega-environments and essential test locations for pearl millet cultivar selection in India. *Crop Science, 53*(6), 2444-2453. https://doi.org/10.2135/cropsci2013.01.0053

Haussmann, B. I. G., Rattunde, H. F. W., Weltzien-Rattunde, E., Traoré, P. S. C., vom Brocke, K., & Parzies, H. K. (2012). Breeding strategies for adaptation of pearl millet and sorghum to climate variability and change in West Africa. *Journal of Agronomy and Crop Science, 198*, 327-339. https://doi.org/10.1111/j.1439-037X.2012.00526.x

IBPGR & ICRISAT. (1993). *Descriptors for Pearl Millet (Pennisetum glaucum (L.) R. Br.)*. IBPGR & ICRISAT, Rome. Retrieved from http://www.bioversityinternational.org/e-library/publications/detail/descriptors-for-pearl-millet-empennisetum-glaucumem-l-r-br

IFAD. (1999). *Farmer participatory testing of technologies to increase sorghum and pearl millet production in the Sahel*. Retrieved from http://www.ifad.org/grants

Ishaq, J., & Meseka, S. (2014). Genetic Stability of Grain Yield and principal component analysis in pearl millet (*Pennisetum glaucum* L.). *Greener Journal of Plant Breeding and Crop Science, 2*(4), 88-92.

Khairwal, I. S., Rai, K. N., Diwakar, D., Sharma, Y. K., Rajpurohit, B. S., Nirwan, B., & Bhattacharjee, R. (2007). *Pearl millet: Crop management and seed production manual* (p. 104). International Crops Research Institute for the Semi-Arid Tropics, Patancheru 502 324, Andhra Pradesh, India.

Lubadde, G., Tongoona, P., Derera, J., & Sibiya, J. (2014). Major pearl millet diseases and their effects on on-farm grain yield in Uganda. *African Journal of Agricultural Research, 9*(39), 2911-2918. https://doi.org/10.5897/AJAR2013.7208

Mashiri, C. E., Chikerema, P., Murewi, C. T. F., Gonzo, L., Maposa, S., Pfupajena, M. H., … Sisito, G. (2014). Estimating environmental effects using the agronomic traits of pearl millet varieties by means of the biplot technique. *International Journal of Advanced Scientific and Technical Research, 2*(4), 72-91.

Matus-Cadiz, M. A., Hucl, P., Peron, C. E., & Tyler, R. T. (2003). Genotype × environment interaction for grain colour in hard white spring wheat. *Crop Science, 43*, 219-226. https://doi.org/10.2135/cropsci2003.2190

Misra, R. C., Das, S., & Patnaik, M. C. (2010). AMMI analysis of stability and adaptability of late duration finger millet (*Eleusine coracana*) genotypes. *World Journal of Agricultural Sciences, 6*(6), 664-669.

Mustapha, M., & Bakari, H. R. (2014). Statistical evaluation of genotype by environment interactions for grain yield in millet (*Pennisetum glaucum* (L.) R. Br). *International Journal of Engineering and Science, 3*(9), 7-16.

Payne, R. W., Murray, D. A., Harding, S. A., Baird, D. B., & Souter, D. M. (2012). *GenStat for windows 14th edition introduction*. VSN International, Hemel Hempstead.

Singh, S. D. (1990). Sources of resistance to downy mildew and rust in pearl millet. *Plant Disease, 74*, 871-874. https://doi.org/10.1094/PD-74-0871

Thakur, R. P., Rajan, S., & Rao, V. P. (2011). Screening techniques for pearl millet diseases. *Information Bulletin No. 89* (p. 56). Patancheru 502 324, Andhra Pradesh, India: International Crops Research Institute for the Semi-Arid Tropics.

Tooley, P. W., & Grau, C. R. (1984). Field characterisation of rate reducing resistance to *Phytophthora megasperma* f.sp. *glycinea* in soybean. *Phytopathology, 74*, 1201-1208. https://doi.org/10.1094/Phyto-74-1201

Vadez, V., Hash, T., Bidinger, F. R., & Kholova, J. (2012). Phenotyping pearl millet for adaptation to drought. *Frontiers of Physiology, 3*, 1-12. https://doi.org/10.3389/fphys.2012.00386

Yan, W., & Kang, M. S. (2003). *GGE Biplot Analysis: A graphical tool for breeders, geneticists, and agronomists*. CRC Press, Boca Raton, FL.

Yan, W., & Racjan, I. (2002). Biplot analysis of test environments and trait relations of soybean in Ontario. *Crop Science, 42*, 11-20. https://doi.org/10.2135/cropsci2002.0011

Yan, W., Kang, M. S., Ma, B., Woods, S., & Cornelius, P. L. (2007). GGE Biplot vs. AMMI Analysis of genotype-by-environment data. *Crop Science, 47*(2), 643-653. https://doi.org/10.2135/cropsci2006.06.0374

15

Biochemical Resistance Traits of Bitter Gourd against Fruit Fly *Bactrocera cucurbitae* (Coquillett) Infestation

Paras Nath[1], A. K. Panday[2], Akhilesh Kumar[2], A. B. Rai[3] & Hemalatha Palanivel[1]

[1] College of Agriculture, Fisheries and Forestry, Fiji National University, Koronivia, Fiji

[2] Department of Entomology and Agricultural Zoology, BHU, Varanasi, India

[3] Indian Institute of Vegetable Research, Varanasi, India

Correspondence: Hemalatha Palanivel, Department of Genetics and Plant Breeding, College of Agriculture, Fisheries and Forestry, Fiji National University, Koronivia, Fiji.
E-mail: hemalatha.palanivel@fnu.ac.fj

Abstract

Host plant resistance is a key factor for management of the melon fruit fly, *Bactrocera cucurbitae* (Coquillett), due to difficulties associated with its chemical and biological control. Various biochemical traits including total sugars, reducing sugars, non-reducing sugars, silica, protein content, ash content, other elements and phenols, and moisture content of fruit were studied on 74 varieties/genotypes of bitter gourd (*Momordica charantia* L.), in relation to resistance against *B. cucurbitae* under field conditions. Seventy-four genotypes of bitter gourd were screened against fruit fly infestation. The correlation coefficients revealed that the larval density and bitter gourd fruit damage (%) had significant positive relationship (r = 0.99). The moisture content had significant positive effect on the fruit damage (r = 0.75) and number of larvae per fruit (0.80). Significant differences were found in tested varieties/genotypes for fruit infestation and larval density per fruit. The nitrogen, phosphorous, potassium and protein content (r = -0.87, -0.90) showed significant negative correlation with fruit fly infestation. The non-reducing, reducing, total sugars, total phenols, silica and ash content had significant impact on the fruit damage and showed significant negative correlation with fruit fly infestation. The ascorbic acid also had significant impact on the fruit damage and showed significant negative correlation with fruit fly infestation (r = -0.79), the chlorophyll 'a', 'b' and total chlorophyll content had non-significant negative effect on the fruit damage and number of larvae per fruit.

Keywords: biochemical traits, bitter gourd, fruit fly resistance, screening

1. Introduction

Vegetables are one of the most important components of Indian horticulture. In recent past tremendous progress has been made and as a result, India has emerged as second largest producer of vegetables next after China (Panday et al., 2009). Among the various vegetable crops, cucurbitaceous vegetables that include bitter gourd, cucumber, melons, pumpkins and various types of gourds are of major importance in the Indian subcontinent and include the largest number of summer and rainy season vegetables. Bitter gourd (*Momordica charantia*) is one of the most popular vegetables in Southeast Asia. Bitter gourd (*Momordica charantia* L., Cucurbitacae), commonly known as balsam pear, or *Karela*, was cultivated throughout the world, especially in the tropical areas (ElBatran et al., 2006). In terms of nutritive value, bitter gourd ranks first among cucurbits, being rich in iron, phosphorus and ascorbic acid (Awasthi & Jaiswal, 1986). A substance with clinical properties of insulin has been isolated from bitter gourd fruits and hence is recommended for consumption to diabetic patients (Baldwa et al., 1977; Aslam & Stockley, 1979; Kedar & Chakraborti, 1982). Bitter gourd has good export potential and its share in export of green vegetable is to the extent of 20 per cent (Anonymous, 1992).

Insect pests are a major constraint for increasing the production and productivity of this crop. Bitter gourds are attacked by several insect pests, among them the fruit fly is one of the most destructive insect-pests (Panday et al., 2008). Melon fruit flies (Diptera: Tephritidae: Dacinae) are economically important pests of the cucurbits and are geographically distributed throughout the tropics and subtropics of the world (Chinajariyawong et al., 2003), especially in most countries of South East Asia (Allwood et al., 1999). The extent of losses varies between 30 and 100% depending on the cucurbit species and the season. (Pareek & Kavadia, 1995; Kapoor, 1993; Panday et al., 2009). The melon fly has been observed on 81 host plants, with watermelon being a highly-preferred host, and has been a major limiting factor in obtaining good-quality fruits and high yield (Nath & Bhushan, 2006).

As a result of the recent efforts, made by the Environmental Protection Agency, to reduce the use of harmful insecticides, especially, organophosphates, organochlorines, some carbamates and pyrethroids, in the agricultural crops, the trend has now shifted towards an integrated pest management (IPM) for the control of tephritid fruit flies (Roger et al., 2010). Integrated pest management (IPM), includes, a combination of chemical, biological and cultural control tactics (Sarfaraz et al., 2002), with insecticides still to continue as an important components of such strategies. But, the larvae of melon fruit flies, like, other fruit flies often pupate either in the soil, inside the fruits or under the fruits, thereby avoiding the exposure to or contact with insecticides, when surface application is practiced. Similarly, the maggots damage the fruits internally. Therefore, it is imperative to explore alternative methods of control of this pest. Hence, the development of varieties resistant to melon fruit fly is an important component for an integrated pest management of this pest (Panda & Khush, 1995).

The development and then the cultivation of fruit fly resistant bitter gourd cultivars has been limited, because of the lack of adequate information on the genetic variability and sources of resistance in the available bitter gourd genotypes and influence of these sources on the pest multiplication (Dhillon et al., 2005). Therefore, it becomes imperative to identify sources of resistance in bitter gourd and get knowledge of their influence on oviposition preference, larval performance and pest multiplication for devising sustainable pest management strategies

The identification of biochemical factors governing resistance is helpful in the development of rapid screening technique. The mechanism of resistance may be antixenosis, antibiosis and tolerance. The bitter gourd varieties having such inhibitory mechanism of resistance to melon fruit fly can be used in transferring the resistance in to the commercially acceptable varieties. The chemical defenses in plants against predators and parasites are known since antiquity. Biochemical studies can only follow a successful discovery of sources of resistance. It is well established that chemical stimuli play a major role in host plant selection by insect for both feeding and oviposition (Maxwell & Jennings, 1980). Antixenosis refers to the potential plant characteristics/traits, either allelochemical or morphological, that impart or alter insect behaviour towards the host (Haldhar et al., 2013). Resistance in a given plant species, as expressed in the field, is a complex phenomenon and there is generally no single chemical to condition it. The extent of losses varies between 30 and 100%, depending on the cucurbit species and the season. As the maggots damage the fruits internally, it is difficult to control this pest with insecticides. Hence, development of bitter gourd varieties/genotypes resistant to the fruit fly is an important component of integrated pest management (Panda & Khush, 1995), but it has been limited in India owing to inadequate information on the sources of plant traits associated with resistance to pest infestations. The present study was designed to identify various morphological biochemical (allelochemical) fruit traits of bitter gourd varieties/genotypes associated with resistance against the fruit fly in terms of fruit infestation and larval density under field conditions.

2. Materials and Methods

2.1 Preliminary Screening

Little is known about the biochemical basis of resistance in bitter gourd against the infestation of fruit fly. If the factors responsible for imparting resistance in fruits of bitter gourd are known they can be used as markers for selecting resistant plant materials within a short period of time. To study the biochemical basis of resistance to the melon fruit fly, estimation of various bio-chemicals were done for this study, sixteen genotypes comprising of two highly resistant, three resistant, three moderately resistant, two susceptible and two highly susceptible were selected from the preliminary screening of 74 bitter gourd genotypes. The seeds were sown in summer season of the year 2009 and 2010 at the vegetable research Farm of Indian Institute of Vegetable Research, Varanasi (25.10° N and 82.52° E) with 3 replicates (blocks) for each variety/genotype in a randomized block design. The pits were dig of 30 cm × 30 cm × 30 cm size at 2 × 1.5 m spacing and form basins and Installed drip system with main and sub-main pipes and place the inline lateral tubes at an interval of 1.5 m. Recommended agronomic practices were followed for raising the healthy crop. The marketable fruits were picked at weekly intervals for observations of fruit fly infestation and number of larvae per fruit. The infested and healthy fruits were counted to know the per cent fruit infestation. The infested fruits were cut and open to count the number of larvae per fruit. All the screened genotypes were grouped in to different categories on the basis of per cent fruit infestation and number of larvae per fruit (Nath, 1966). The pericarp was air dried for five days and milled with an aid of electrical grinder. The ethanol extract of each sample was prepared by soaking 150 g of dried powder in one litre of ethanol for 24 hours as incubation period. The extract was filtered with Whatman's filter paper number 42. The observations on biochemical traits were recorded on five randomly selected marketable size fruits in three replications. The moisture content was calculated by the following formula,

$$Per\ cent\ moisture\ content = \frac{Fresh\ weight\ of\ sample - Dry\ weight\ of\ sample}{Fresh\ weight\ of\ samples} \times 100 \qquad (1)$$

The different biochemical constituents were estimated by different standard methods, Chlorophyll 'a', 'b' and total chlorophyll were estimated by the method suggested by Mahlberg and Venketeswaran (1966), ascorbic acid was determined by 2, 6 dichlorophenol indophenol titration method (AOAC, 1960), total nitrogen was determined by Microkjeldahl's method (AOAC, 1985). The nitrogen percentage was multiplied by 6.25 to get total protein percentage. Phosphorous (Jackson, 1973) potassium (Tewatia, 1994), Reducing sugars were estimated by the method suggested by Paleg (1960). Total sugars were estimated by the method of Yemn and Willis (1954). Non reducing sugar was estimated by subtracting reducing sugars from total sugars. Total phenol content of samples was estimated by the method of Swain and Hills (1959). Silica percentage was determined as per AOAC (1980), by the formula given below.

$$Silica\ (\%) = \frac{Weight\ of\ silica + (Filter\ paper - Weight\ of\ Filter\ paper)}{Weight\ of\ sample} \times 100 \qquad (2)$$

The per cent fruit damage and number of larvae were correlated with different biochemical traits of bitter gourd fruits to know the role of different biochemical traits in imparting resistant or susceptibility in bitter gourd fruits against melon fruit fly.

3. Results and Discussion

3.1 Screening of Bitter Gourd Genotypes for Fruit Fly Infestation

Seventy-four genotypes of bitter gourd were screened against fruit fly infestation in Agricultural Research Farm, Banaras Hindu University (24°56′ N to 82°12′ E). On the basis of per cent fruit infestation and the average number of larvae per damaged fruit, the genotypes were categorized in to different groups i.e. (highly resistant, resistant, moderately resistant, susceptible, and highly susceptible). The genotypes of different categories showed that the lowest fruit infestation was recorded in the genotypes IC 248282, Kerala collection -1, VRBT-4 and DRAR-1, IC 68314 and these were grouped as resistant genotypes while 61 genotypes as moderately resistant, 5 genotypes as susceptible and 3 genotypes highly susceptible (Table 1). There were significant differences in per cent fruit infestation and larval density per fruit among the genotypes tested. The fruit infestation during 2009 and 2010 summer season (average of two years) ranged from 13.64% to 81.57% and larval density per fruit ranged from 2.75 to 7.88 larvae per fruit (Table 2). The correlation coefficients revealed that the larval density and bitter gourd fruit damage (%) had significant positive relationship (r = 0.98) (Table 4). Dhillon et al. (2005) was also reported that the genotypes with low fruit fly infestation had low larval numbers in the fruits and there was positive correlation (r = 0.96) between percentage fruit infestation and number of larvae per fruit. Plant varieties/ genotypes possess physiological and biochemical variations due to the environmental stress or genetic makeup, which alter the nutritional values for herbivores (Rafiq et al., 2008). The percent fruit infestation and larval density were significantly lower in resistant varieties/genotypes and higher in susceptible varieties/ genotypes of bitter gourd.

Table 1. Grouping of bitter gourd germplasms/lines/genotypes in to different categories on the basis of per cent fruit infestation during summer season (average of two years)

S. No.	Fruit infestation (%)	No. of larvae/fruit	Reaction	No. of genotypes	Genotypes
1	1-10	4.3	Highly resistant	0	-
2	11-20	4.9	Resistant	5	IC-248282, Kerala collection -1, VRBT-4, DRAR-1, IC-68314
3	21-50	5.0	Moderately resistant	61	VRBT 6, VRBT-7, VRBT 11, VRBT 14, VRBT 28, VRBT 29, VRBT 31, VRBT 32, VRBT 35, VRBT 37, VRBT 39, VRBT 41, VRBT 46, VRBT 54, VRBT 63, VRBT 68, VRBT 69, VRBT 71, VRBT 72, VRBT 73, VRBT 75, VRBT 76, VRBT 77, VRBT 83, VRBT 85, VRBT 86, VRBT 87, VRBT 90, VRBT 91, VRBT 92, VRBT 94, VRBT 95, VRBT 96, VRBT 98, VRBT 99, VRBT 100, VRBT 103, VRBT 107, VRBT 113, VRBT 115, VRBT 128, VRBT 135, VRBT 139, VRBT145, VRBT147, VRBT 178, VRBT 179, VRBT 187, VRBT 188, DVBTG-1, DVBTG-2, DVBTG 5, DVBTG 7, NDBT-1, Long green, Pusavishesh, Hirkani, US-6201, Jhalribaramasi, BL-237, Konkan Tara
4	51-75	6.1	Susceptible	5	VRBT 21, VRBT 22, VRBT 38, VRBT 93, VRBT 175
5	76-100	8.1	Highly Susceptible	3	Jaunpuri, Arka harit, Pusa Do Mausmi

Table 2. Per cent fruit infestation and larval densities of fruit fly in different genotypes of bitter gourd during summer season (Average of the 2 years 2009-2010)

S. No.	Genotypes	Fruit infestation (%)	No. of larvae/fruit	Reaction
1	IC 248282	13.64 (21.53)	2.75 (1.78)	Resistant
2	Kerala Collection -1	15.68 (23.17)	2.85 (1.81)	Resistant
3	DRAR-1	18.12 (25.12)	3.26 (1.91)	Resistant
4	VRBT 4	18.76 (25.49)	3.40 (1.95)	Resistant
5	Konkan Tara	24.80 (29.73)	4.13 (2.13)	Moderately resistant
6	Hirkani	30.91 (33.71)	4.59 (2.24)	Moderately resistant
7	DVBTG-1	44.16 (41.60)	5.65 (2.47)	Moderately resistant
8	NDBT-1	40.11 (39.07)	5.23 (2.37)	Moderately resistant
9	VRBT 35	46.49 (43.11)	5.61 (2.45)	Moderately resistant
10	VRBT 93	55.11 (48.16)	6.19 (2.57)	Susceptible
11	VRBT 38	57.33 (49.58)	6.09 (2.55)	Susceptible
12	VRBT 21	67.01 (55.32)	6.77 (2.69)	Susceptible
13	VRBT 175	65.54 (54.57)	6.71 (2.67)	Susceptible
14	Jaunpuri	76.21 (61.31	7.47 (2.81)	Highly Susceptible
15	Arkaharit	78.17 (62.95)	7.79 (2.87)	Highly Susceptible
16	Pusa Do Mausami	81.57 (64.84)	7.88 (2.88)	Highly Susceptible

3.2 Biochemical Fruit Traits and Correlation Analysis

From each category of genotypes, 3-4 genotypes were selected for biochemical analysis (Table 3). The moisture content of the fruits of all the selected genotypes varied significantly from each other, however it varied from 86.27% (IC 248282 and DRAR-1) to 95.37% (VRBT 175). There was a significant increase in fruit fly infestation and number of larvae per fruit with an increase in moisture content of the fruits. In general, the moisture content in the resistant genotypes is lower in comparison with the highly susceptible genotypes. The moisture content had significant positive effect on the per cent fruit damage (r = 0.75) and number of larvae (r = 0.80) per fruit. All the tested genotypes of bitter gourd differed significantly from each other in nitrogen, phosphorous, potassium, protein, sugars, phenol, ascorbic acid, chlorophyll, silica and ash content (Table 3 and 4). The relationship with ash content with the infestation was illustrated in Figure 1.Nitrogen, phosphorous, potassium and protein content ranged from 1.93% to 2.85%, 0.32% to 0.59%, 1.80% to 4.63% and 12.08% to 17.81%, respectively, being minimum in Pusa Do Mausami and maximum in IC 248282. The nitrogen, phosphorous, potassium and protein content of the fruit of bitter gourd were found highest in resistant genotypes and lowest in susceptible genotypes and showed, nitrogen (r = -0.87, -0.90), phosphorus (r = -0.68, -0.69), potassium (r = - 0. 83, -0.88) and protein content (r = - 0.87, -0.90) significant negative correlation with fruit fly infestation and number of larvae per fruit. Benepal and Hall (1967) reported that feeding of squash bug (*Anasa tristis* De Geer) was not affected by protein content of plants of *Cucurbita pepo* L. Protein as a whole may not affect the feeding of insects, whereas, particular amino acid may be responsible for imparting resistance. The non-reducing, reducing, total sugar, total phenol, silica and ash content ranged from 0 .81% to 1.45%, 1.77% to 2.90%, 2.61% to 4.41. 71%, 93.78 mg to 153.64 mg/100 g, 0.83% to 1.77% and 9.63% to 13.40% respectively, being minimum in Pusa Do Mausami and maximum in IC 248282. The non-reducing sugar (r = -0.82, -0.82), reducing sugar (r = -0.86, -0.85), total sugar (r = -0.87, -0.87), total phenol (r = -0.95, -0.92), silica (r = -0.93, -0.92) and ash content (r = -0.92, -0.94), had significant impact on the fruit damage and showed significant negative correlation with fruit fly infestation and number of larvae per fruit (Tables 3 and 4).

Table 3. Biochemical constituents of the fruits of different germplasms/lines/genotypes of bitter gourd during summer season of the years 2009-2010

Varieties/Genotypes	Moisture content (%)	Nitrogen (%)	Phosphorus (%)	Potassium content (%)	Protein (%)	Non-reducing Sugars (%)	Reducing sugars (%)	Total Sugars (%)	Ascorbic acid (mg/100 g)	Total Phenol (mg/100 g)	Chlorophyll (a) (mg/100 g)	Chlorophyll (b) (mg/100 g)	Total Chlorophyll (mg/100 g)	Ash (%)	Silica (%)
IC 248282	86.27	2.85	0.59	4.63	17.81	1.45	2.90	4.41	185.69	153.64	3.30	1.17	4.66	13.4	1.77
Kerala Collection -1	87.64	2.79	0.47	4.06	17.46	1.32	2.83	4.19	188.71	152.46	3.69	1.69	5.38	13.17	1.76
DRAR-1	86.27	2.77	0.58	3.81	17.33	1.30	2.76	4.09	174.55	153.34	3.58	1.75	5.36	12.70	1.72
VRBT 4	88.86	2.64	0.57	3.62	16.48	1.27	2.72	4.02	160.13	152.30	3.56	1.66	5.24	12.57	1.70
Konkan Tara	89.27	2.62	0.52	3.47	16.38	1.36	2.70	4.10	152.22	152.18	3.64	1.79	5.88	12.15	1.66
Hirkani	93.12	2.37	0.49	2.84	14.79	1.11	2.68	3.81	132.45	150.75	3.69	1.71	5.18	11.63	1.63
DVBTG-1	93.03	2.34	0.42	2.47	14.63	1.10	2.63	3.76	119.10	150.09	2.53	1.31	3.92	10.97	1.60
NDBT-1	93.76	2.32	0.44	2.39	14.48	1.15	2.61	3.79	112.97	146.56	2.89	1.51	4.44	10.83	1.50
VRBT 35	94.07	2.36	0.43	2.44	14.73	1.17	2.64	3.84	135.51	140.02	2.94	1.62	4.57	10.66	1.34
VRBT 93	95.27	2.31	0.40	2.37	14.44	1.01	2.53	3.59	143.48	132.75	3.17	1.55	4.13	10.23	1.20
VRBT 38	94.95	2.33	0.34	2.26	14.54	0.95	2.55	3.53	119.08	131.37	3.30	1.44	4.34	10.34	1.25
VRBT 21	95.27	2.21	0.33	2.13	13.83	0.83	2.37	3.23	113.76	127.69	3.11	1.21	4.37	10.01	0.97
VRBT 175	95.37	1.93	0.32	2.11	12.08	0.93	2.26	3.22	132.86	120.10	3.42	1.44	4.90	10.18	1.15
Jaunpuri	91.27	2.37	0.59	2.98	14.83	1.23	2.53	3.80	139.79	111.27	3.58	1.64	5.28	10.20	1.18
Arkaharit	92.83	2.11	0.33	1.91	13.17	0.92	2.15	3.10	102.74	102.92	1.14	0.62	1.80	10.52	1.31
Pusa Do Mausami	94.87	1.98	0.32	1.80	12.35	0.81	1.77	2.61	79.63	93.78	3.47	1.72	5.21	9.63	0.83
SEM ±	1.35	0.44	0.05	0.46	2.76	0.25	0.42	0.46	7.31	1.61	0.70	0.51	0.87	0.95	0.30
CD (0.05)	2.74	0.90	0.11	0.94	5.62	0.52	0.85	0.93	14.89	3.28	1.43	1.04	1.77	1.93	0.60

Table 4. The correlation coefficients of the fruit damage, larval density and biochemical traits of the fruits of different germplasms/lines/genotypes of bitter gourd during summer season of the years 2009-2010

Biochemical traits	Fruit damage (%)	Larvae density (fruit⁻¹)	Moisture content (%)	Nitrogen (%)	Phosphorus (%)	Potassium (%)	Protein (%)	Non-reducing sugars (%)	Reducing sugars (%)	Total sugars (%)	Ascorbic acid (mg/100 g)	Phenol (mg/100 g)	Chlorophyll 'a' (mg/100 g)	Chlorophyll 'b' (mg/100 g)	Total chlorophyll (mg/100 g)	Ash (%)	Silica (%)
Fruit damage (%)	1.00																
Larval density (fruit⁻¹)	0.99**	1.00															
Moisture content (%)	0.75**	0.80**	1.00														
Nitrogen (%)	-0.87**	-0.90**	-0.90**	1.00													
Phosphorus (%)	-0.68**	-0.69**	-0.82**	0.82**	1.00												
Potassium (%)	-0.83**	-0.88**	-0.94**	0.95**	0.85**	1.00											
Protein (%)	-0.87**	-0.90**	-0.90**	1.00**	0.82**	0.95**	1.00										
Non-reducing sugars (%)	-0.82**	-0.82**	-0.86**	0.90**	0.90**	0.91**	0.90**	1.00									
Reducing sugars (%)	-0.86**	-0.85**	-0.64**	0.86**	0.73**	0.79**	0.86**	0.84**	1.00								
Total sugars	-0.87**	-0.87**	-0.77**	0.92**	0.84**	0.88**	0.92**	0.94**	0.97**	1.00							
Ascorbic acid (mg/100 g)	-0.79**	-0.83**	-0.82**	0.88**	0.75**	0.92**	0.88**	0.85**	0.83**	0.87**	1.00						
Phenol (mg/100 g)	-0.95**	-0.92**	-0.56**	0.78**	0.60*	0.69**	0.78**	0.73**	0.91**	0.87**	0.70**	1.00					
Chlorophyll 'a' (mg/100g)	-0.38	-0.40	-0.29	0.38	0.44	0.47	0.38	0.37	0.33	0.36	0.46	0.36	1.00				
Chlorophyll 'b' (mg/100 g)	-0.35	-0.33	-0.19	0.29	0.40	0.29	0.29	0.34	0.26	0.30	0.30	0.34	0.86**	1.00			
Total chlorophyll (mg/100 g)	-0.43	-0.44	-0.37	0.41	0.51	0.49	0.41	0.46	0.33	0.40	0.45	0.38	0.96**	0.90*	1.00		
Ash (%)	-0.92**	-0.94**	-0.92**	0.92**	0.73**	0.94**	0.92**	0.86**	0.77**	0.84**	0.86**	0.77**	0.32	0.20	0.37	1.00	
Silica (%)	-0.93**	-0.92**	-0.79**	0.84**	0.71**	0.81**	0.84**	0.85**	0.85**	0.88**	0.75**	0.86**	0.16	0.16	0.23	0.91**	1.00

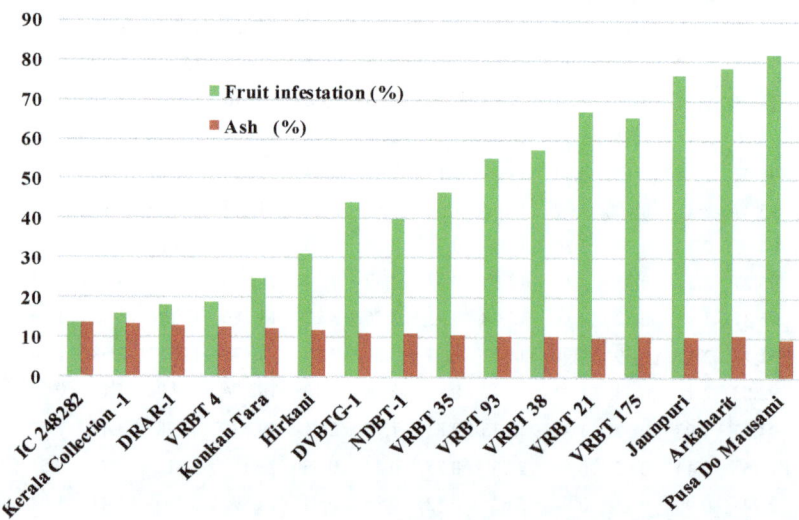

Figure 1. Relationship between ash content and fruit infestation percentage

The total sugars, reducing sugars and non - reducing sugars were higher in resistant genotypes as compare to susceptible genotypes. However, in an earlier study, Sharma and Hall (1971) reported positive correlation between spotted cucumber beetle (*Diabrotica undecimpunctata.* Howardi) feeding and total sugars concentration of various cucurbitaceous crops. Ingoley et al. (2005) they reported that the cucumber fruits of susceptible and highly susceptible genotypes revealed higher amount of total sugar as compared to moderately resistance genotypes. In earlier study, Chelliah and Sambandam (1971) also reported that the resistance to melon fruit fly (*D. cucurbitae*) in *Cucumis callosus* appeared to be associated with high silica content of the fruits. It is concluded that higher amounts of tannin, flavonol, total phenol and silica contents of fruits of resistance genotypes may be responsible for imparting resistance against melon fruit fly in bitter gourd. The bitter gourd genotypes differed significantly from each other in ascorbic acid content. The ascorbic acid content differed from 79.63 mg/100 g to 188.71 mg/100 g of fresh weight. The highest ascorbic acid content was recorded in the genotype Kerala collection -1 (188.71 mg/100 g), and lowest was recorded in the genotype Pusa Do Mausami (79.63 mg/100 g). The ascorbic acid also had significant impact on the fruit damage and showed significant negative correlation with fruit fly infestation (r = -0.79) and number of larvae per fruit (r = -0.83). Chelliah (1970) suggested that perception of chemical stimuli was well developed in *B. cucurbitae*. Melon fruit fly infestation and larval density per fruit increased with increase in moisture level; while ascorbic acid, reducing, non-reducing and total sugars, nitrogen, protein, phosphorous and potassium contents were greater in resistant genotypes when compared with susceptible ones. Similar findings have also been reported by Tewatia et al. (1998). The significant difference was observed for chlorophyll 'a', 'b', and total chlorophyll content in the fruits of various genotypes of bitter gourd and it was ranged from 1.14 to 3.69 mg/100 g, 0.62 to 1.79 mg/100 g and 1.80 to 5.88 mg/100 g fresh weight respectively. The chlorophyll content was recorded highest in the fruits of genotype Konkantara (5.88 mg/100 g), and lowest in the genotype Arkaharit. The chlorophyll 'a' (r = - 0. 38, -0.40), chlorophyll 'b' (r = - 0. 35, -0.33) and total chlorophyll (r= - 0. 43, -0.44) had non-significant negative effect on the fruit damage (%) by fruit fly and number of larvae per fruit. The allelochemical compounds of fruit were significantly different among the tested muskmelon varieties/genotypes. In brinjal and okra crop, the biochemical characters such as total sugar and crude protein were positively correlated with fruit borer infestation, whereas total phenols were negatively correlated (Jat & Pareek, 2003; Sharma & Singh, 2010). Very little information is available on correlation of the biochemical traits. Biochemical characters such as total sugar and crude protein were positively correlated whereas total phenols were negatively correlated with fruit borer nfestation (Haldhar et al., 2013). Phenolic heteropolymers play a central role in plant defense against insects and pathogens (Barakat et al., 2010). Phenols also play an important role in cyclic reduction of reactive oxygen species such as superoxide anion and hydroxide radicals, H_2O_2, and singlet oxygen, which in turn activate a cascade of reactions leading to the activation of defensive enzymes (Maffei et al., 2007).

4. Conclusion

The silica and ash content were playing an important role in pest resistance. It is very well documented silica has clear prophylactic effect with wide range of insect feeding guilds, including lepidopteran borers, folivores, phloem feeding insects and other plant feeders. Hence we suggest that reduction in fruit fly infestations on resistant varieties/genotypes could be due to prophylactic effects of Silica and ash content and biochemical/Allelochemical traits. Certain biochemical traits (*e.g.*, Total sugar, phenols and ascorbic acid, described related to resistance of bitter gourd against *B. cucurbitae* and, therefore, can be used as marker traits in plant breeding programs to select resistant varieties/genotypes.

References

Allwood, A. J., Chinajariyawong, A., Drew, R. A. I., Hamacek, E. L., Hancock, D. L., Hengsawad, C., ... Vijasegaran, S. (1999). Host plant records for fruit flies (Diptera: Tephritidae) in Southeast Asia. *The Raffles Bulletin of Zoology, 7*, 1-92.

Anonymous. (1992). *Report on infrastructure for export of agricultural commodities and processed food* (p. 44). Govt. of India, Planning commission (Agriculture Division), Yojana Bhawan, New Delhi.

AOAC. (1960). *Official methods of Analysis, Association of Agricultural Chemists*. Washington, D. C.

AOAC. (1980). *Official methods of Analysis, Association of Agricultural Chemists*. Washington, D. C.

AOAC. (1985). *Official methods of Analysis, Association of Agricultural Chemists*. Washington, D. C.

Aslam, M., & Stockley, I. H. (1979). Interaction between curry ingredient (Karela) and drug (chlorpropamide). *Lancet, 1*(8116), 607. https://doi.org/10.1016/S0140-6736(79)91028-6

Awasthi, C. P., & Jaiswal, R. C. (1986). Biochemical composition and nutritional quality of fruits of bitter gourd grown in Uttar Pradesh. *Progressive Horticulture, 18*, 265-269.

Baldwa, V. S., Bhandari, C. M., Pangaria, A., & Goyal, R. K. (1977). Chemical trial in patients with diabetes mellitus of an insulin like compounds obtained from plant source. *Upsala Journal of Medical Sciences, 82*, 39-41. https://doi.org/10.3109/03009737709179057

Barakat, A., Bagniewska-Zadworna, A., Frost, C. J., & Carlson, J. E. (2010). Phylogeny and expression profiling of CAD and CAD-like genes in hybrid Populus (*P. deltoids* × *P. nigra*): Evidence from herbivore damage for sub functionalization and functional divergence. *BMC Plant Biology, 10*, 1-11. https://doi.org/10.1186/1471-2229-10-100

Benepal, P. S., & Hall, C. V. (1967). The influence of mineral nutrition of varieties of *Cucurbita pepo* L. on the feeding response of squash bug *Anasa tristis* De Geer. *Proceedings of the American Society of Horticultural Science, 90*, 304-312.

Chelliah, S. (1970). Host influence on the development of melon fly, *Dacus cucurbitae* Coquillett. *Indian Journal of Entomology, 32*, 381-383.

Chelliah, S., & Sambandam, C. N. (1971). Role of certain mechanical factors in *Cucumis callosus* (Rottl.) Cogn. in imparting resistance to *Dacus cucurbitae*. *Auara, 3*, 48-53.

Chinajariyawong, A., Kritsaneepaiboon, S., & Drew, R. A. I. (2003). Efficacy of protein bait sprays in controlling fruit flies (Diptera: Tephritidae) infesting angled luffa and bitter gourd in Thailand. *The Raffles Bulletin of Zoology, 51*(1), 7-15.

Dhillon, M. K., Singh, R., Naresh, J. S., & Sharma, N. K. (2005). Influence of physio-chemical traits of bitter gourd, *Momordica charentia* L. on larval density and resistance to melon fruit fly, *Bactrocera cucurbitae* (Coquillet). *Journal of Applied Entomology, 129*(7), 395-399. https://doi.org/10.1111/j.1439-0418.2005.00911.x

El-Batran, S. A. E. S., El-Gengaihi, S. E., & El-Shabrawya, O. A. (2006). Some toxicological studies of *Momordica charantia* L. on albino rats in normal and alloxan diabetic rats. *Journal of Ethnopharmacology, 108*, 236-242. https://doi.org/10.1016/j.jep.2006.05.015

Haldhar, S. M., Bhargava, R., Choudhary, B. R., Pal, G., & Kumar, S. (2013). Allelochemical resistance traits of muskmelon (*Cucumis melo*) against the fruit fly (*Bactrocera cucurbitae*) in a hot arid region of India. *Phytoparasitica, 41*, 473-481. https://doi.org/10.1007/s12600-013-0325-x

Ingoley, P., Mehta, P. K., Chauhan, Y. S., Singh, N., & Awasthi, C. P. (2005). Evaluation of cucumber genotype for resistance to fruit fly under mid hill condition of Himachal Pradesh. *Journal of Entomological Research, 29*(1), 57-60.

Jackson, M. L. (1973). Vandomolybdate phosphoric yellow colour method for determination of phosphorus. *Soil Chemical Analysis*. Prentice Hall of India, New Delhi.

Jat, K. L., & Pareek, B. L. (2003). Biophysical and biochemical factors of resistance in brinjal against *Leucinodes orbonalis* (Guen). *Indian Journal of Entomology, 65*, 252-258.

Kapoor, V. C. (1993). Indian fruit flies. *Oxford and IBH Publications* (p. 228). New Delhi, India.

Kedar, P., & Chakraborti, C. H. (1982). Effect of bitter gourd and glibenclamide in streptozotocin induced diabetes Mellitus. *Indian Journal of Experimental Biology, 28*, 232-235.

Maffei, M. E., Mithöfer, A., & Boland, W. (2007). Insects feeding on plants: Rapid signals and responses preceding the induction of phytochemical release. *Phytochemistry, 68*, 2946-2959. https://doi.org/10.1016/j.phytochem.2007.07.016

Mahlberg, P., & Venketeswaran, G. S. (1966). Pigment analysis of normal and proliferated genetical strains of Nicotiana under cultural conditions. *Botanical Gazette, 127*, 114-9. https://doi.org/10.1086/336351

Maxwell, P. G., & Jennings, P. R. (Eds.). (1980). *Breeding plants resistance to insects* (p. 683). John Wiley and Sons, New York, USA.

Nath, P. (1966). Varietal resistance of gourds to the fruit flies. *Indian Journal of Horticulture, 23*(2), 69-79.

Nath, P., & Bhushan, S. (2006). Screening of cucurbit crops against fruit fly. *Annals of Plant Protection Science, 14*, 472-473.

Paleg, L. G. (1960). Physiological effects of gibberellic acid on carbohydrate metabolism and amylase activity of barley endosperm. *Plant Physiology, 35*, 293-299. https://doi.org/10.1104/pp.35.3.293

Panda, N., & Khush, G. S. (1995). *Host plant resistance to insects* (p. 435). Wallingford, UK: CAB International.

Panday, A. K., Nath, P., & Rai, A. B. (2008). Efficacy of some eco-friendly insecticides, poisons baits and their combinations against bitter gourd infestation by melon fruit fly (*Bactrocera cucurbitae* Coquillet). *Vegetable Science, 35*(2), 152-155.

Panday, A. K., Nath, P., Rai, A. B., & Kumar, A. (2009). Screening of some bitter gourd varieties/germplasms on the basis of some biological and biometrical parameters of melon fruit fly (*Bactrocera cucurbitae* Coquillet) *Vegetable Science, 36*(Suppl. 3), 399-400.

Pareek, B. L., & Kavadia, V. S. (1995). Screening of muskmelon varieties against melon fruit fly, *Dacus cucurbitae* Coquillett under field conditions. *Indian Journal of Entomology, 57*, 417-420.

Rafiq, M., Ghaffar, A., & Arshad, M. (2008). Population dynamics of whitefly (*Bemisia tabaci*) on cultivated crop hosts and their role in regulating its carry-over to cotton. *International Journal of Agricultural Biology, 9*, 68-70.

Ronald, J., Prokopy, N. W., Miller, J. C., Piñero, L. O., Nancy, C., Hannah, R., & Roger, I. V. (2004). How effective is gf-120 fruit fly bait spray applied to border area sorghum plants for control of melon flies (Diptera: Tephritidae)? *Florida Entomologist, 87*(3), 354-360. https://doi.org/10.1653/0015-4040(2004)087 [0354:HEIGFF]2.0.CO;2

Sarfaraz, A., Ansari, S. H., & Porchezhian, E. (2002). Antifungal activity of alcoholic extracts of *Ziziphus vulgaris* and *Acacia concinna. Hamdard Medicus* (pp. 14-15, pp. 42-45). Bait Al-Hikmah, Karachi, Pakistan.

Sharma, B. N., & Singh, S. (2010). Biophysical and biochemical factors of resistance in okra against shoot and fruit borer. *Indian Journal of Entomology, 72*, 212-216.

Sharma, G. C., & Hall, C. V. (1971). Influence of cucurbitacins, sugars and fatty acids on cucurbit susceptibility to spotted cucumber beetle. *American Society of Horticultural Science, 96*, 675-680.

Swain, T., & Hills, H. (1959). Phenolic constituents of *Prunus domestica.* Quantitative analysis of phenolic constituents. *Journal of the Science of Food and Agriculture, 10*, 63-68. https://doi.org/10.1002/jsfa.27 40100110

Tewatia, A. S. (1994). *Resistance studies in bitter gourd against melon fruit fly* (p. 73, Ph.D. thesis). Chaudhary Charan Singh Haryana Agriculture University, Hisar, India.

Tewatia, A. S., Dhankhar, B. S., & Singh, R. (1998). Evaluation of bitter gourd (*Momordica charentia* L.) cultivars for resistance to melon fruit fly (*Bactrocera cucurbitae* Coq.). *Haryana Journal of Horticultural Science, 27*(4), 266-271.

Yemm, E. W., & Willis, A. J. (1954). The estimation of carbohydrate in the plant extract by anthrone reagent. *Journal of Biochemistry, 57*, 508-514. https://doi.org/10.1042/bj0570508

16

C-Repeat Binding Factor and Dehydrin Genes are Induced Co-Ordinately in Drought Tolerance Response of Wheat Cultivars

Csilla Deák[1], Katalin Jäger[2], Veronika Anna Nagy[1], Réka Oszlányi[1], Beáta Barnabás[2] & István Papp[1]

[1] Department of Plant Physiology and Plant Biochemistry, Faculty of Horticultural Science, Szent István University, Budapest, Hungary

[2] Agricultural Institute, Centre for Agricultural Research, Hungarian Academy of Sciences, Martonvásár, Hungary

Correspondence: István Papp, Department of Plant Physiology and Plant Biochemistry, Faculty of Horticultural Science, Szent István University, H-1118 Budapest, Villányi út 29-43, Hungary.
E-mail: papp.istvan@kertk.szie.hu

Abstract

Four bread wheat genotypes with contrasting drought stress tolerance were studied. Expression levels of dehydrin (*Wdhn13*) and C-repeat binding factor (*Cbf14*, *Cbf15*) genes were investigated in leaves of two drought tolerant (Plainsman V, Mv Emese) and two sensitive (GK Élet, Cappelle Desprez) cultivars by semi-quantitative RT-PCR during drought treatment at anthesis. Coordinate induction of *Cbf14*, *Cbf15* and *Wdhn13* genes occurred at a late stage of stress treatment in all cultivars except the most sensitive Cappelle Desprez, where no induction was evident. The most pronounced late induction of genes was observed in the tolerant Mv Emese genotype. *Cbf14*, *Cbf15* and *Wdhn13* showed largely parallel changes of expression in stressed adult plants. The mRNA level of the same set of genes was measured in leaves of non-stressed seedlings with qRT-PCR method. Expression level of *Wdhn13* was high and low in seedlings of tolerant and sensitive cultivars, respectively. *Cbf15* specific transcript was barely detectable in leaves of non-stressed seedlings. In order to shed light on any potential difference in hormone responsiveness, seedlings were subjected to ABA treatment *in vitro*. At low hormone concentrations (10 and 20 µM ABA) consistently weaker ABA induced root growth retardation of GK Élet was found in comparison with the other three cultivars. Results highlight pronounced and late induction of a set of defence genes and low ABA sensitivity as features appearing in drought tolerant and sensitive responses, respectively. Data is discussed in the light of multifactorial determination of the complex phenotype of drought tolerance in wheat.

Keywords: wheat (*Triticum aestivum* L.), drought stress, abscisic acid, dehydrin genes, CBF genes

1. Introduction

C-repeat binding factor (*Cbf*) genes are AP2/ERF transcription factors implicated both in cold stress response (Galiba et al., 2009) and in ABA mediated drought stress tolerance (Mizoi et al., 2012; Knight et al., 2004; Xiao et al., 2006; Kidokoro et al., 2015). *Cbf14* and *Cbf15* genes of wheat were first identified in *Triticum monococcum* L. and mapped to a major locus determining frost tolerance in this species (Miller et al., 2006). *Cbf14* and *Cbf15* of *Triticum aestivum* were functionally tested by transformation of barley, proving the role of these genes in freezing tolerance (Soltész et al., 2013). Along with alleviation of consequences of freezing stress, transcriptional activation of some dehydrin genes was also noted in the transgenic barley lines. Dehydrins are protective proteins against cellular damage in freezing and osmotic stress with multiple ways of functioning (Battaglia et al., 2008; Hara, 2010), complementing other molecular defences (Hegedűs et al., 2004) activated by these stress factors. Fifty-four wheat genes have been selected as candidates for coding dehydrins, most of them induced by dehydration (Wang et al., 2014). *Wdhn13* is a relatively well characterized member of this group. Its ABA inducibility has been suggested by some authors (Kurahashi et al., 2009), but has not been proposed by others (Ohno et al., 2003). This gene was found to exhibit variable expression/inducibiliy in divergent cultivars in response to dehydration (Rampino et al., 2006).

Connection between abscisic acid (ABA) signalling and drought tolerance has been well established (*e.g.* Lee & Luan, 2012; Golldack et al., 2014). ABA sensitivity may be manifested at any of the divergent processes this hormone regulates, such as germination, growth retardation or other stress responses leading to tolerance. Although these responses probably do not use the same set of signalling components, ABA sensitivity at germination for instance was found to be associated with enhanced drought tolerance in a number of cases (Cutler et al., 1996; Hugouvieux et al., 2001; Papp et al., 2004). Measuring retardation of root elongation at the seedling stage is a relatively simple way to characterize responsiveness to exogenously applied ABA (Thole et al., 2014). This trait was found to be associated with high drought stress tolerance levels in a set of synthetic hexaploid wheat lines derived from *Aegilops tauschii* (Kurahashi et al., 2009).

In this study, expression levels of genes potentially associated with a drought stress response were investigated in four bread wheat genotypes of contrasting drought tolerance. The cultivars were already characterized earlier and were shown to exhibit tolerance (Plainsman V, Mv Emese) or sensitivity (GK Élet, Cappelle Desprez) against drought stress (Guóth et al., 2009; Jäger et al., 2014). We hypothesized that known regulators of water stress tolerance and/or effector genes might contribute to this contrasting behaviour. As candidates, a dehydrin gene and *CBF* transcription factors were characterized under drought conditions at anthesis. The transcription pattern of these genes was also established in non-stressed seedlings of the same cultivars. Further on we set out to compare ABA sensitivity of root elongation at the seedling stage to explore ABA responsiveness of the cultivars.

2. Materials and methods

2.1 Growth of Plant Material, Sampling and Drought Stress Treatment

Seeds of four winter wheat (*Triticum aestivum* L.) cultivars (Plainsman V, Mv Emese, GK Élet, Cappelle Desprez) were imbibed in distilled water for 1 hour at 22 °C. All subsequent steps of the procedure were performed at this temperature. Imbibition was followed by surface sterilization in sodium hypochlorite with 2% available chlorine for 8 min and by rinsing 4 times with sterile distilled water. Imbibed seeds were germinated in glass Petri dishes on a filter paper moistened with distilled water in the darkness. The seedlings were subjected to ABA treatment immediately after the emergence of the radicle.

For water stress treatment plants of the winter wheat varieties were planted in a soil-sand-peat mixture (3:1:1, v/v/v) after 7 weeks of vernalisation at a temperature of 4 °C, and grown in PGV-15 growth chambers (Conviron, Winnipeg, Canada) using the spring climatic program T1 (Tischner et al., 1997). The min/max temperature rose from 12.5/5.5 °C to 21/14 °C until anthesis. Irrigation was carried out regularly in the morning at a rate of 150 ml/pot/day. In order to compensate for the delay in heading and flowering of Cappelle Desprez compared to the other three cultivars (according to preliminary experiments), plants of the former variety were planted 2 weeks earlier to align developmental differences. Drought stress was generated by total water withholding starting at the time of inflorescence emergence (Zadoks' growth stage 53) to complete anthesis (Zadoks' growth stage 67) until the volumetric water content of the soil dropped below 10 % in the pots. Leaf samples were collected from both drought-stressed and control plants. Soil volumetric water content (VWC) of 20 pots per genotype and treatment was measured using an HH2 moisture meter (Delta-T Devices Ltd., Cambridge, UK) at field capacity and during treatments. Relative water content (RWC) of leaves was determined on whole flag leaves of three plants per genotype and treatment at the end of water withholding by measuring fresh weight (FW) at excision, saturated weight (SW) after 24 h rehydration on distilled water at 4 °C in the dark, and dry weight (DW) after oven drying for 48 h at 80 °C. The leaf RWC (%) was calculated using the following equation:

$$RWC\ (\%) = [(FW - DW)/(SW - DW)] \times 100 \qquad (1)$$

For sample preparation of non-stressed young seedlings, plants were planted as above and were grown at constant 21 °C temperature under 10/14 hours light/dark illumination cycle. At Zadok's growth stage 13 leaves of seedlings were harvested for RNA preparation (Zadoks et al., 1974).

2.2 RNA Extraction, RT-PCR, qRT-PCR, DNA Fragment Isolation and Sequencing

Leaf samples were grinded in liquid nitrogen and RNA was isolated with Tri Reagent (Molecular Research Center, Cincinnati, OH, USA) according to the instructions of the manufacturer. First strand cDNA synthesis kit (Fermentas, Vilnius, Lithuania) was used for reverse transcription of the RNA samples. RT-PCR was performed with GoTaq Flexi DNA Polymerase from Promega (Madison, WI, USA) according to the instructions of the manufacturer, applying a total of 30 cycles. An equal use of cDNA templates was confirmed by RT-PCR with control primers TalKa and TalKb, amplifying sequence Ta2776, as suggested by Paolacci et al. (2009). The gene-specific primers for the dehydrin gene were designed with the Primer Premier software (Premier Biosoft, Palo Alto, CA, USA) and were purchased from Biomi Kft (Gödöllő, Hungary) or Sigma (St Luis, MI, USA). Primers specific for the CBF genes were kindly provided by Gábor Galiba (Agricultural Institute of HAS, Martonvásár, Hungary). All primers used are listed in Table 1. qRT-PCR was performed in a Rotor Gene 6000 instrument (Corbett Research, Australia), by applying conditions recommended by the manufacturer and 60 °C annealing temperature. Quality of products was confirmed with melting curve analysis by heating samples from 66 °C to 99 °C in 0.5 °C increments. Evaluation of data was done with the help of Comparative Quantitation Analysis program (McCurdy et al., 2008). RT-PCR and qRT-PCR were performed at least two times on different biological samples with a minimum of two technical repetitions in each case. RT-PCR products were purified from agarose gels using an Illustra GFX PCR DNA and gel band purification kit (GE Healthcare, Uppsala, Sweden). Primary DNA sequences were determined by Biomi Kft (Gödöllő, Hungary). Differences of genes expression levels were tested with one-way ANOVA.

2.3 Measuring Root Growth Retardation of Seedlings in Response to ABA Treatment

Synchronously germinated seedlings were transferred into Petri dishes with filter paper wetted with distilled water or with 10, 20 or 50 µM ABA (Sigma) solution. At least 40 seedlings per treatment and genotype were incubated for further one week in the darkness and the lengths of their primary roots were recorded. The experiment was repeated three times. Differences in root length were tested with one-way ANOVA and a subsequent Dunnett's T3 (T3) test (Dunnett, 1955).

3. Results

3.1 Drought Stress Treatment

As a consequence of water deprivation soil volumetric water content declined below 10% in all cultivars by the end of the treatment (Figure 1). There was no consistent, significant difference in the dynamics of water loss among the cultivars based on repetitions of the experiment. Relative water content of leaves remained significantly higher in the drought tolerant cultivars (Figure 2).

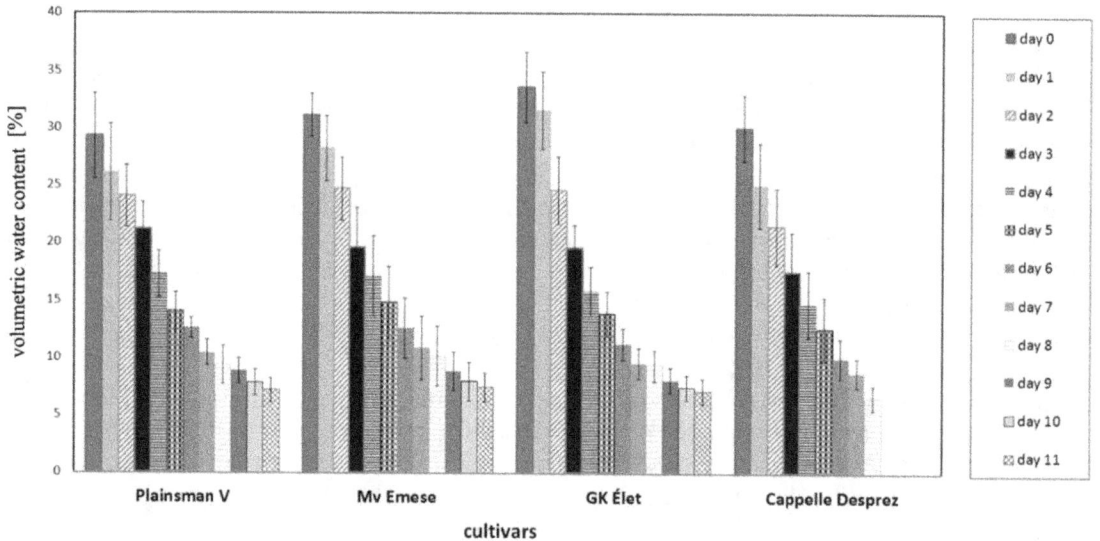

Figure 1. Volumetric water content (VWC) of soil during drought stress. Number of days of the treatment is indicated. Means ± standard deviations are shown

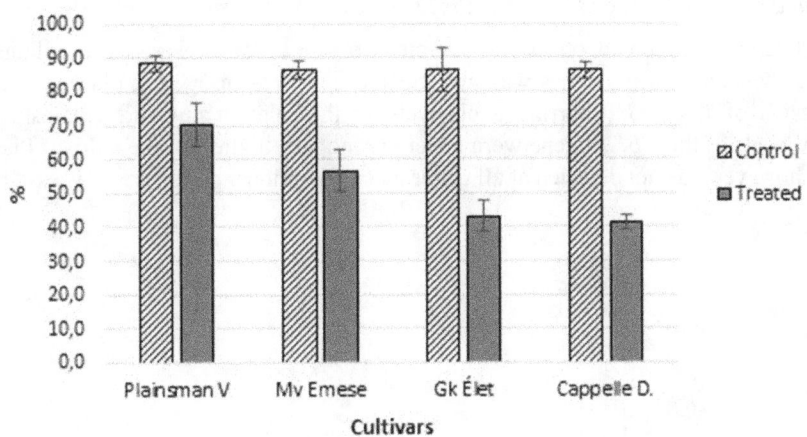

Figure 2. Relative water content (RWC) of leaves at the end of the treatment. Means ± standard deviations are shown

3.2 Transcriptome Analysis in Adult Plants under Water Deprivation

Gene specific RT-PCR product from *Wdhn13* was most induced in Mv Emese at 6th-10th days of water deprivation (Figure 3). mRNA of this gene was barely detectable in the other three genotypes, with low level of induction in Plainsman V and Gk Élet. Declining signal in semi quantitative RT-PCRs of all studied genes was followed by a wave of induction around the 8th day of treatment, except for the most sensitive Cappelle Desprez, where this induction did not occur. Parallel changes in expression of *Cbf14*, *Cbf15* and *Wdhn13* genes were noted, but this effect was not stringent. In order to confirm identity of the PCR products representative RT-PCR fragments of *Cbf14*, *Cbf15* and *Wdhn13* were extracted from agarose gels and sequenced directly.

Table 1. Gene specific and control primers used in RT-PCRs and qRT-PCR

Primer name	Primer sequence
Cbf14a	5'-CCAAACCAGTGTCATTCAA-3'
Cbf14b	5'-TTGTCTCAACTTCGCCACT-3'
Cbf15a	5'-GTGTCTCAACTTCGCCGACT-3'
Cbf15b	5'-ATGTGTCCAGGTCCATTTTCC-3'
Wdhn13a	5'-ATTCTGCAAAGTAGCGGGTC-3'
Wdhn13b	5'-AGAACCAGTGTCAGATTTCCCT-3'
TalKa	5'-GTAGCATTATGTTTGTGCCTTG-3'
TalKb	5'-GGAGAGCCAGTCAAGACCCTCG-3'

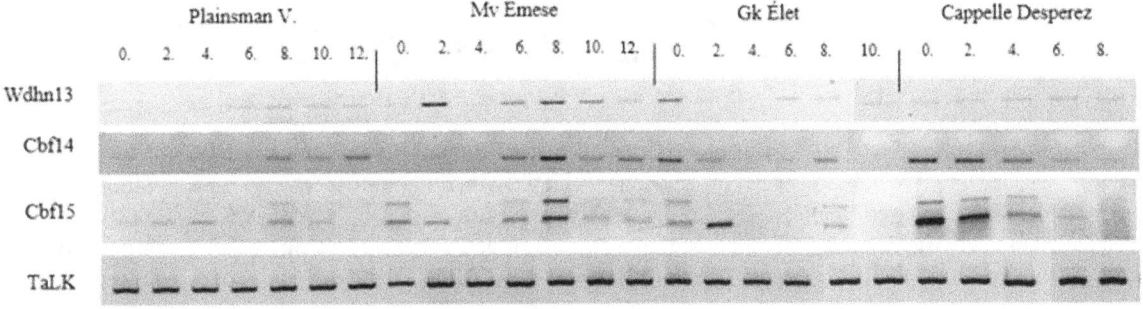

Figure 3. RT-PCR products for the studied genes in the four genotypes during drought stress treatment. Numbers indicate days of treatment

3.3 Transcript Abundance of Cbf14, Cbf15 and Wdhn13 Genes in Non-Stressed Seedlings

Quantitative RT-PCR was performed on cDNA samples prepared from unstressed seedlings of the cultivars studied. Transcript abundance of the genes was investigated in leaves of 3 week old non-stressed plants. Data show high expression of *Wdhn13* occurring exclusively in the tolerant cultivars (Plainsman V and Emese) (Figure 4). mRNA levels of the *CBF14* gene were generally modest, highest in the cultivar Plainsman V. mRNA of the *Cbf15* gene however was not detected at all under these conditions, or its level was extremely low.

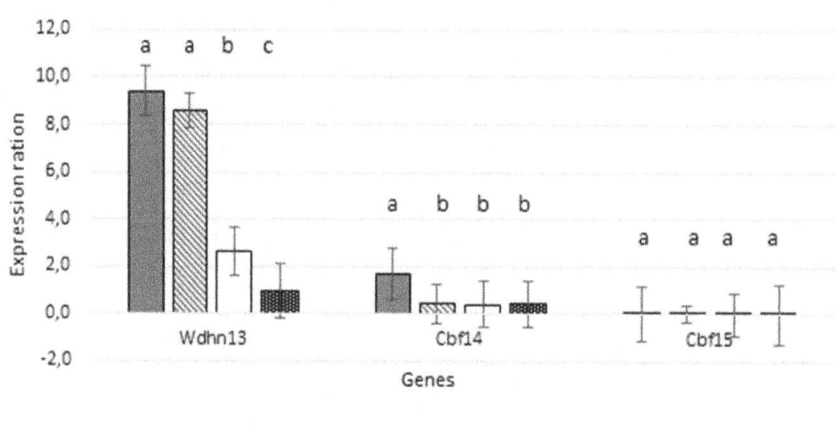

Figure 4. Expression levels of the genes, measured by qRT-PCR, in unstressed seedlings. One-way ANOVA with post-hoc Tukey HSD test was used to test for statistical differences. Means ± standard deviations are shown. Different letters indicate differences among the cultivars for each gene separately at $p < 0.05$ of probability

3.4 ABA Sensitivity of Seedlings

Exogenous application of ABA at low concentrations retarded root elongation to a different degree. Cultivar GK Élet showed significantly the weakest response to 10 and 20 µM ABA applications (Figure 5).

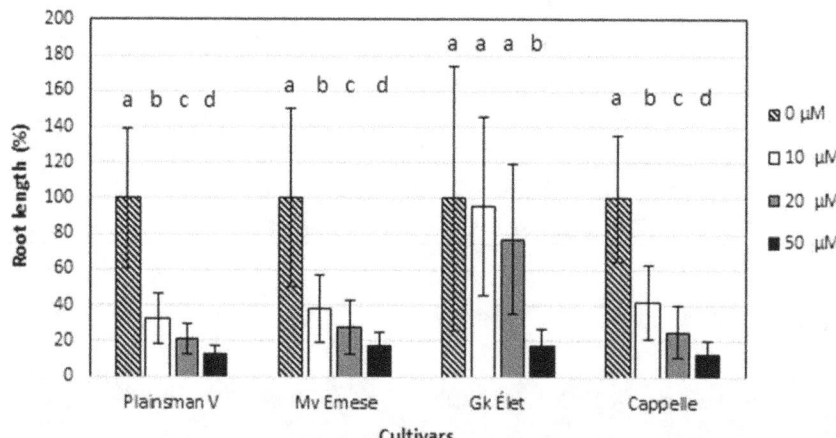

Figure 5. Relative root lengths of seedlings after one week growth at the indicated ABA concentrations (10, 20 and 50 µM) were compared to untreated controls. Welch's ANOVA test ($F_{(3, 125,9)} = 158,5$; $p = 0,000$) was used to test for statistically significant differences among data sets of genotypes and hormone concentrations. Means ± standard deviations are shown. Different letters indicate differences at $p < 0.05$ of probability

4. Discussion

Drought tolerance is a complex phenotype with multi factorial determination (Ashraf, 2010). Studies presented here focused on expression of potentially stress related genes as well as ABA sensitivity in wheat genotypes with contrasting ability to tolerate water stress. The investigated attributes may indicate differences in stress response mechanisms with potential contribution to drought hardiness.

This work is an extension of earlier experiments (Jäger et al., 2014) where several traits potentially contributing to different drought tolerance level of four selected bread wheat cultivars have already been identified. Smaller leaf area and lower stomatal number pro leaves were found in the tolerant varieties, while a thin and water permeable cuticle was described in the sensitive cv. Cappelle Desprez. Besides morphological traits, stress induced morpho-physiological changes also differentiated the cultivars especially after repeated exposures to water deficit. Loss of cellular integrity (indicated by increased electrolyte leakage), more severe ultrastructural damage to the chloroplasts and faster decline of relative water content of leaves were characteristic to the sensitive cultivars during repeated drought cycles (Jäger et al., 2014). Apparently the two tolerant genotypes exhibited stronger defencesagainst oxidative damage at the cellular level. This coincided with higher glutathione synthase and glutathione peroxidase activity in cv. Plainsman V as was established by Gallé and co-workers (2009). Therefore, several facets of the more effective drought tolerance response of Plainsman V and Mv Emese have been already uncovered. Different responses to water shortage were confirmed in our present experiments, as tolerant cultivars exhibited significantly higher relative water content values in leaves already at the end of one water deprivation period applied (Figure 2).

Given the high complexity of the drought tolerant phenotype we hypothesized that further relevant physiological differences could still be established among these cultivars. In a search for further clues explaining differences, expression levels of potential stress tolerance related genes were studied.

Dehydrin genes are well appreciated targets of *Cbf* genes when responding to cold. Parallel expression of *DHN* and *Cbf* genes has been found in cold stress (*e.g.* Kume et al., 2005). *Cbf* and other cold responsive genes were found activated by ABA independent and ABA dependent pathways as well (Knight et al., 2004). More recently *Cbf* gene expression has been also linked to induction of a dehydrin gene during the drought tolerance response of *Brachypodium distachyon* (Ryu et al., 2014). *Wdhn13* is probably subject to regulation by multiple transcription factors, including *Cbf* genes. The *Wcbf2* wheat transcription factor has been shown to activate the promoter of *Wdhn13* in tobacco (Takumi et al., 2008). Transcriptional regulation of *Wdhn13* involving *Cbf* genes under drought conditions fits well with our data. In this study dehydrin and *Cbf* gene specific RT-PCR reactions were performed on leaf samples from drought stressed plants of the four genotypes around anthesis. Data indicated that *Wdhn13*, *Cbf14* and *Cbf15* genes showed coordinate induction in Mv Emese during a late stage of the water limitation period. This induction of dehydrin and *CBF* genes occurred at approximately one week after total water deprivation and was most pronounced in Mv Emese, a genotype with high tolerance and yield under drought stress conditions. Based on this expression pattern in drought stressed adult plants *Wdhn13* (and probably other dehydrins) may be regulated by both *Cbf* genes. In order to confirm association between drought, stress-induced *CBF*/dehydrin gene expression at anthesis and drought tolerance, a larger number of genotypes ought to be examined. Transcript profiles of *Wdreb2* and other potential regulators as well as more dehydrin genes should be tested in this respect.

We presumed that a clearer picture could be obtained about interactions among dehydrin and *Cbf* genes by examining plants exposed to as low a level of stress as possible. To this end the same set of genes was investigated in leaves of non-stressed young seedlings. According to observations of Rorat and co-workers (2004) in five Solanaceae species and barley, leaves of young, developing seedlings exhibited high dehydrin protein levels even in the absence of stress. Similarly, light induced expression of a dehydrin-encoding gene, *HaDhn1*, was found in sunflower seedlings in the absence of environmental stresses (Natali et al., 2007). This made us speculate that wheat seedlings may also express at least some *DHN* and/or *Cbf* genes spontaneously (*i.e.* probably as a result of the unavoidable, very low level of stress received during cultivation). In order to minimize stress factors as much as possible, wheat seedlings were well watered and a constant temperature was applied to exclude cold induced gene expression. Target gene expression was investigated by quantitative qRT-PCR. Cultivar specific expression of the genes was found, which didn't match with that found around anthesis (day 0 of drought treatment). This confirms that transcription of *Cbf* and dehydrin genes may be subject to development specific regulation. In seedlings of some cultivars *Cbf14* and *Cbf15* transcripts were found expressed at moderate and at very low levels, respectively. This contrasted with parallel expression of *Cbf14* and *Cbf15* in adult stage plants under drought stress implying that the two genes may be regulated by different mechanisms. Independent regulation of the two genes was also shown earlier by Knox et al. (2008) in cold stress.

The dehydrin gene *Wdhn13* was found highly expressed in unstressed seedlings of the two tolerant varieties but at a lower level in the sensitive cultivars. A potential connection between dehydrin expression early in plant development and drought tolerance of adult plants should be further investigated on a larger set of genes and cultivars. It is worth noting however that high levels of a desiccation induced dehydrin protein at the seedling stage werealso correlated with adult stage drought tolerance in wheat (Lopez et al., 2001).

ABA sensitivity of root growth has been associated with drought tolerance in wheat (Kurahashi et al., 2009). As ABA is also known to activate expression of some dehydrin genes, we performed an in vitro experiment to this end. In our study ABA induced root growth retardation of GK Élet was found to be consistently weaker at low external hormone concentrations (10 and 20 µM ABA applied) in comparison with the other three cultivars. This data suggest that ABA insensitivity may contribute to the low level of drought tolerance of GK Élet, and confirm that ABA responsiveness may be a factor of drought sensitivity in some wheat cultivars. No direct correlation was found however between inducibility of the genes studied and ABA sensitivity of the cultivars. Stress induced modification of root growth in cvs Mv Emese and GK Élet has been also characterized by Tari et al. (2010). Smaller osmotically (PEG) induced reduction of root growth rate was found in GK Élet, which corresponds well with our findings. In the experiment of Tari et al. (2010) endogenous ABA content of the root tips remained unchanged in GK Élet, but decreased in Mv Emese. Our data about ABA insensitivity can explain why the relatively higher ABA content could still not have led to efficient growth retardation in Gk Élet root tips. ABA responsiveness of root elongation can be measured at the seedling stage, making this method especially feasible for early pre-screening drought sensitivity of breeding lines.

We conclude that pronounced and coordinate induction of *Cbf* and dehydrin genes (*Cbf14*, *Cbf15* and *Wdhn13*) in a late stage of water stress in a tolerant genotype (Mv Emese) at anthesis may represent a module in the multi facet phenotype of drought hardiness. We propose that the other tolerant cultivar involved in our studies (Plainsman V) probably uses alternative strategies to avoid detrimental effects of drought. Morphological traits may support lower level of water loss in this cultivar (Jäger et al., 2014). Evading water shortage by accelerated development of the root system may also be a part of an acclimation process in Plainsman V. This aspect of the water stress response however, was not investigated in our studies. A further basic factor of drought tolerance is ABA responsiveness, which is apparently weak in the cultivar Gk Élet, probably contributing to its sensitivity. Our experiments also uncovered expression of the *Wdhn13* dehydrin gene in leaves of unstressed seedlings in the tolerant cultivars. Whether this implies differences in sensitivity of signalling steps leading to defencegene expression needs further investigations.

Acknowledgements

The work was supported by the TÁMOP 4.2.1/B-09/1/KMR/-2010-0005, NKTH-OTKA CK80211, NKTH-OTKA CK80274 and NKTH K108644 grants from the National Development Agency of Hungary. The authors would like to thank Dr Gábor Galiba for providing primers for *Cbf* genes, Dr László Gáspár and Dr András Ittzés for help in data management and statistics, Erika Gondos and András Miskó for excellent technical assistance and David J Cleary for language editing. Csilla Deák and Réka Oszlányi were partially supported by the Doctoral School of Horticultural Science of Szent István University.

References

Ashraf, M. (2010). Inducing drought tolerance in plants: Recent advances. *Biotechnology Advances, 28*(1), 169-183. https://doi.org/10.1016/j.biotechadv.2009.11.005

Battaglia, M., Olvera-Carrill, O. Y., Garciarrubio, A., Campos, F., & Covarrubias, A. A. (2008). The enigmatic LEA proteins and other hydrophilins. *Plant Physiol., 148*(1), 6-24. https://doi.org/10.1104/pp.108.120725

Cutler, S., Ghassemian, M., Bonetta, D., Cooney, S., & McCourt, P. (1996). A protein farnesyl transferase involved in abscisic acid signal transduction in Arabidopsis. *Science, 273*(5279), 1239-41. https://doi.org/10.1126/science.273.5279.1239

Dunnett, C. W. (1955). A multiple comparison procedure for comparing several treatments with a control. *Journal of the American Statistical Association, 50*, 1096-1121. https://doi.org/10.1080/01621459.1955.10501294

Galiba, G., Vágújfalvi, A., Li., C., Soltész, A., & Dubcovsky, J. (2009). Regulatory genes involved in the determination of frost tolerance in temperate cereals. *Plant Science, 176*, 12-19. https://doi.org/10.1016/j.plantsci.2008.09.016

Gallé, A., Csiszár, J., Secenji, M., Guóth, A., Cseuz, L., Tari, I., Györgyey, J., & Erdei, L. (2009). Glutathione transferase activity and expression patterns during grain filling in flag leaves of wheat genotypes differing in drought tolerance: Response to water deficit. *J. Plant Physiol., 166*(17), 1878-91. https://doi.org/10.1016/j.jplph.2009.05.016

Golldack, D., Li, C., Mohan, H., & Probst, N. (2014). Tolerance to drought and salt stress in plants: Unraveling the signaling networks. *Front Plant Sci., 5*, 151. https://doi.org/10.3389/fpls.2014.00151

Guóth, A., Tari, I., Gallé, Á., Csiszár, J., Pécsváradi, A., Cseuz, L., & Erdei, L. (2009). Comparison of the drought stress responses of tolerant and sensitive wheat cultivars during grain filling: Changes in flag leaf photosynthetic activity, ABA levels, and grain yield. *J. Plant Growth Regul., 28*, 167-176. https://doi.org/10.1007/s00344-009-9085-8

Hara, M. (2010). The multifunctionality of dehydrins, An overview. *Plant Signaling & Behavior, 5*(5), 503-508. https://doi.org/10.4161/psb.11085

Hegedűs, A., Erdei, S., Janda, T., Tóth, E., Horváth, G., & Dudits, D. (2004). Transgenic tobacco plants overproducing alfalfa aldose/aldehyde reductase show higher tolerance to low temperature and cadmium stress. *Plant Science, 166*(5), 1329-1333. https://doi.org/10.1016/j.plantsci.2004.01.013

Hugouvieux, V., Kwak, J. M., & Schroeder, J. I. (2001). An mRNA cap binding protein, ABH1, modulates early abscisic acid signal transduction in Arabidopsis. *Cell., 106*(4), 477-87. https://doi.org/10.1016/S0092-8674 (01)00460-3

Jäger, K., Fábián, A., Eitel, G., Szabó, L., Deák, C., Barnabás, B., & Papp, I. (2014). A morpho-physiological approach differentiates bread wheat cultivars of contrasting tolerance under cyclic water stress. *J. Plant Physiol., 171*, 1256-1266. https://doi.org/10.1016/j.jplph.2014.04.013

Kidokoro, S., Watanabe, K., Ohori, T., Moriwaki, T., Maruyama, K., Mizoi, J., ... Yamaguchi-Shinozaki, K. (2015). Soybean DREB1/CBF-type transcription factors function in heat and drought as well as cold stress-responsive gene expression. *Plant J., 81*(3), 505-18. https://doi.org/10.1111/tpj.12746

Knight, H., Zarka, D. G., Okamoto, H., Thomashow, M. F., & Knight, M. R. (2004). Abscisic acid induces CBF gene transcription and subsequent induction of cold-regulated genes via the CRT promoter element. *Plant Physiol., 135*(3), 1710-7. https://doi.org/10.1104/pp.104.043562

Knox, A. K., Li, C., Vágújfalvi, A., Galiba, G., Stockinger, E. J., & Dubcovsky, J. (2008). Identification of candidate CBF genes for the frost tolerance locus Fr-Am2 in Triticum monococcum. *Plant Mol. Biol., 67*(3), 257-70. https://doi.org/10.1007/s11103-008-9316-6

Kume, S., Kobayashi, F., Ishibashi, M., Ohno, R., Nakamura, C., & Takumi, S. (2005). Differential and coordinated expression of Cbf and Cor/Lea genes during long-term cold acclimation in two wheat cultivars showing distinct levels of freezing tolerance. *Genes Genet Syst., 80*(3), 185-97. https://doi.org/10.1266/ggs.80.185

Kurahashi, Y., Terashima, A., & Takumi, S. (2009). Variation in dehydration tolerance, ABA sensitivity and related gene expression patterns in D-genome progenitor and synthetic hexaploid wheat lines. *Int. J. Mol. Sci., 10*, 2733-2751. https://doi.org/10.3390/ijms10062733

Lee, S. C., & Luan, S. (2012). ABA signal transduction at the crossroad of biotic and abiotic stress responses. *Plant Cell Environ., 35*(1), 53-60. https://doi.org/10.1111/j.1365-3040.2011.02426.x

Lopez, C. G., Banowetz, G., Peterson, C. J., & Kronstad, W. E. (2001). Differential accumulation of a 24-kd dehydrin protein in wheat seedlings correlates with drought stress tolerance at grain filling. *Hereditas., 135*(2-3), 175-81. https://doi.org/10.1111/j.1601-5223.2001.00175.x

McCurdy, R. D., McGrath, J. J., & MacKay-Sim, A. (2008). Validation of the comparative quantification method of real-time PCR analysis and a cautionary tale of housekeeping gene selection. *Gene Therapy and Molecular Biology., 12*(1), 15-24.

Miller, A. K., Galiba, G., & Dubcovsky, J. (2006). A cluster of 11 CBF transcription factors is located at the frost tolerance locus Fr-Am2 in Triticum monococcum. *Mol. Genet. Genomics., 275*(2), 193-203. https://doi.org/10.1007/s00438-005-0076-6

Mizoi, J., Shinozaki, K., & Yamaguchi-Shinozaki, K. (2012). AP2/ERF family transcription factors in plant abiotic stress responses. *Biochim Biophys Acta., 1819*(2), 86-96. https://doi.org/10.1016/j.bbagrm.2011.08.004

Natali, L., Giordani, T., Lercari, B., Maestrini, P., Cozza, R., Pangaro, T., ... Cavallini, A. (2007). Light induces expression of a dehydrin-encoding gene during seedling de-etiolation in sunflower (*Helianthus annuus* L.). *J. Plant Physiol., 164*(3), 263-73. https://doi.org/10.1016/j.jplph.2006.01.015

Ohno, R., Takumi, S., & Nakamura, C. (2003). Kinetics of transcript and protein accumulation of a low-molecular-weight wheat LEA D-11 dehydrin in response to low temperature. *J. Plant Physiol., 160*(2), 193-200. https://doi.org/10.1078/0176-1617-00925

Paolacci, A. R., Tanzarella, O. A., Porceddu, E., & Ciaffi, M. (2009). Identification and validation of reference genes for quantitative RT-PCR normalization in wheat. *BMC Mol. Biol., 10*, 11. https://doi.org/10.1186/1471-2199-10-11

Papp, I., Mur, L. A., Dalmadi, A., Dulai, S., & Koncz, C. (2004). A mutation in the Cap Binding Protein 20 gene confers drought tolerance to Arabidopsis. *Plant Mol. Biol., 55*(5), 679-86. https://doi.org/10.1007/s11103-004-1680-2

Rampino, P., Pataleo, S., Gerardi, C., Mita, G., & Perrotta, C. (2006). Drought stress response in wheat: Physiological and molecular analysis of resistant and sensitive genotypes. *Plant Cell Environ., 29*(12), 2143-52. https://doi.org/10.1111/j.1365-3040.2006.01588.x

Rorat, T., Grygorowicz, W. J., Irzykowski, W., & Rey, P. (2004). Expression of KS-type dehydrins is primarily regulated by factors related to organ type and leaf developmental stage during vegetative growth. *Planta, 218*(5), 878-85. https://doi.org/10.1007/s00425-003-1171-8

Ryu, J. Y., Hong, S. Y., Jo, S. H., Woo, J. C., Lee, S., & Park, C. M. (2014). Molecular and functional characterization of cold-responsive C-repeat binding factors from Brachypodium distachyon. *BMC Plant Biol., 14*, 15. https://doi.org/10.1186/1471-2229-14-15

Soltész, A., Smedley, M., Vashegyi, I., Galiba, G., Harwood, W., & Vágújfalvi, A. (2013). Transgenic barley lines prove the involvement of TaCBF14 and TaCBF15 in the cold acclimation process and in frost tolerance. *J. Exp. Bot., 64*(7), 1849-62. https://doi.org/10.1093/jxb/ert050

Takumi, S., Shimamura, C., & Kobayashi, F. (2008). Increased freezing tolerance through up-regulation of downstream genes via the wheat CBF gene in transgenic tobacco. *Plant Physiol. Biochem., 46*(2), 205-11. https://doi.org/10.1016/j.plaphy.2007.10.019

Tari, I., Guóth, A., Benyó, D., Kovács, J., Poór, P., & Wodala, B. (2010). The roles of ABA, reactive oxygen species and nitric oxide in root growth during osmotic stress in wheat: comparison of a tolerant and a sensitive variety. *Acta Biol. Hung., 61*, 189-96. https://doi.org/10.1556/ABiol.61.2010.Suppl.18

Thole, J. M., Beisner, E. R., Liu, J., Venkova, S. V., & Strader, L. C. (2014). *Abscisic Acid Regulates Root Elongation through the Activities of Auxin and Ethylene in Arabidopsis thaliana.* G3 (Bethesda). May 15. pii: g3.114.011080.

Tischner, T., Kőszegi, B., & Veisz, O. (1997). Climatic programmes used in the Martonvásár phytotron most frequently in recent years. *Acta Agron Hung., 45*, 85-104.

Wang, Y., Xu, H., Zhu, H., Tao, Y., Zhang, G., Zhang, L., ... Ma, Z. (2014). Classification and expression diversification of wheat dehydrin genes. *Plant Sci., 214*, 113-20. https://doi.org/10.1016/j.plantsci.2013.10.005

Xiao, H., Siddiqua, M., Braybrook, S., & Nassuth, A. (2006). Three grape CBF/DREB1 genes respond to low temperature, drought and abscisic acid. *Plant Cell Environ., 29*(7), 1410-21. https://doi.org/10.1111/j.1365-3040.2006.01524.x

Zadoks, I. C., Chang, T. T., & Konzak, C. F. (1974). A decimal code for the growth stages of cereals. *Weed Res., 14*, 415-421. https://doi.org/10.1111/j.1365-3180.1974.tb01084.x

17

Different Methods for Overcoming Integumental Dormancy during *in vitro* Germination of Red Araza Seeds

Cassio G. Freire[1], João P. P. Gardin[2], César M. Baratto[2] & Renato L. Vieira[3]

[1] University Alto Vale do Rio do Peixe, St. Victor Baptista Adami, Center, Caçador, SC, Brazil

[2] University of West of Santa Catarina, Unoesc, Videira, SC, Brazil

[3] Agricultural Research and Rural Extension Company of Santa Catarina, Epagri, Experimental Station of Caçador, SC, Brazil

Correspondence: Cassio G. Freire, University Alto Vale do Rio do Peixe, 800, St. Victor Baptista Adami, Center, Caçador, Santa Catarina, Zip Code: 89500-000, Brazil. Email: cassio.geremia@uniarp.edu.br

The research is financed by UNIEDU Postgraduate Program of the State of Santa Catarina, Unoesc, & Epagri.

Abstract

Red Araza, or Red Strawberry Guava (*Psidium cattleianum* Sabine) is a native Brazilian Atlantic Forest species of the Myrtaceae family, whose seeds exhibit integumental dormancy. Due to its importance to different industries worldwide, recent research efforts are seeking to expand this species' micropropagation processes using *in vitro* seedling germination, especially since *in vitro* micropropagation of adult plant material has, so far, been limited. This research effort evaluated different methods of overcoming integumental dormancy during *in vitro* germination of the Red Araza, so as to allow future micropropagation of the species. The seeds' emergence and vigor were evaluated based on mechanical and acid scarification, using different substrates and immersions in solutions with different levels of gibberellic acid (GA_3), and on the influence of the pre-immersion of seeds in water and sulfuric acid. The mechanical and acid scarification of the seeds, combined or separate, resulted in higher *in vitro* germination percentages and a higher germination rate index (GRI). Pre-immersion in distilled water (20 hours) also proved to be efficient for the germination of the Red Araza seed, with 76.2% of the seeds germinating and a higher speed of emergence (GRI = 0.18). When compared to a Murashige and Skoog (MS-zero) medium, sowing in a hydrophilic cotton substrate showed greater emergence and vigor, with approximately 70% of the seeds germinating. Treating the seeds by pre-immersing them in GA_3 turned out to be unnecessary. The methods used for overcoming integumental dormancy during *in vitro* germination of Red Araza seeds proved to be efficient, and could be used to develop micropropagation protocols of seminal origin for this species.

Keywords: acid scarification, Araçá, gibberelic acid, mechanical scarification, myrtaceans, red strawberry guava, tegumentar dormancy

1. Introduction

A species of the Myrtaceae family, Red Araza, or, hereinafter, Red Strawberry Guava (*Psidium cattleianum* Sabine) is native to the Brazilian Atlantic Forest, but also found in different tropical and subtropical ecosystems (Tng et al., 2015). It's an arboreal species that produces a fleshy fruit with peculiar taste, possessing a rich chemical composition (Galho, Lopes, Bacarin, & Limac, 2007), a large variety of bioactive compounds, such as phenolics and carotenoids (Silva, Rodrigues, Mercadante, & De Rosso, 2014), as well as essential oils and other volatile compounds of pharmacological interest (Marin et al., 2008). Previous studies have already determined that aqueous and ketone extracts from the Red Strawberry Guava generate various antioxidant activities, have an anti-proliferative effect on human cancer cells, and acts as an antimicrobial agent against *Salmonella enteritidis* (Medina et al., 2011).

Due to its heightened importance to the agro-industrial and pharmacological fields, many studies have been done with the intent to broaden the knowledge base concerning this species' variability and genetic selection, its use in biotechnological processes, details regarding its ecological management and the establishment of commercial orchards/plant husbandries (Kinupp, 2011). Studies that focus on the *in vitro* propagation of Red Strawberry Guava can promote important mechanisms for the sustainable exploitation of this species; mechanisms that, up until now, have been rather scarce (Pasqual, Chagas, Soares, & Rodrigues, 2012).

To date, tests for the *in vitro* introduction and multiplication of these Myrtaceae using herbaceous branches have proved ineffective, mainly due to the high percentage of *in vitro* phenol oxidation and microbial contamination (Rodríguez, 2013; Freire, Oliveira, & Vieira, 2014). Consequently, protocols developed from *in vitro* germinated seedlings are being actively investigated as a feasible alternative to species micropropagation, especially considering that they have contributed to producing healthy explants and allowed for the continuity of *in vitro* propagation processes.

Red Strawberry Guava seeds, however, are known to show integumental dormancy and other studies about its *in vitro* germination can be developed under different conditions so as to optimize and homogenize seedling emergence (Da Silva, Perez, & De Paula, 2011). With that in mind, this study set out to optimize the *in vitro* germination of red Strawberry Guava (*Psidium cattleianum* Sabine, Myrtaceae), using different methods to overcome its inherent integumental dormancy, in order to obtain healthy seedlings that can later contribute to the development of micropropagation processes.

2. Materials and Methods

2.1 Obtaining the Seeds

The seeds were obtained in 2014 from ripe fruits collected from 9-year-old Red Strawberry Guava trees (26°49′06″S-50°59′29″W and 26°46′15″S-51°02′09″W) located in Caçador, a city in the southern state of Santa Catarina, in Brazil. The seeds were completely extracted from the pulp, in running water, placed on paper towels, at room temperature and without direct sunlight, and kept under these conditions for five days, until it was time to start the experiments.

2.2 Experimental Conditions

The experiments were performed according to ISTA rules (ISTA, 1999), in the state of Santa Catarina, Brazil, at EPAGRI's (Portuguese acronym for State of Santa Catarina Agricultural, Livestock, and Rural Extension Research Company) Plant Tissue Culture Laboratory. The experiments were set up in a growth chamber where the flasks were then exposed to a 16-hour photoperiod sourced by cold, white fluorescent lamps, with intensity set at 75 μmol m^{-2} s^{-1}, and a temperature of 25±2 °C. The following two types of substrate mediums were used: a cotton medium (2.05 grams flask^{-1} of hydrophilic cotton moistened with 20 mL of distilled water and autoclaved for 25 minutes, at 121 °C and 1.2 atm) and a complete MS-zero medium (25 mL flask^{-1}) (Murashige & Skoog, 1962). The substrates were placed in closed flasks with the following dimensions and capacity: height = 95 mm, diameter = 65 mm, and a capacity of 230 mL. Operating within a laminar flow hood, a standard asepsis method was used on the seeds by immersing them for one minute in 70% v/v ethanol, followed by a 15 minutes immersion in a NaClO solution with 1.5% an active principle and containing Tween 20® detergent (10 drops L^{-1}), and washing them three times with sterile distilled water. The seeds that exhibited root protrusions equal to or greater than 2.0 mm were considered to have been germinated (Borghetti & Ferreira, 2004). Two were the factors assessed during the experiments: the germination rate index (GRI) and the germination percentage. According to Maguire (1962), the GRI was calculated using Equation (1) based on every other day assessments. The germination percentage was calculated after periods of time stipulated for each experiment.

$$GRI = (G1/N1) + (G2/N2) + (G3/N3) + (Gn/Nn)$$ (1)

Where,

G1, G2, G3, ... Gn = number of seeds germinated in the first, second, third and thru to the last count;

N1, N2, N3, ... Nn = number of days from the time of sowing to the first, second, third and thru to the last count.

Every time the terms standard asepsis method, cotton substrate and MS-zero substrate are mentioned in this paper, they shall be consistent with the descriptions contained in this section.

2.3 Determining Moisture Level

The seeds' moisture level was determined according to Brasil (2009); in other words, in three groups of 50 seeds, each using the oven-dried method, at 105±3 °C, for a period of 24 hours.

2.4 Mechanical and Acid Scarification of the Integument

Eight treatments were arranged in a 2 × 4 factorial design, containing two levels of mechanical scarification (sanded and unsanded seeds), four levels of acid scarification (immersion in a 9 mol L^{-1} H_2SO_4 solution for zero, five, ten, and 20 minutes) and five replications, each containing 10 seeds. For the mechanical scarification, an autoclaved metal sandpaper was used, scouring both sides of the seeds' integument adjacent to the micropyle. Subsequently, the acid scarification was performed by immersing the different seed samples in an acid solution under constant agitation. The seeds were then drained of the acid and subjected to a standard asepsis method, with cotton as the substrate. The GRI and the germination percentage were assessed, with the latter being assessed after 62 days *in vitro*.

2.5 Substrate and Gibberellic Acid (GA₃) Concentrations Tests

Using a laminar flow hood, the seeds were immersed in distilled water for 20 hours, and in a 9 mol L^{-1} H_2SO_4 solution for ten minutes, washed in sterile distilled water, and then subjected to a standard asepsis method. The treatments were arranged in a 2 × 4 factorial design, with two substrates (cotton and MS-zero medium), and four concentrations of GA_3 (Sigma ®) (0 mg L^{-1}, 250 mg L^{-1}, 500 mg L^{-1}, and 1000 mg L^{-1}). Six replications, each containing five seeds, were performed per treatment so the GRI and germination percentage could be assessed (after 80 days).

2.6 Influence of Water Immersion and Acid Scarification

Four different treatments were tested (T1, T2, T3, and T4), with seven replications, each containing six seeds, as described in Table 1.

Cotton was used as the substrate, and the GRI and germination percentage were assessed (70 days after sowing).

Table 1. Different treatments for the *in vitro* germination of Red Strawberry Guava seeds

Treatments	Description
T1	Control – only "Standard Asepsis Method"[a]
T2	"Water Immersion"[b] associated with "Standard Asepsis Method"
T3	"Water Immersion" + "Acid Scarification"[c] and "Standard Asepsis Method"
T4	"Water Immersion"[b] associated with "Reduced Asepsis"[d]

Note. [a] – One minute in 70% v/v ethanol, followed by 15 minutes in a 1.5% NaClO solution, with an active principal containing 10 drops L^{-1} of Tween 20® detergent. [b] – Immersion for 20 hours in distilled water. [c] – Immersion for 10 minutes in a 9 mol L^{-1} H_2SO_4 solution. [d] – Same as the standard aseptic method, but with immersion time reduced to 3 minutes in NaClO.

2.7 Statistics and Data Analysis

All tests were performed in a randomized design. The results were submitted to the Shapiro-Wilk normality test ($p < 0.05$) and for analysis of variance (ANOVA), and, using the Scott-Knott test ($p < 0.05$), their group means were later separated into qualitative variables and regression studies for quantitative variables. When outside the expected normality, the data was transformed into $\sqrt{(x+0.1)}$.

3. Results

Under the experimental conditions described, the average moisture percentage for the Red Strawberry Guava seeds was stipulated at 9.29%±0.036%.

3.1 Mechanical and Acid Scarification of the Integument

By itself, mechanical scarification, achieved by sanding or not sanding the seed integument, did not significantly change the seeds' germination percentages ($p = 0.8783$), resulting in average values of 43.0% and 43.5%, respectively (see Figure 1). However, acid scarification ($p < 0.0001$) by itself and the interaction of both acid and mechanical scarification ($p = 0.0013$) did significantly change *in vitro* germination percentages (see Figure 1). As immersion times in a 9 mol L^{-1} H_2SO_4 solution increased, mechanically scarified seeds yielded positive quadratic germination increments, while the non-scarified seeds showed positive linear germination increments. Using regression analysis (see Figure 1), optimum immersion times for acid scarification were determined to be 20 minutes for unsanded seeds and 15 minutes for sanded seeds.

As was the case with germination percentages, only acid scarification ($p < 0.0001$) by itself and the interaction of both acid and mechanical scarification ($p = 0.0013$) yielded significant changes in germination rate indexes (GRIs) (see Figure 1). For seeds that were not mechanically scarified, results showed that germination vigor was highest when the acid scarification immersion time in a 9 mol L^{-1} H_2SO_4 solution was 20 minutes. GRIs for mechanically scarified seeds, on the other hand, did not differ ($p < 0.05$) for immersion times of five, ten and 20 minutes, which indicates that, under these conditions, a five-minute immersion is enough to obtain the highest GRI value.

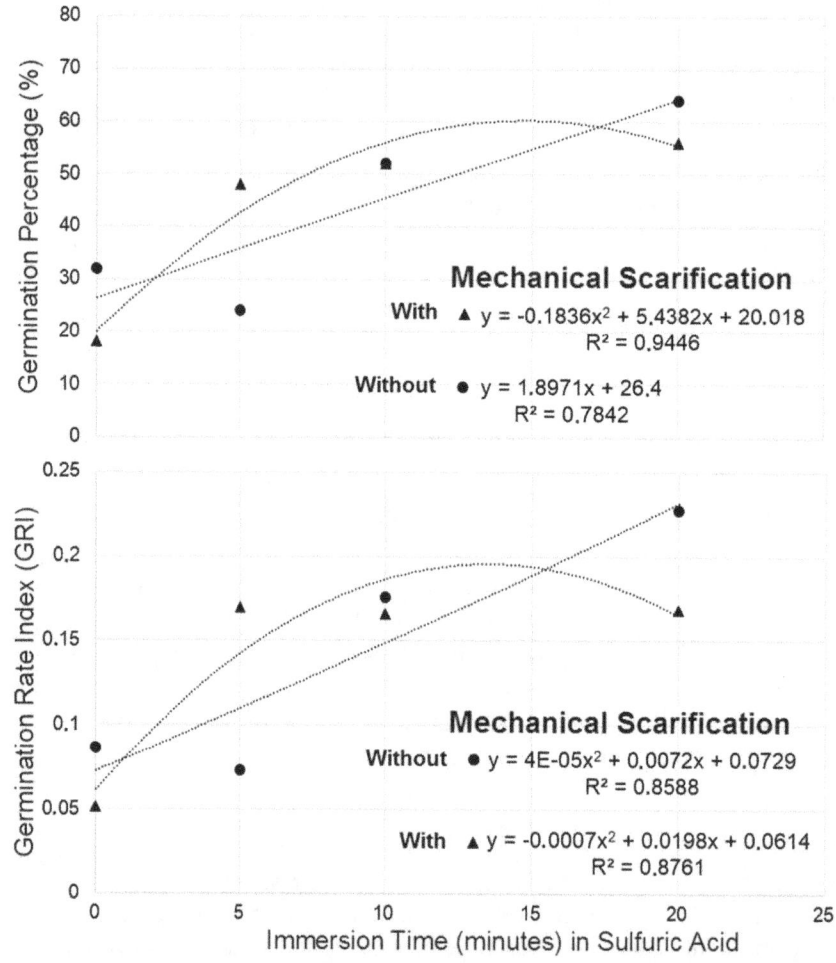

Figure 1. Germination percentages and germination rate index (GRI) for Red Strawberry Guava seeds (*Psidium cattleianum*) submitted or not to mechanical scarification (using sandpaper), and acid scarification with different immersion times in a 9 mol L^{-1} H_2SO_4 solution, 62 days after *in vitro* sowing ($p < 0.05$)

3.2 Substrate and Gibberellic Acid (GA₃) Concentrations Tests

Based on the data shown in Figure 2, one can see that the *in vitro* germination percentages of seeds placed on hydrophilic cotton were significantly higher ($p < 0.05$) than for those placed in an MS-zero medium, with averages of 70.14% and 15.28%, respectively. In spite of this, one can also see that both substrates exhibited very similar behavior. As GA_3 concentrations increased, polynomial regressions indicated a reduction in germination percentage when the concentration increased above 100 mg L^{-1}, increasing again as it went above 700 mg L^{-1} (see Figure 2). Also note that, of the four concentrations used in the experiment, the 0 mg L^{-1}, 250 mg L^{-1}, and 1000 mg L^{-1} did not differ between them and provided higher germination percentages than the 500 mg L^{-1} concentration ($p < 0.05$), regardless of the substrate used (see Figure 2).

Pertaining to GRIs, after 94 days *in vitro*, significant effects ($p < 0.0001$) were detected only with regards to the type of substrate used, showing no GRI variations for the different concentrations of GA_3 used ($p = 0.0690$) nor for the interaction between the two factors ($p = 0.9045$). Just like for the germination percentages (see Figure 2), the GRI obtained was higher for the cotton medium (on average, 7.32 times higher) ($p < 0.0001$) than for the MS-zero medium.

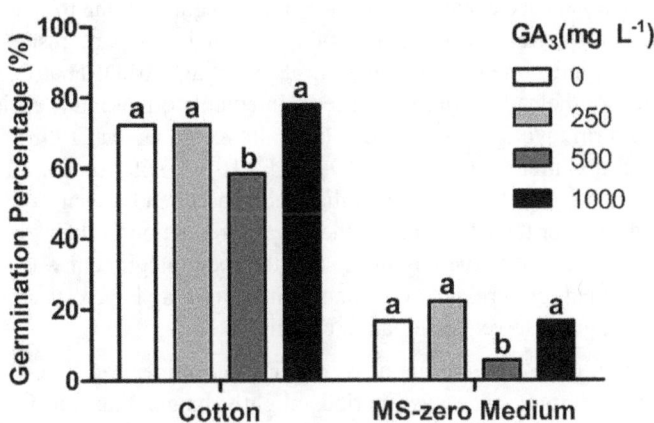

Figure 2. Germination percentages for Red Strawberry Guava seeds (*Psidium cattleianum*) submitted to different concentrations of GA_3 on a moistened hydrophilic cotton substrate and a solidified MS-zero medium (6 g L^{-1} agar), 94 days after *in vitro* sowing

Note. Original values shown; for statistical analysis, values were transformed into $\sqrt{(x+0.1)}$, ($p < 0.05$).

3.3 Influence of Water Immersion and Acid Scarification

The experiment showed that germination percentages did not vary significantly ($p = 0.0911$) as a result of the different treatments. However, when compared to the control group (T1), there was a definite increase in the number of seeds germinated when they were immersed in distilled water (T2), with 76.19% of seeds being germinated (Table 2).

Table 2. Germination rate index (GRI) and germination percentages for Red Strawberry Guava seeds (*Psidium cattleianum*) submitted to different treatments, 70 days after *in vitro* sowing

Treatments	GRI	Germination (%)
T1	0.10[b]	52.38[ns]
T2	0.18[a]	76.19
T3	0.08[b]	52.38
T4	0.14[a]	69.05
Variation coefficient (%)	37.88	32.79

Note. Within a column, the means with the same superscript letter do not differ statistically from the Scott-Knott test at 5%. [ns] = not significant.

In terms of GRIs, the treatments yielded significant effects ($p = 0.0062$), with treatments T2 and T4 showing greater seed vigor (Table 2). For Treatment 2, the first germinated seed was observed on the 12th day; for Treatment 4, it happened on the 16th day; and, for Treatment 3, it happened only on the 20th day. These results indicate that a 20-hour immersion in distilled water (done for treatments T2, T3 and T4) reduces the time needed for the onset of germination, except when associated with acid scarification of ten minutes in a 9 mol L^{-1} H_2SO_4 solution (treatment T3). In Table 2, one can see that treatment T2 yielded a GRI almost twice that of the control group (T1). The table also shows that, when acid scarification was used (T3), the GRI decreases significantly (approximately 55%) when compared to treatment T2.

4. Discussion

The acid scarification of Red Strawberry Guava seeds with H_2SO_4 proved to be more effective in increasing *in vitro* germination percentages and GRIs than mechanical scarification (Figure 1). Da Silva (2009) had already tested different H_2SO_4 immersion times in *ex vitro* germination experiments with the same species and, as was the case in this study, he noticed a quadratic germination decreasing trend when the seeds were subjected to immersion times greater than 15 minutes.

It is well known that *P. cattleianum* seeds have an impermeable integument due to its rocklike consistency (*testa petrea*) (Cisneiro, Matos, Lemos, Reis, & Queiroz, 2003), which is, in part, responsible for the seed's low imbibition rate and consequent integumental dormancy (Da Silva et al., 2011). That fact led to the reasoning that mechanical or acid scarification helps to 'injure' the seed's integument, making it easier for water and gases to enter the embryo, which, in turn, favors its germination (Bertalot & Nakagawa, 1998). Using that reasoning, this experiment found that increasing immersion times in a 9 mol L^{-1} H_2SO_4 solution contributed to reducing the time needed for germination of the seeds. However, when the seeds had already been sanded and were subjected to acid scarification for periods greater than 15 minutes, there was a reduction in their germination percentages and GRIs (Figure 1). This could be related to the possibility that the seed integument was previously deteriorated, in which case, the immersion in acid for periods longer than 15 minutes may have caused chemical or physiological damage to the embryo, resulting in a lower germination capacity.

C. C. Baskin and J. M. Baskin (2014) claim that acid scarification immersion times for seeds with integument dormancy should be carefully tested, since even periods slightly longer than the ideal time may damage the embryo and prevent it from germinating. Furthermore, these authors also say that the addition of concentrated acids may change the pH levels of the solutions surrounding the embryo and, because of the specific pH levels required for germination, it would interfere with the germination process.

With respect to the *in vitro* germination substrates, this research effort found that sowing seeds in cotton increased the rate of germination and the final number of germinated seeds when compared to the MS-zero medium solidified with agar. Passos, Tavarez, and Alves (2007) also used moistened hydrophilic cotton for the germination of Sabia (*Mimosa caesalpiniifolia*) seeds and obtained a higher germination percentage with this substrate than with any of the others that were tested. In this study, the difference in how *in vitro* germination was affected by the different substrates became evident during experiment evaluations. On the 56th day after the *in vitro* sowing, approximately 40% of the seeds placed in the cotton substrate had already germinated, whereas none had germinated in the MS-zero medium.

Unlike what was observed in this study, other authors obtained good *in vitro* germination percentages for myrtaceans, using saline culture mediums such as MS-zero, LPm and WPM. Cid, Machado, Carvalheira and Brasileiro (1999), for example, obtained good germination percentages for the *Eucalyptus* spp. in an MS-zero medium; similar results were obtained for the Cagaita (*Eugenia dysenterica*), using the same medium and yielding almost 90% germination (Martinotto et al., 2007). Rodríguez (2013) obtained a 68.00% germination result for the Red Strawberry Guava (*P. cattleianum*) using an LPm medium (Von Arnold & Erikson, 1981). Also, Souza, Fior, Souza and Schwarz (2011) achieved an average of 70% *in vitro* germination when placing Guabijuzeiro (*Myrcianthes pungens*) seeds in a WPM medium.

Several substrate factors may affect a seed's germination quality. Since water and oxygen are essential to many metabolic processes that occur during a developing embryo's germination (Taiz & Zeiger, 2013), one of these factors is the substrate's ability to maintain good water availability and aeration for the seed (Popinigis, 1985).

According to Gulliver and Heydecker (1973), up to a certain limit, the more water available to the seeds, the greater the speed of imbibition and absorption, resulting in faster germination of the seeds. Carvalho and Nakagawa (2000) also claim that water is the most significant factor influencing the germination process, especially considering that the embryo's development and growth depend on water imbibition by the seed and its subsequent absorption by the tissues, mechanisms that enhance breathing and other metabolic activities necessary to achieve seedling emergence and root protrusions.

The positive influence of high water availability in the substrate had already been established for the germination of different species, such as the white spruce (*Picea glauca* [Moench.] Voss.) (Downie et al., 1998), the *Mesua ferrea* (Joshi, Phartyal, Khan, & Arunkumar, 2015), and also myrtaceans, such as the Guavira [*Campomanesia adamantium* (Cambess.) O. Berg.] (Dresch, Scalon, & Kodama, 2011), the Brazilian Strawberry Guava (*Psidium guineense* Swartz) (Gonçalves et al., 2009) and the Uvaia (*Eugenia pyriformis*) (Scalon & Jeromine, 2013).

However, the high salt concentration of germination substrates and the solidification of the same using agar, for example, have been reported as negative factors to the seedling emergence process (Grattapaglia & Machado, 1998). The solidification of the substrate increases the colloidal state of the same (Stoltz, 1971) and, as well as saline ions, increase the effectiveness of intermolecular interactions with water, which increases the osmotic pressure of the medium where the seeds are inserted (Doneen & MacGilliwray, 1943). These factors increase the retention of water by the substrate, reducing the water absorption/imbibition by the seed, and directly affecting their physiological responses (Carvalho & Nakagawa, 2000).

Accordingly, in this study, the difference in the availability of water between the cotton and the MS-zero mediums is believed to have contributed to the better germination percentages observed in the moistened cotton substrate (Figure 2).

This study also indicated that different concentrations of gibberellic acid (GA_3), in which the Red Strawberry Guava seeds were immersed, were not efficient enough to promote an increase in germination percentage, when compared to the control group. It's possible that this lack of efficiency occurred because the GA_3 didn't properly stimulate the supply of nutrients to the embryo, or didn't contribute to the seed's production of endogenous gibberellins, a plant hormone that plays an essential role in the germination process. Other research efforts for the same species showed conflicting results regarding the effects of GA_3 solutions on germination. Rodríguez (2013), for example, not only noted that there was no significant difference in germination when using GA_3 concentrations of 0 mg L^{-1}, 10 mg L^{-1}, 20 mg L^{-1} and 50 mg L^{-1}, but also found that the control group had a higher germination percentage. Tomaz et al. (2011), on the other hand, observed that, when previously immersed in a 500 mg L^{-1} GA_3 solution, a higher number of Red Strawberry Guava seeds successfully germinated. Interestingly, the concentration used by Tomaz et al. (2011) is the same that, in this study, resulted in lower germination percentages and vigor.

Exogenous applications of GA_3 don't always have the expected effect on germination (Kermode, 2005) because it's dependent on other factors such as the endogenous concentration of abscisic acid and other inhibiting compounds present in the seed (Taiz & Zeiger, 2013; Carvalho & Nakagawa, 2000; Khan, 1971), or whether or not they are associated to beneficial microorganisms (Dalal & Kulkarni, 2015). In addition, seed responses to exogenous applications of GA_3 are quite specific, varying among different species of the same genus, or even within the same species (Kumar et al., 2012). As an example, for the genus *Psidium*, immersion in a GA_3 solution promoted a significant increase in germination percentages of *P. guajava* seeds, when compared to the control group, to the group submitted to acid scarification with HCl and H_2SO_4 treatments, and even to the group that got immersed in distilled water (Chandra & Govind, 1990). In contrast, germination of *P. guineense* seeds was not stimulated by GA_3 solution immersions, yielding lower germination percentages than both the control group and the group immersed in distilled water (Dresch, Scalon, Neves, & Masetto, 2014).

In terms of GRIs, results obtained in this study indicated that pre-immersion of the Red Strawberry Guava seeds in water for a period of 20 hours (T2 in Table 2) increases their germinating vigor, with GRI values almost double that of the control group. Other research efforts also showed that pre-immersion in water decreases the time to the onset of germination in many other species, for instance, the *Tamarindus indica* (Azad, Nahar, & Matin, 2015), the *Acrocomia aculeata* (Rodrigues Junior et al., 2013), and the *Acacia origena* (Aref, Atta, Shahrani & Mohamed, 2011). Acid scarification after the period of pre-immersion in water (T3 in Table 2), however, significantly reduced both the GRI and the *in vitro* germination percentages. Such results may have been caused by acidity induced chemical and/or physiological damages to the embryo, which would decrease the seed's germination capacity. Similar results were attained by Tavares, Lucca Filho, and Kersten (1995) when they immersed Guava (*Psidium guajava*) seeds in a H_2SO_4 solution, yielding a significant reduction in GRI, when compared to the control group.

It is important to note that acid scarification of seeds may trigger oxidative processes and deregulate pH levels in the regions surrounding the embryo (C. C. Baskin & J. M. Baskin, 2014). This, in turn, may have negatively interfered in the germination process of Red Strawberry Guava seeds, possibly reducing physiological responses that affect the embryo's development.

5. Conclusion

Results led to the determination that immersion of non-mechanically scarified Red Strawberry Guava seeds in a 9 mol L^{-1} H_2SO_4 solution for 20 minutes yields higher GRIs and *in vitro* germination percentages. When the seeds are mechanically scarified, 5- and 15-minute immersions in a 9 mol L^{-1} H_2SO_4 solution yielded better results for GRI and germination percentages, respectively. The experiments also revealed that pre-immersing the seeds in solutions of GA_3, with concentrations ranging from 0 mg L^{-1} to 1000 mg L^{-1}, was irrelevant. Also, sowing the seeds in a moistened hydrophilic cotton substrate is more efficient and promotes higher rates of emergence and vigor, when compared to a complete MS-zero medium. Furthermore, pre-imbibition of the Red Strawberry Guava in distilled water for 20 hours increases both the *in vitro* germination speed and percentages.

In conclusion, this research effort showed that optimization processes for the *in vitro* germination of Red Strawberry Guava seeds are efficient, and can be used to obtain healthy *in vitro* seedlings, making it possible to further develop micropropagation protocols of seminal origin for this species.

References

Aref, I. M., Atta, H. A. E., Shahrani, T. A., & Mohamed, I. A. (2011). Effects of seed pretreatment and seed source on germination of five *Acacia* spp. *African Journal of Biotechnology, 10*(71), 15901-15910. https://doi.org/10.5897/AJB11.1763

Azad, M. S., Nahar, N., & Matin, M. A. (2013). Effects of variation in seed sources and pre-sowing treatments on seed germination of Tamarindus indica: A multi-purpose tree species in Bangladesh. *Forest Science and Practice, 15*(2), 121-129. https://doi.org/10.1007/s11632-013-0211-0

Baskin, C. C., & Baskin, J. M. (2014). *Seeds: Ecology, biogeography and evolution of dormancy and germination* (2nd ed., p. 1585). Academic Press.

Bertalot, M. J., & Nakagawa, J. (1998). Superação da dormência em sementes de *Leucaena diversifolia* (Schlecht.), *Revista Brasileira de Sementes, 20*(1), 39-42. https://doi.org/10.17801/0101-3122/rbs.v20 n1p39-42

Borghetti, F., & Ferreira, A. F. (2004). Interpretation of results germination. In A. G. Ferreira, & F. Borghetti (Eds.), *Germination: from basic to applied* (pp. 209-222). Porto Alegre: Artmed.

Carvalho, N. M., & Nakagawa, J. (2000). *Sementes: ciência, tecnologia e produção* (4th ed., p. 588). Jaboticabal: FUNEP.

Chandra, R., & Govind, S. (1990). Gibberellic acid, thiourea, ethrel and acid treatments in relation to seed germination and seedling growth in guava (*Psidium guajava* L.). *Progressive Horticulture, 22*(1), 40-43.

Cid, L. P. B., Machado, A. C. M. G., Carvalheira, S. B. R., & Brasileiro, A. C. M. (1999). Plant regeneration from seedling explants of *Eucalyptus grandis* × *E. urophylla. Plant Cell, Tissue and Organ Culture, 56*(1), 17-23. https://doi.org/10.1023/A:1006283816625

Cisneiro, R. A., Matos, V. P., Lemos, M. A., Reis, O. V., & Queiroz, R. M. (2003). Qualidade fisiológica de sementes de araçazeiro durante o armazenamento. *Revista Brasileira de Engenharia Agrícola e Ambiental, 7*(3), 513-518. https://doi.org/10.1590/S1415-43662003000300018

Da Silva, A. (2009). *Morfologia, conservação e ecofisiologia da germinação de sementes de Psidium cattleianum Sabine.* Unpublished dissertation in partial fulfillment of the requirements for the degree of Master in Ecology and Natural Resources. Universidade Federal de São Carlos, UFSCar, SP.

Da Silva, A., Perez, S. C. J. G. A., & De Paula, R. C. (2011). Qualidade fisiológica de sementes de *Psidium cattleianum* Sabine acondicionadas e armazenadas em diferentes condições. *Revista Brasileira de Sementes, 33*(2), 197-206. https://doi.org/10.1590/S0101-31222011000200001

Dalal, J., & Kulkarni, N. (2015). Effect of endophytic treatments on plant growth performance and disease incidences in soybean (*Glycine max* (L.) Merril) Cultivar JS-335 against challenge inoculation with *R. solani. American Journal of Agricultural and Biological Sciences, 10*(2), 99-110. https://doi.org/10.3844/ajabssp.2015.99.110

Downie, B., Coleman, J., Scheer, G., Wang, B. S. P., Jensen, M., & Dhir, N. (1998). Alleviation of seed dormancy in white spruce (*Picea glauca* [Moench.] Voss.) is dependent on the degree of seed hydration. *Seed Science and Technology, 26*, 555-569.

Dresch, D. M., Scalon, S. P. Q., & Kodama, F. M. (2011). Crescimento inicial de mudas de *Campomanesia adamantium* (Camb.) O. Berg em diferentes substratos e disponibilidades hídricas. In XIII Congresso Brasileiro de Fisiologia Vegetal-XIV Reunião Latino-Americana de Fisiologia Vegetal, 2011, Buzios. Livro resumo, *Periodical Brazilian Journal of Plant Physiology*, Londrina.

Dresch, D. M., Scalon, S. P. Q., Neves, E. M. S., & Masetto, T. E. (2014). Effect of pre-treatments on seed germination and seedling growth in *Psidium guineense* Swartz. *Agrociencia Uruguay, 18*(2), 33-39. Retrieved April 1, 2016, from https://www.researchgate.net/publication/271325658_Effect_of_Pre-treatments_on_Seed_Germination_and_Seedling_Growth_in_Psidium_guineense_Swartz

Freire, C. G., Oliveira, L. P. de, & Vieira, R. L. (2014). *Tratamento fungicida e diferentes assepsias na introdução in vitro de araçazeiro-vermelho (Psidium cattleianum)* (pp. 14-14). In X Simpósio Florestal Catarinense, 2014, Curitibanos/SC. Anais do X Simpósio Florestal Catarinense Florestas Produtivas, Cenários e Perspectivas.

Galho, A. S., Lopes, N. F., Bacarin, M. A., & Limac, M. G. S. (2007). Chemical composition and growth respiration in *Psidium cattleyanum* Sabine fruits during the development cycle. *Revista Brasileira de Fruticultura, 29*(1), 61-66. https://doi.org/10.1590/S0100-29452007000100014

Gonçalves, C. A. R. L., Kodama, F. M., Dresch, D. M., Scalon, S. P. Q., & Pereira, Z. V. (2009). Germinação de Araza (*Psidium guineense* Swartz) em diferentes substratos e regimes hídricos. XVI Congresso Brasileiro de Sementes, Qualidade: Desafio Permanente. *Informativo ABRATES, 19*(2), 237.

Grattapaglia, D., & Machado, M. A. (1998). Micropropagação. In A. C. Torres, L. S. Caldas, & J. A. Buso (Eds.), *Tissue culture and genetic transformation of plants* (Vol. 1, pp. 183-260). Brasília: SPI/Embrapa, CNPH.

Gulliver, R. L., & Heydecker, W. (1973). Establishment of seedlings in a changeable environment. In W. Heydecker (Ed.), *Seed Ecology* (pp. 433-462). London, Butterworth.

ISTA. (1999). International rules for seed testing. *Seed Science Technology, 27*(Suppl.).

Joshi, G., Phartyal, S. S., Khan, M. R., & Arunkumar, A. N. (2015). Recalcitrant morphological traits and intermediate storage behaviour in seeds of *Mesua ferrea*, a tropical evergreen species. *Seed Science and Technology, 43*, 121-126. https://doi.org/10.15258/sst.2015.43.1.13

Kermode, A. R. (2005). Role of abscisic acid in seed dormancy. *Journal of Plant Growth Regulation, 24*, 319-44. https://doi.org/10.1007/s00344-005-0110-2

Khan, A. A. (1971). Cytokinins: permissive role in seed germination. *Science, 171*, 853-859. https://doi.org/10.1126/science.171.3974.853

Kinupp, V. F. (2011). Native food species from southern Brazil. In L. Coradin, A. Siminski, & A. Reis (Eds.), *Native species of flora of current or potential economic value: Plans for the future* (p. 934). South Brasília: MMA.

Kumar, R., Misra, K. K., Misra, D. S., & Brijwal, M. (2012). Seed germination of fruit crops: A review. *HortFlora Research Spectrum, 1*(3), 199-207.

Maguire, J. D. (1962). Speed of germination aid in selection and evaluation for seedling emergence and vigor. *Crop Science, 2*(1), 176-177. https://doi.org/10.2135/cropsci1962.0011183X000200020033x

Marin, R., Apel, M. A., Limberger, R. P., Raseira, M. C. B., Pereira, J. F. M., Zuanazzi, J. A. S., & Henriques, A. T. (2008). Volatile components and antioxidant activity from some myrtaceous fruits cultivated in southern Brasil. *Latin American Journal of Pharmacy, 27*(2), 172-177.

Martinotto, C., Paiva, R., Santos, B. R., Soares, F. P., Nogueira, R. C., & Silva, A. A. N. (2007). Effect of scarification and light on *in vitro* seed germination of (*Eugenia dysenterica* DC.). *Ciência e Agrotecnologia, 31*(6), 1668-1671. https://doi.org/10.1590/S1413-70542007000600010

Medina, A. L., Haas, L. I. R., Chaves, F. C., Salvador, M., Zambiazi, R. C., Da Silva, W. P., ... Rombaldi, C. V. (2011). Araza (*Psidium cattleianum* Sabine) fruit extracts with antioxidant and antimicrobial activities and antiproliferative effect on human cancer cells. *Food Chemistry, 128*, 916-922. https://doi.org/10.1016/j.foodchem.2011.03.119

Ministério da Agricultura, Pecuária e Abastecimento, Brasil. (2009). *Regras para análise de sementes. [Rules for seed testing]* (p. 399). Secretaria de Defesa Agropecuária. Brasília: Mapa/ACS.

Murashige, T., & Skoog, F. (1962). A revised médium for rapid growth and bioassay with tobacco tissue cultures. *Physiologia Plantarum, 15*, 473-497. https://doi.org/10.1111/j.1399-3054.1962.tb08052.x

Pasqual, M., Chagas, E. A., Soares, J. D. R., & Rodrigues, F. A. (2012). Tissue culture techniques for native Amazonian fruit trees. In A. Leva, & L. M. R. Rinaldi (Eds.), *Recent advances in plant in vitro culture* (p. 220). Intech. https://doi.org/10.5772/52211

Passos, M. A., Tavares, K. M. P., & Alves, A. R. (2007). Dormancy and development in "sabiá" (*Mimosa caesalpiniifolia* Benth.) seeds. *Revista Brasileira de Ciências Agrárias, 2*(1), 51-56.

Popinigis, F. (1985). *Fisiologia da semente* (2nd ed., p. 289). Brasília, DF: [s.n].

Rodrigues Junior, A. G., Oliveira, T. G. S., Souza, P. P., & Ribeiro, L. M. (2013). Water uptake and pre-germination treatments in macaw palm (*Acrocomia aculeata*-Arecaceae) seeds. *Journal of Seed Science, 35*(1), 99-105. https://doi.org/10.1590/S2317-15372013000100014

Rodríguez, E. A. G. (2013). *Contribuições à propagação de araçazeiro (Psidium cattleianum Sab.) e grumixameira (Eugenia brasiliensis Lam.).* Unpublished dissertation in partial fulfillment of the requirements for the degree of Master in Plant Science with emphasis on horticulture. Universidade Federal do Rio Grande do Sul, Porto Alegre, RS.

Scalon, S. P. Q., & Jeromine, T. S. (2013). Substrate and water levels on the germinative potential of seeds of uvaia. *Revista Árvore, 37*(1), 49-58. https://doi.org/10.1590/S0100-67622013000100006

Silva, N. A., Rodrigues, E., Mercadante, A. Z., & De Rosso, V. V. (2014). Phenolic compounds and carotenoids from four fruits native from the Brazilian Atlantic Forest. *Journal of Agricultural and Food Chemistry, 62*, 5072-5084. https://doi.org/10.1021/jf501211p

Souza, L. S., Fior, C. S., Souza, P. V. D., & Schwarz, S. F. (2011). Disinfestation of seeds and *in vitro* multiplication of guabijuzeiro from apical segments juveniles (*Myrcianthes pungens* O. Berg) D. Legrand. *Revista Brasileira de Fruticultura, 33*(3), 691-697. https://doi.org/10.1590/S0100-29452011005000081

Stoltz, L. P. (1971). Agar restriction of the growth of excised mature Iris embryos. *Journal of the American Society for Horticultural Science, 96*, 681-684.

Taiz, L., & Zeiger, E. (2013). *Plant Physiology* (5th ed., p. 918). In A. M. D. Junior et al. (trad.), P. L. de Oliveira (rev. tecn.). Porto Alegre: Artmed.

Tavares, M. S. W., Lucca Filho, O. A., & Kersten, E. (1995). Germination and vigor of guava seeds (*Psidium guajava* L.) submitted to different methods to supress dormancy. *Ciência Rural, 25*(1), 11-15. https://doi.org/10.1590/S0103-84781995000100003

Tng, D. Y. P., Goosem, M. W., Paz, C. P., Preece, N. D., Goosem, S., Fensham, R. J., & Laurance, S. G. W. (2015). Characteristics of the *Psidium cattleianum* invasion of secondary rainforests. *Austral Ecology.*

Tomaz, Z. F. P., Galarça, S. P., Lima, C. S. M., Betemps, D. L., Gonçalves, M. A., & Rufato, A. R. (2011). Tratamentos pré-germinativos em sementes de araçazeiro (*Psidium cattleyanum* Sabine L.). *Revista Brasileira de Agrociência, Pelotas, 17*(1-4), 60-65.

Von Arnold, S., & Erikson, T. (1981). *In vitro* studies of adventitious roots formation in *Pinus contorta. Canadian Journal of Botany, 59*, 870-874. https://doi.org/10.1139/b81-121

18

Lactobacillus plantarum Exopolysaccharides Induce Resistance against Tomato Bacterial Spot

Juliane Mendes Lemos Blainski[1], Argus Cesar da Rocha Neto[1], Caroline Luiz[1], Márcio José Rossi[1] & Robson Marcelo Di Piero[1]

[1] Center of Agricultural Sciences, Federal University of Santa Catarina, Florianópolis, Santa Catarina, Brazil

Correspondence: Juliane Mendes Lemos Blainski, Center of Agricultural Sciences, Federal University of Santa Catarina, Rodovia Admar Gonzaga, 1346-Itacorubi 88034-000 Florianópolis, Santa Catarina, Brazil. E-mail: juliane.lemos@yahoo.com.br

The research is financed by CAPES.

Abstract

Lactic acid bacteria produce several exopolysaccharides (EPS) that may have antimicrobial action and/or induce defense responses in plants. This work aims to evaluate the potential of EPS produced by *Lactobacillus plantarum* in the protection of tomato plants against bacterial spot caused by *Xanthomonas gardneri*, as well as to predict the possible mechanisms of action. The EPS were characterized through FTIR and applied at 0; 0.5; 1.5 and 3.0 mg mL^{-1} in tomato plants with five expanded leaves, followed by the pathogen inoculation after 3 or 7 days. Antimicrobial activity of the biopolymer (1.5 or 10.0 mg mL^{-1}) was evaluated in bioassay when EPS was incorporated into culture medium or embedded in antibiogram disk. The defense mechanisms *i.e.*, total phenolic compounds and flavonoids content, phenylalanine ammonia-lyase (PAL), glutathione reductase (GR) and lipoxygenase (LOX) activities, were measured in tomato plants treated with EPS (1.5 mg mL^{-1}), inoculated or not with *X. gardneri*. EPS reduced bacterial spot symptoms by up to 72.0% compared to the control. There were no direct effects of EPS on the *in vitro* growth of *X. gardneri*. The spectrophotometric profile, ascorbic and ellagic acid concentrations were change in tomato plants after EPS application, in plants challenged with the pathogen. Increases in PAL, GR and LOX activities were observed in plants treated with EPS. Thus, the application of *L. plantarum* exopolysaccharides can be considered as an effective alternative for controlling bacterial spot in tomato plants. This paper also discusses how these exopolysaccharides reduced the severity of the disease.

Keywords: *Xanthomonas gardneri*, *Solanum lycopersicum*, resistance induction, lactic acid bacteria

1. Introduction

Tomatoes (*Solanum lycopersicum*) are very important for human nutrition because of its high levels of mineral salts, vitamins and soluble sugars (Raiola et al., 2014). This crop contributes economically because its chain production employs a great number of people, in conventional and organic food production (Dimitri & Oberholtzer, 2009; FAO, 2014). On the other hand, tomato plants are highly affected by several diseases that may restrict their yield, depending on the level of genetic resistance. In this sense, bacterial leaf spot caused by *Xanthomonas* sp. (*X. euvesicatoria*, *X. gardneri*, *X. perforans* and *X. vesicatoria*) is one of the major diseases affecting tomato plant growth, development and overall productivity (Jones et al., 2004; Quezado-Duval et al., 2007).

Bacterial leaf spot develops mainly in hot and rainy climate, since water favors the spread of the phytopathogen, as well as its infection and the colonization of plant tissues, reducing up to 50% of tomato crop yield (Quezado-Duval et al., 2007; Potnis et al., 2015). In favorable conditions, the progression of this disease is hard to control, even when specific strategies are adopted, *e.g.*, the use of bacteria-free seeds or seedlings, the elimination of alternative hosts and the use of chemical control (Obradovic et al., 2004; Araujo et al., 2013; Potnis et al., 2015). Among the chemicals currently used for the pathogen control, antibiotics and copper-based compounds have variable efficiency, impairing plant development, causing damages to the environment and selecting bacteria resistant to the active ingredient (Silva & Fay, 2006; Abbasi et al., 2014; Potnis et al., 2015).

One alternative to the conventional control of plant diseases may reside in the induction of resistance, where elicitors are applied in plants to activate genes that promote the synthesis of defense compounds, as the enzyme phenylalanine ammonia-lyase - PAL (Tian et al., 2006; Ge et al., 2013; Ebrahim et al., 2011; Salazar et al., 2013). The PAL is the first enzyme activated in the phenylpropanoid pathway, being responsible to the deamination of L-phenylalanine, transforming it into trans-cinnamic acid and ammonia. The trans-cinnamic acid can then be incorporated in different phenolic compounds, which are present in the formation of esters, coumarins, flavonoids, lignins and salicylic acid, among others. For example, while salicylic acid is normally associated with the induction of resistance by acting as a signal, ellagic and ascorbic acid act as antioxidants, protecting cell against oxidative damage (Gao et al., 2015; Tian et al., 2006; Zhang & Zhou, 2010).

Similarly to the phenolic compounds, glutathione is involved in the process of cell protection against oxidative damage, being considered as an oxidative stress biomarker, and found as reduced glutathione (GSH) and oxidized glutathione (GSSG). The reduction of GSSG to GSH occurs by the activity of glutathione reductase (GR), where the increase of enzymatic activity indicates that the plant is under some stress and that the "glutathione apparatus" will provide greater tolerance to oxidative stress (Sharma & Dubey, 2007; Sharma et al., 2012).

The jasmonic acid (JA) is also an important indicator related to stress, responsible to trigger several defense responses in plants, such as the synthesis of lipoxygenase (LOX). LOX is a key-enzyme that catalyzes the polyunsaturated fatty acids to hydro peroxides, resulting in reactive molecules, such as H_2O_2. The products of this pathway act in the plant development, in its defense against herbivores and pathogens, in detoxification reactions, among others (Feussner & Wasternack, 2002).

In the search for compounds that can induce resistance and integrate the alternative control of plant diseases, polysaccharides such as β-glucans, chitins and chitosans present in the cell walls of fungi and yeasts, as well as exopolysaccharides (EPS) secreted by bacteria, are of interest because they are easily obtainable natural compounds (Wang et al., 2008; Mahapatra & Banerjee, 2013; Leemhuis et al., 2013). Furthermore, plants are genetically prepared to sense microbial molecular signatures, called microbe-associated molecular patterns (MAMPs), activating immune responses (Zhang & Zhou, 2010).

In general, lactic acid bacteria produce a large variety of EPS whose physical-chemical properties are unique (Notararigo et al., 2013). Species of the genus *Lactobacillus*, particularly *L. plantarum*, have been reported as important probiotic agents, bringing several benefits to human health, being also generally recognized as safe (GRAS) by the USFDA (Laws et al., 2001), and non-toxic to the environment (Seo et al., 2015). Exopolysaccharides can be recognized by plants, triggering different responses of defense. Moreover, the bioprocess involved to their obtainment is generally cheap and generates low impact to the environment. In this sense, the use of a microbiological approach to induce resistance in plants has a strong technological appeal and a high commercial potential due to the possibility of an accurate control of the production process, resulting in a stable, standardized and high-quality raw materials that are easily extracted and that can bring environmental advantages (Baque et al., 2012; Huang & Mcdonald, 2009). In this study, we evaluated the efficacy of EPS extracted from *L. plantarum* to control bacterial spot and elicit defense mechanisms in tomato plants.

2. Method

2.1 Lactobacillus Strain, Growth Conditions and Extraction of Exopolysaccharides

L. plantarum (CCT 0580, ATCC 8014) was obtained from André Tosello Tropical Culture Collection (Campinas, Brazil). The bacterial suspension was prepared by adding 2 mL of an initial inoculum stored in sterile glycerol 2% into 20 mL of nutrient broth. Subsequently, the nutrient broth containing the inoculum was added into 200 mL of bean curd whey (BCW, Tofutura Indústria de Alimentos Ltda). BCW was supplemented with glucose (11.5 g L^{-1}), yeast extract (1 g L^{-1}), sodium citrate (3 g L^{-1}), Tween 80 (1 mL L^{-1}), KH_2PO_4 (1 g L^{-1}), K_2HPO_4 (1.4 g L^{-1}), $MgSO_4$ (0.2 g L^{-1}), $MnSO_4$ (0.05 g L^{-1}). Each step of the experiment was performed at 30 °C for 18 h. Furthermore, the production of *L. plantarum* EPS was performed in BCW in a 5-L airlift bioreactor. The production was conducted for 30 h and operated with a specific airflow rates, ranging from 0.2 to 1.3 vvm (air volume per medium volume per minute). Temperature and pH were automatically maintained (30 °C, pH 6.0). Polypropylene glycol (0.4 mL L^{-1}) was used as an antifoaming agent.

The resultant microbial biomass was separated from the culture medium by centrifugation (2000 rpm, 20 min), and the cell-free filtrate used to obtain the EPS through nanofiltration, condensation in a rotary-evaporator and lyophilization (Camelini et al., 2013). The EPS were analyzed by FTIR spectroscopy (ABD Bomem Inc. FTLA 2000), using KBr pellets. Protein content was determined by Bradford method, using BSA as reference.

2.2 Xanthomonas Strain and Growth Conditions

X. gardneri was provided by Sakata Seed Sudamerica LTDA and identified at the Centro Nacional de Pesquisa de Hortaliças (CNPH, EMBRAPA, Brazil) using BOX-PCR and the primer 5'-CTACGGCAAGGCGAC GCTGACG-3'. The bacteria were maintained at 25 °C in phosphate buffer (8.6 mM K_2HPO_4; 7.4 mM KH_2PO_4). Periodic subcultures were replicated in Nutrient Agar (NA) culture medium [Composition (g/L): meat peptone 5.0; meat extract 3.0; agar 12.0] (Merck, Darmstadt, Germany) and the plates incubated at 25 °C, for 48 h. The bacterial suspension was obtained by adding distilled water to the growth medium and spreading them with the aid of Drigalski spatula. The concentration of the suspensions was adjusted to 0.6 (OD_{600nm}) (Coqueiro & Di Piero, 2011; Luiz et al., 2012).

2.3 Disease Severity Assay in Tomato Plants

Commercial tomato seeds of Santa Cruz Kada cultivar (Paulista) were provided by Isla Sementes Ltda and sown in polystirene trays containing substrate Plantmax®. Fifteen days after sowing, two plants were transferred to 2 L pots, filled with organic compost. Every 15 days, 20 mL of a solution containing 4.0 g of urea and 3.8 mL of Eurofit® per liter of distilled water was added to each pot. The experiments were conducted inside a greenhouse.

The EPS were sprayed on five-leaf tomato plants at the concentrations of 0.5, 1.5 and 3 mg mL^{-1}, 3 days before the inoculation with X. gardneri. All leaves of each plant were sprayed with 10 mL of a suspension, using an HVLP paint gun (maximum pressure = 58 psi, 0.7 mm nozzle, manufactured by Grifo, Italy) coupled to an air compressor (Schulz, Brazil; pressure = 25 lbf/in2; power = 180 W; air flow = 105 mL min^{-1}). After inoculation, the plants remained in a moist chamber for 60 h, in order to favor the bacterium development.

After the most effective EPS concentration was established (1.5 mg mL^{-1}), the time interval between plant treatment and inoculation was evaluated, being investigated the intervals of 3 and 7 days. Two experiments were conducted in greenhouse, under different environmental conditions: the average temperature of the first experiment was 28 ± 2 °C, whereas in the second, it was 19.8 ± 1 °C. All plants were treated and inoculated as described previously, and all experiments were set up under a completely randomized design, with seven replications per treatment, where a pot containing two plants represented a repetition. The assessment of disease severity was performed 15 days after the inoculation, with the aid of a diagrammatic scale for bacterial spot described by Mello et al. (1997). For these experiments, distilled water and the commercially available resistance inducer, Acibenzolar-S-Methyl (ASM), 0.05 mg mL^{-1}, were used as negative and positive control, respectively. ASM was obtained from the commercial product Bion® (Syngenta Proteção de Cultivos Ltda., Brazil).

2.4 Determination of Antibiotic Activity

The EPS were prepared and incorporated into NA culture medium (1.5 mg mL^{-1}). The mixture was placed in 8 cm Petri dishes, and 100 µL of the bacterial suspension (0.1 OD_{600nm}, diluted to 1/1000) pipetted over the surface and spread with the aid of a Drigalski spatula. NA without EPS was used as control. The plates were incubated (25 ± 1 °C, 48 h) and the evaluation performed counting the number of colony-forming units (CFU) (adapted from Luiz et al., 2012).

An antibiogram bioassay was also performed. For this, 50 µL of a bacterial suspension (0.3 OD_{600nm}, diluted to 1/1000) were spread in NA plates, and left drying. Then, paper discs (5 mm diameter) were soaked with 10 µL of a suspension containing EPS (10 mg mL^{-1}), distilled water or the antibiotic oxytetracycline (10 mg mL^{-1}). Five replications were made for each treatment, where a replication was represented by a single plate containing four disks. The plates were incubated at 25 ± 1 °C for 48 h and the evaluation performed by analyzing the formation of a bacterial growth inhibition halo.

2.5 Biochemical Analyses

EPS (1.5 mg mL^{-1}), ASM (0.05 mg mL^{-1}) or distilled water (control) were applied on tomato plants with five true leaves. At the 3th day after application, the inoculation was performed with bacterial suspension of X. gardneri (OD 0.6; 600 nm). Leaf samples were collected at 3, 5 and 7 days after spraying the treatments. The second, third and fourth leaves were sampled. Five replications for each treatment were made. The collected samples were stored in transparent plastic bags, put in contact with liquid nitrogen, and subsequently stored in ultra-freezer (-80 °C), until processing time. The samples were processed for evaluation of spectrophotometric profile, quantification of total phenolic compounds and flavonoids, identification of phenolic compounds through HPLC, as well as for the determination of enzymatic activity of phenylalanine ammonia-lyase (PAL), glutathione reductase (GR) and lipoxygenase (LOX).

2.5.1 Spectrophotometric Profile, Quantification of Total Phenolic Compounds and Flavonoids and Identification of Phenolic Compounds through HPLC

For the spectrophotometric profile, leaf samples (250 mg) were homogenized in 5 mL ethanol-toluene (1:1; v/v), with the aid of a mortar for 3 min, and left standing for 15 min. The extracts were diluted in ethanol-toluene (1:5; v/v) and the absorbance scanned spectrophotometrically (250 to 750 nm). A scan was performed for each sample and the results were expressed by the mean absorbance values of three replications.

In order to quantification of total phenolic compounds, flavonoids and identification of phenolic compounds through HPLC, 200 mg of the foliar tissue of each sample were crushed in a porcelain pestle with liquid nitrogen, and homogenized with 3 mL of 80% acidified methanol (methanol: HCl = 80:1, v/v). The resulting mixture was incubated in dark for 1 h, at room temperature, and subsequently centrifuged (3000 g, 5 min), recovering the supernatant for the subsequent analysis.

The phenolic compounds were performed according of McCue et al. (2000), with modifications. Firstly, 0.5 mL of the obtained extract was mixed with 0.5 mL of methanol (95%). Then, 1 mL of ethanol (95%), 1 mL of distilled water and 0.5 mL of Folin-Ciocalteu were added to this mixture. After 5 min, 1 mL of Na_2CO_3 (5%) was added, and the sample was incubated in the dark for another 1 h. The absorbance of the final solution was measured at 725 nm using a spectrophotometer and the quantification of phenolic compounds calculated based on a gallic acid standard curve (0.0-100 µg). The results were expressed in µg of gallic acid equivalents per gram of fresh weight (µg EAG·g·FW^{-1}).

For the flavonoid quantification, 0.5 mL of the crude extract was added with 2.5 mL of ethanol (99%) and 0.5 mL of methanol solution of aluminum chloride (2%), and left 1 h in darkness. The absorbance was measured at 420 nm and flavonoid content was expressed in µg of quercetin equivalents per g of fresh weight (µg EQ·g·FW^{-1}).

For the identification of phenolic compounds through HPLC, aliquots (10 µL) of the crude extract were injected into liquid chromatography equipment (Thermo Scientific Dionex UltiMate 3000) equipped with a reverse phase column (C_{18} reverse phase; Phenomenex LC-18, 250 mm × 4.6 mm, 5 µm Ø inner; 40 °C) and a spectrophotometric detector UV-visible ($\lambda = 280$ nm). A solution of acidified water (pH 3.0) and methanol PA (85:15 v/v), in a flow of 1 mL min^{-1}, was used as mobile phase. The identification of the compounds of interest was performed through the comparison between the samples retention times with the retention times of standard compounds (ascorbic acid, ellagic acid, gallic acid, epicatechin, gallocatechin, caffeic acid, rutin). The quantification of the phenolic acids was performed using the standard curves of the major compound (ascorbic acid). The results were expressed in µg per g of fresh weight (µg·g·FW^{-1}) and refer to the average of three consecutive injections for each sample (n = 3).

2.5.2 Determination of Phenylalanine Ammonia-lyase Activity

Sampled leaf tissue (100 mg) was homogenized in sodium borate buffer (25 mM, pH 8.8) containing ethylene diamine tetra-acetic acid (EDTA, 1 mM) and polyvinylpyrrolidone (PVP, 0.5%) for determining phenylalanine ammonia-lyase activity. The obtained solution was centrifuged (20,000 g, 30 min, 4 °C) and the supernatant (protein extract) recovered. The enzymatic activity was determined according to Falcón et al. (2008), with modifications. Fifty mM of phenylalanine was added to 100 mM sodium borate buffer (pH 8.8). Then, a total of 250 µL of protein extract was added to 250 µL to the previous mixture, and incubated at 40 °C for 1 h. The reaction was interrupted by the addition of 200 µL of 5 N HCl and ice bath for 5 min. Subsequently, 300 µL of distilled water were added, and the absorbance of the final solution was measured at 290 nm. The results were expressed in nmol of trans-cinnamic acid formed per mg of protein per minute of reaction (nmol trans-cinnamic acid min^{-1}·mg protein).

2.5.3 Glutathione Reductase and Lipoxygenase Activity

Leaf samples (100 mg) were macerated in liquid nitrogen, and added 1.5 mL of Tris-HCl buffer (50 mM) with $CaCl_2$ (20 mM), pH 8.0. The resultant extract was centrifuged (5.500 g, 10 min, 4 °C), and the supernatant recovered. The glutathione reductase (GR) activity was determined according to the method described by Calrberg e Mannervick (1985) while lipoxygenase activity (LOX) of linoleic acid was determined according to the method described by Axelrod et al. (1981).

The GR activity was determined by adding 50 µL of the obtained extract to 250 µL of reaction buffer (Tris-HCl buffer; 0.10 M, pH 7.5), magnesium chloride (0.3 mM), glutathione oxidized (GSSG; 1.0 mM), and NADPH (0.20 mM). The decrease in the absorbance was measured over the first minute of the reaction (340 nm, 28 °C) and the results expressed in GR units min^{-1}·mg·protein^{-1}, where 1 unit of GR was considered the quantity of the enzyme necessary to reduce 0.01 absorbance units.

For LOX activity, 5 µL of the obtained extract was added with 20 µL of sodium linoleate (10 mM), previously prepared in 250 µL phosphate buffer (50.0 mM, pH 6.5). The rate of the reaction was determined at every 20 seconds interval (234 nm, 2 min, 40 °C). The results were expressed in LOX units min^{-1}·mg·protein^{-1}, where 1 unit of LOX was considered the quantity of enzyme required to reduce 0.01 absorbance units.

For all samples, the amount of total protein was evaluated accordingly to the method described by Bradford (1976).

2.6 Statistical Analyses

The statistical analysis of variance (*one-way* or factorial ANOVA) and Tukey`s test were performed to verify the differences between the means of the analysed variables in the experiments. When necessary, the regression analysis was performed to verify the effect of treatment doses (quantitative factor). The analyses were performed using the statistical software Statistica 8.0 (Statsoft, 2007) and SISVAR (Ferreira, 2003).

3. Results

3.1 Characterization of the EPS, Their Protective and Antimicrobial Effects

The infrared spectroscopy allowed the identification of the typical functional groups of the studied biopolymer (Figure 1). It showed an intense band in the region of 3.425 cm^{-1}, relative to the O-H stretch, while in the region of 2.931 cm^{-1} the band refers to the C-H stretch. The band in the region of 1.650 cm^{-1} showed a potential presence of C=C groups in the structure of the carbohydrate. Additionally, the bands in the regions of 1.413, 1.338 and 1.020 cm^{-1} indicate a C-C stretch, a C-O bend, and a C-O-C stretch, respectively.

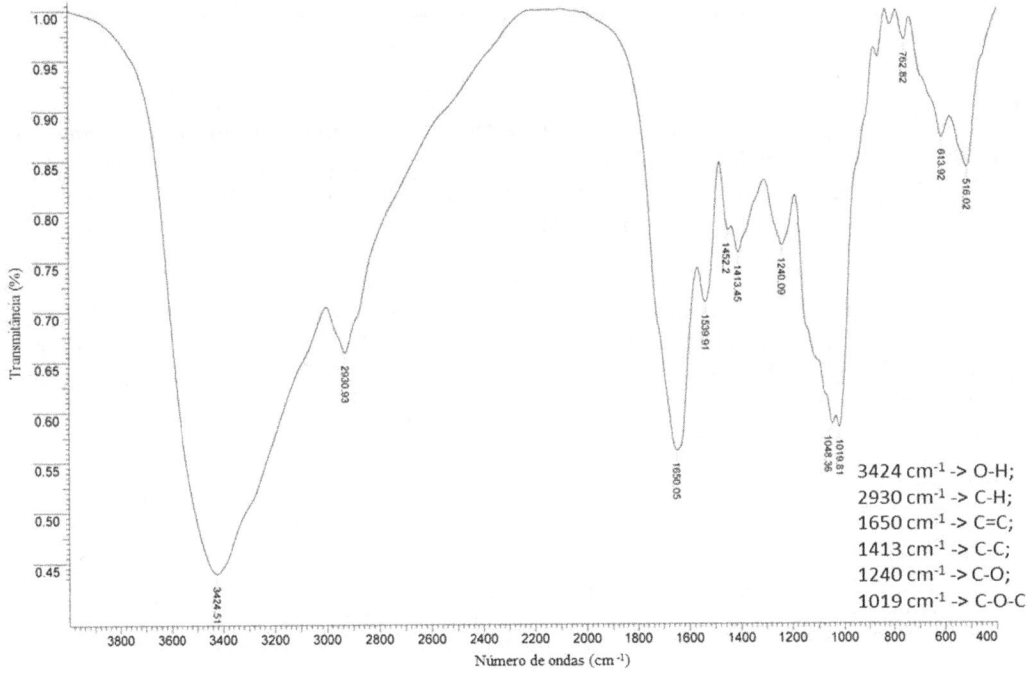

Figure 1. Infrared spectrum of the EPS produced by *Lactobacillus plantarum*, in the region of 4.000-400 cm^{-1}, using KBr pellets

It was observed an effect of EPS doses against the severity of bacterial spot in tomato plants 3 days after the treatments application (Figure 2). Nine days after *X. gardneri* inoculation, a reduction of 52%, 72% and 63% on the severity of the disease was observed when the EPS was applied on the leaves at concentrations 0.5, 1.5 and 3.0 mg mL^{-1}, respectively. Eighteen days after the inoculation, the protection levels reached 42%, 54% and 58%, respectively. Distinctively, ASM decreased the symptoms of bacterial spot by approximately 93.0 %.

After verified the best concentration of EPS against bacterial spot in tomato plants, an assay was performed to determine the best time interval between the application of the treatments and the inoculation with *X. gardneri* (3 or 7 day-interval) (Table 1). It was observed an interaction between the variables in both experiments. The EPS (1.5 mg mL^{-1}) controlled the disease when applied 3 days before the inoculation of the pathogen. However, this efficiency was reduced when a 7 day interval was used. Furthermore, it was observed that the ASM provided higher disease control independent of time interval between the application of the product and the pathogen inoculation, in both experiments.

The EPS did not inhibit the *in vitro* growth of *X. gardneri* (Table 2). In the assays where EPS were incorporated to the culture medium (NA), no significant difference was observed in the number of colonies of *X. gardneri*, compared to control. This result was further confirmed by the disk diffusion test (antibiogram), which showed that the EPS did not reduce the growth of *X. gardneri*. An inhibition halo (1 cm) only appeared in the bactericide (tetracycline) used as control.

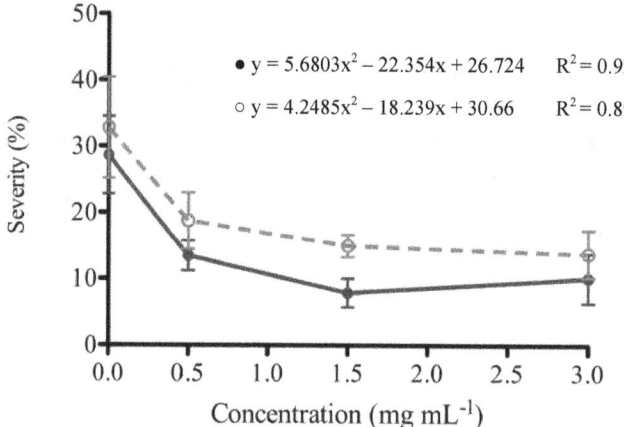

Figure 2. Bacterial leaf spot severity in tomato plants treated with the EPS of *Lactobacillus plantarum* (0; 0.5; 1.5 and 3.0 mg mL^{-1}) 3 days before the inoculation with *Xanthomonas gardneri*. The assessment of disease severity was performed at 9 and 18 days after inoculation (1st and 2nd assessment). There was a significant effect of doses according to the F-test ($p < 0.05$). Bars represent the mean ± standard deviation

Table 1. Bacterial leaf spot severity in tomato plants submitted to different time intervals (3 and 7 days) between the application of the EPS of *Lactobacillus plantarum* (EPS Lac), Acibenzolar-S-Methyl (ASM) or distilled water, and the inoculation with *Xanthomonas gardneri*. Assessments performed at 15 days after inoculation

| | Severity (%) | | | |
| | Experiment 1 | | Experiment 2 | |
	3 day	7 day	3 day	7 day
Water	40.23 ± 2.91 Aa	40.62 ± 6.63 Aa	21.58 ± 2.01 Aa	20.00 ± 1.26 Aa
EPS Lac	19.76 ± 5.19 Bb	35.69 ± 6.31 Aa	13.67 ± 2.16 Bb	18.75 ± 3.06 Aa
ASM	7.14 ± 4.06 Ca	6.71 ± 4.23 Ba	0.85 ± 0.53 Ca	0.71 ± 0.33 Ba

Note. Means followed by the same capital letters in the column and lower case letters in the line, do not differ statistically by the Tukey test ($p < 0.05$).

Table 2. Effect of the EPS of *Lactobacillus plantarum* incorporated to a culture medium or added in antibiogram disks on the growth of *Xanthomonas gardneri*. Assessment of the number of colony forming units (CFU) and diameter of the bacterial growth inhibition halo after 48 h of incubation

	Number of CFU		Growth inhibition halo (cm)	
	Experiment 1	**Experiment 2**	**Experiment 1**	**Experiment 2**
Water	$120,5 \pm 3,9^{ns}$	$159,3 \pm 2,1^{ns}$	0	0
EPS Lac	$124,5 \pm 3,2$	$154,9 \pm 3,5$	0	0
Oxytetracycline	-	-	1,0	1,0

Note. ^{ns} Not significative statistically ($p < 0.05$).

3.2 Changes in the Spectrophotometric Profiles, Phenolic Compounds and Flavonoid Content in Plants Treated with the EPS

Changes in the spectrophotometric profiles of plants treated at different time intervals were observed (Figure 3).

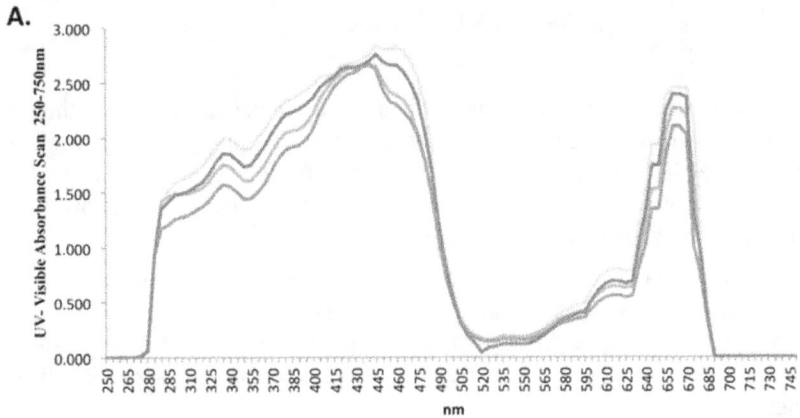

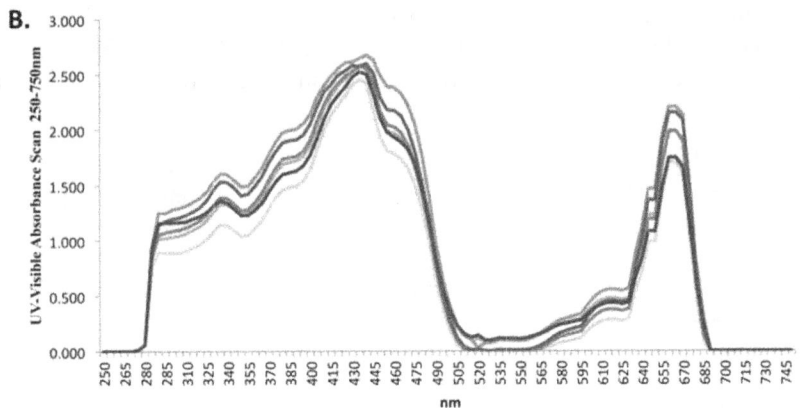

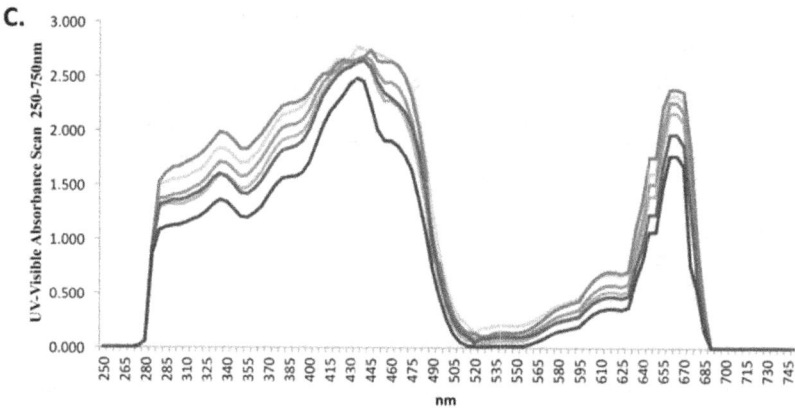

Figure 3. The spectrophotometric profile of ethanol-toluene extract of tomato plants treated with the EPS of *Lactobacillus plantarum (EPS Lac)*, Acibenzolar-S-Methyl (ASM) and distilled water, from plants inoculated (Inoc) or not (NI) with *Xanthomonas gardneri*, at 3 (A), 5 (B) and 7 (C) days after the treatments (dat). A. (———) Water, (———) EPS Lac and (———) ASM; B. and C. (   ) Water Inoc, (   ) EPS Lac Inoc, (———) ASM Inoc, (———) Water NI, (———) EPS Lac NI and (———) ASM NI

At the 3rd and 7th days after the treatments, a significant decrease in the absorbance in the bands 285-325 nm in plants treated with the EPS, inoculated or not, was observed when compared to control. At 5 days after the treatments (dat), plants that have been sprayed with ASM and inoculated, showed a slight increase in the absorbance within this range. At 7 dat, plants treated with ASM, inoculated or not, showed significant reduction in absorbance, similar to those treated with EPS (Table 3).

No significative variations were observed in the content of total phenolic compounds over time in plants treated with the EPS, distilled water or ASM. Even after the inoculation of the pathogen, the evaluated parameters remained unchanged (Figure 4A). In a different manner, the flavonoids content increased at 7 dat, after the inoculation with *X. gardneri*, independently of the treatment, whereas plants previously sprayed with ASM showed minor changes (Figure 4B).

The HPLC results revealed a similar profile for all samples. Two compounds were identified and quantified (ascorbic acid and ellagic acid), being observed a significant difference in their concentration between treatments over time (Figure 5). At 5 dat, it was observed that among the plants not inoculated, those treated with ASM showed the highest concentrations of ascorbic acid, *i.e.*, approximately 30% more than the control. In this same period, EPS and ASM decreased by 30 and 34%, respectively, the concentration of ascorbic acid in inoculated plants (Figure 5A).

Table 3. Average values of absorbance in the range between 285-325 nm in tomato plants treated with the EPS of *Lactobacillus plantarum* (EPS Lac), Acibenzolar-S-Methyl (ASM) or distilled water, from plants inoculated (Inoc) or not (NI) with *Xanthomonas gardneri*, at 3, 5 and 7 days after treatment (dat)

Treatments	Absorbance 285-325 nm				
	3 dat (before Inoc)	5 dat (2DAI)		7 dat (4DAI)	
		N I	Inoc	N I	Inoc
Water	1.44±0.08 A	1.09±0.01 Ab	0.90±0.06 Aa	1.60±0.05 Aa	1.51±0.02 Aa
EPS Lac	1.25±0.02 B	1.19±0.05 Ab	1.05±0.02 Aa	1.33±0.05 Ba	1.30±0.14 Ba
ASM	1.44±0.06 A	1.15±0.04 Aa	1.27±0.04 Bb	1.12±0.02 Ca	1.39±0.02 Bb

Note. 5 and 7 dat correspond to 2 and 4 days after inoculation (DAI). Means followed by the same capital letters in the column and lower case letters in the line, do not differ statistically by the Tukey test (p < 0.05) on time interval.

The concentration of ascorbic acid was also different at 7 dat (Figure 5B). Plants sprayed with EPS or ASM, both non-inoculated, showed a reduction in the phenolic concentration (50 and 48%, respectively), compared to the respective control; after the inoculation with *X. gardneri*, plants previously treated with EPS or ASM showed an increase (16 and 35%, respectively) in ascorbic acid concentration, when compared to the inoculated control (Figure 5B).

Moreover, the concentration of ellagic acid was altered at 5 dat, in plants sprayed with EPS non-inoculated, showing a decreased of the ellagic acid concentration around 28% while plants treated with ASM showed increase of 54% compared to the respective control (Figure 5C). After the inoculation of the causal agent of bacterial spot, an increase of ellagic acid concentration was observed in plants treated with EPS or ASM (19 and 15%, respectively), when compared to the control (Figure 5C). Seven days after the treatments, the concentration of ellagic acid had a similar behavior to the ascorbic acid. In this case, EPS or ASM, both applied in non-inoculated plants, decreased 21 and 30%, respectively, the concentration of this phenol. In the same period, plants previously treated with ASM and inoculated with *X. gardneri* showed an increase in the ellagic acid concentration, about 27% higher than inoculated control (Figure 5D).

3.3 Effects of EPS on Phenylalanine Ammonia-lyase, Glutathione Reductase and Lipoxygenase Activities

The EPS and ASM didn't change PAL activity at the 3rd day after treatment, i.e., before inoculation (Figure 6A). Nevertheless, it was observed a reduction in PAL activity in non-inoculated plants but treated with ASM or EPS at 5 dat. In this same period, a significant increase of 2.5 times in PAL activity was observed in plants sprayed with ASM and inoculated with *X. gardneri* compared to inoculated control (Figura 6B). At 7 dat, PAL activity increased more than 2.5 times in plants previously treated with EPS and challenged with *X. gardneri* (Figure 6C).

A similar trend occurred for GR activity. Significative changes at the 3rd day after the treatments were not observed (Figure 7A). At 5 dat, plants sprayed with ASM and challenged with the pathogen showed an increase in the enzymatic activity compared to the other treatments (Figure 7B). Nonetheless, at 7 dat, plants sprayed with EPS followed by inoculation showed a higher GR activity (Figure 7C).

The EPS and ASM reduced the LOX activity at the 3 days after treatments (Figure 8A). After 5 days of the treatments, plants inoculated with the pathogen showed a significant increase in the enzymatic activity, regardless the treatment (Figure 8B). Finally, it was observed that among the inoculated plants, the plants previously sprayed with EPS (7 dat) had a higher LOX activity (Figure 8C).

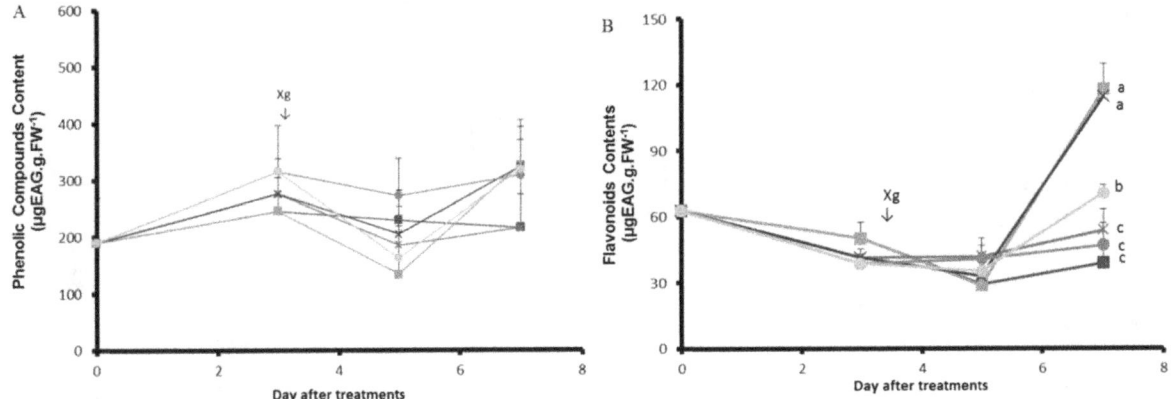

Figure 4. Content of phenolic compounds (A) and flavonoids (B) in tomato plants treated with the EPS of *Lactobacillus plantarum* (EPS Lac), Acibenzolar-S-Methyl (ASM) and distilled water, from plants inoculated (Inoc) or not (NI) with *Xanthomonas gardneri*. "Xg" indicates the inoculation with *X. gardneri* (OD_{600} = 0.6). Bars represent the mean ± standard deviation. Bars followed by different letters indicate significant difference at 5% significance by the Tukey test on time interval. (■) Water NI, (■) Water Inoc, (▲) EPS Lac NI, (▼) EPS Lac Inoc, (●) ASM NI and (●) ASM Inoc

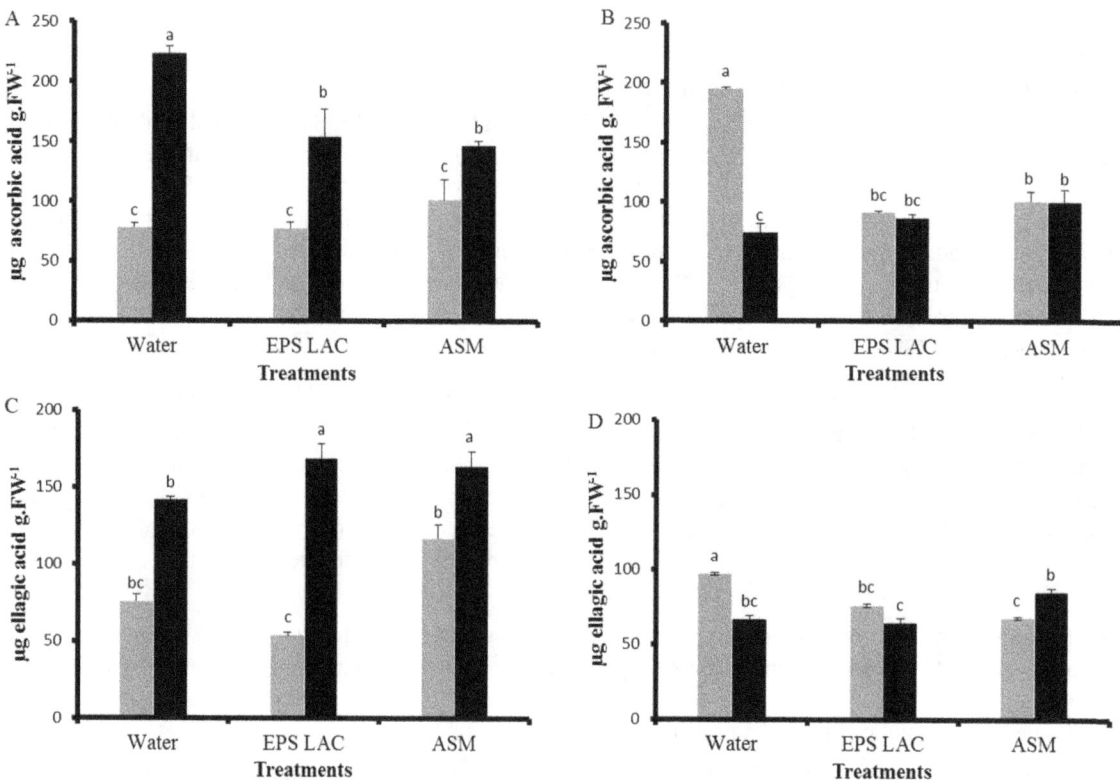

Figure 5. Concentration of ascorbic (A and B) and ellagic (C and D) acid determined by HPLC on tomato plants treated with EPS of *Lactobacillus plantarum* (EPS Lac), Acibenzolar-S-Methyl (ASM) or distilled water, from plants inoculated (Inoc) or not (NI) with *Xanthomonas gardneri*, at 5 (A and C) and 7 (B and D) days after treatment (dat). 5 and 7 dat correspond to 2 and 4 days after inoculation. Bars represent the mean ± standard deviation. Means followed by the same letters do not differ statistically by the Tukey test (p < 0.05). (■) NI and (■) Inoc

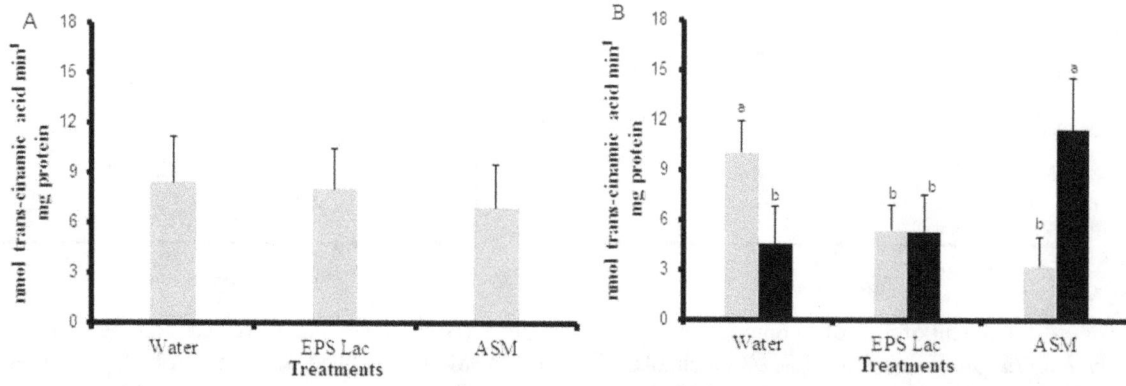

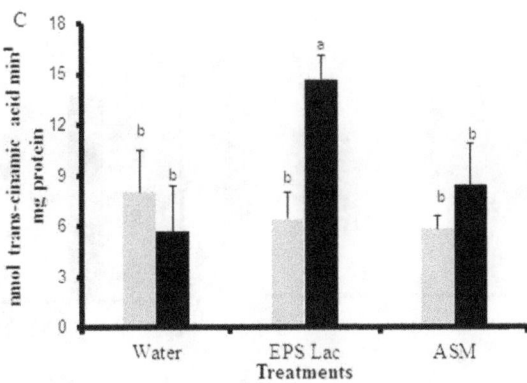

Figure 6. Phenylalanine ammonia-lyase (PAL) activity in tomato plants treated with the EPS of *Lactobacillus plantarum* (EPS Lac), Acibenzolar-S-Methyl (ASM) or distilled water, from plants inoculated (Inoc) or not (NI) with *Xanthomonas gardneri*, 3 (A), 5 (B) and 7 (C) days after the treatments (dat). 5 and 7 dat correspond to 2 and 4 days after inoculation. Bars represent the mean ± standard deviation. Means followed by the same letters do not differ statistically by the Tukey test (p < 0.05). (▪) NI and (■) Inoc

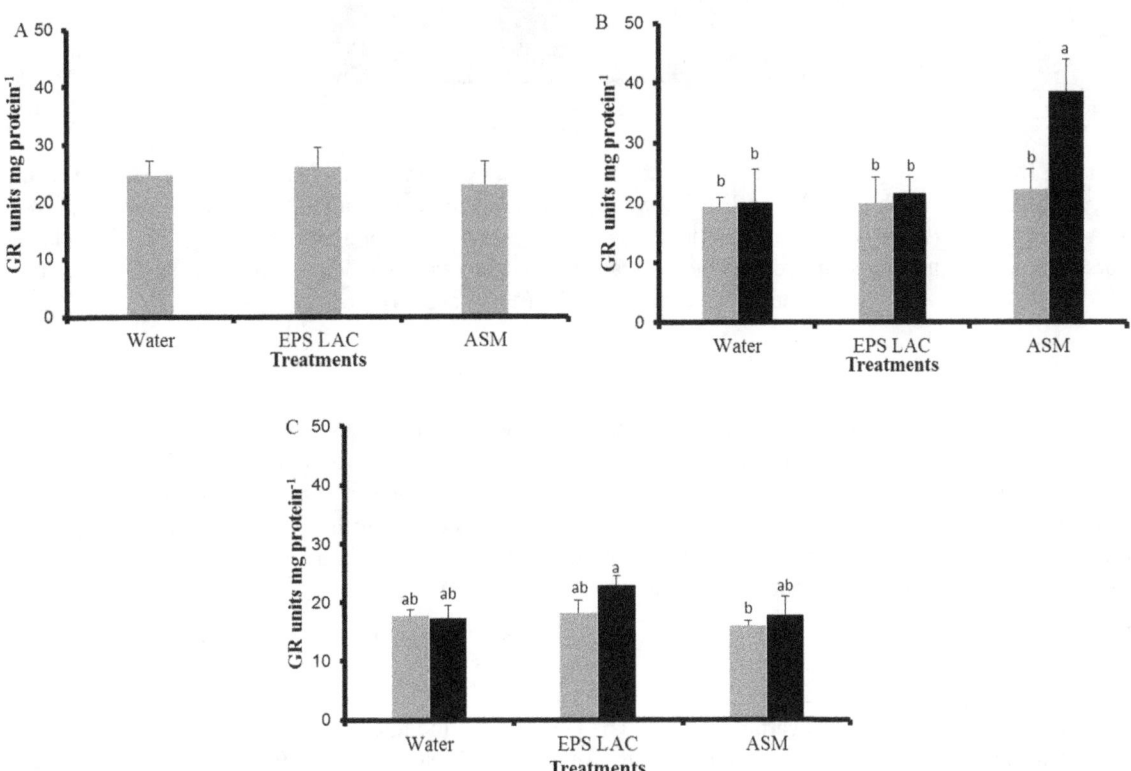

Figure 7. Glutathione redutase (GR) activity in tomato plants treated with the EPS of *Lactobacillus plantarum* (EPS Lac), Acibenzolar-S-Methyl (ASM) or distilled water, from plants inoculated (Inoc) or not (NI) with *Xanthomonas gardneri*, 3 (A), 5 (B) and 7 (C) days after the treatments (dat). 5 and 7 dat correspond to 2 and 4 days after inoculation. Bars represent the mean ± standard deviation. Means followed by the same letters do not differ statistically by the Tukey test (p < 0.05). (▪) NI and (■) Inoc

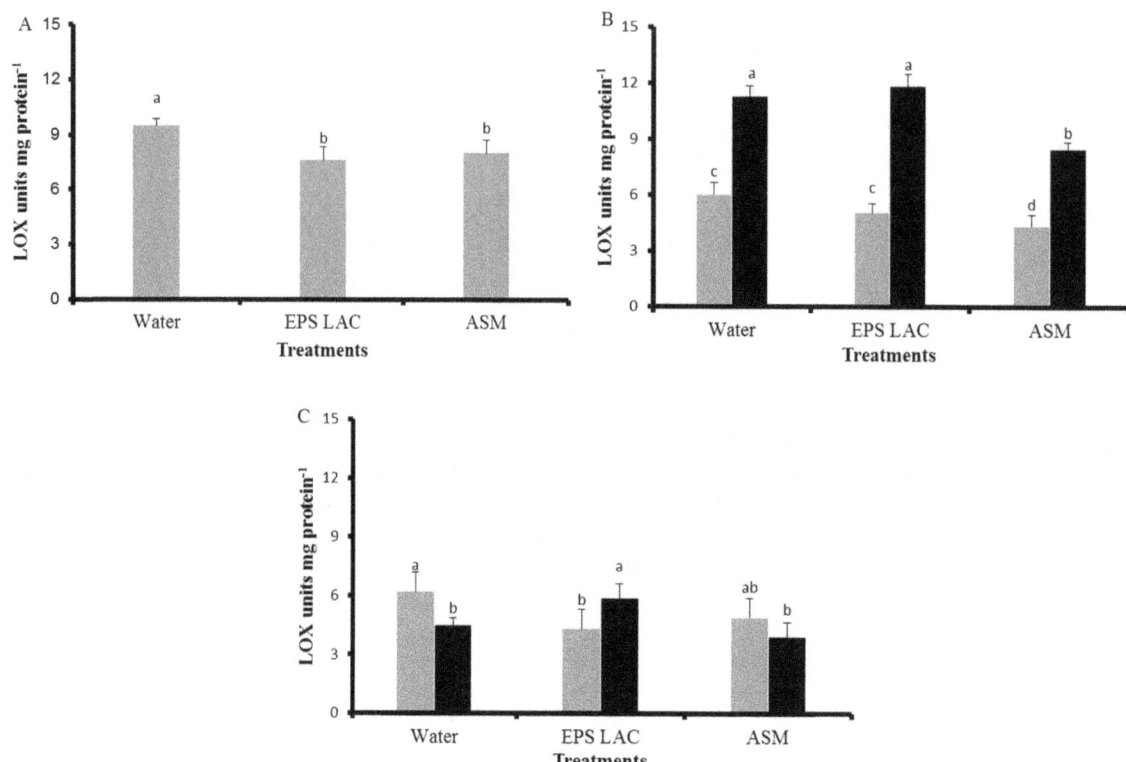

Figure 8. Lypoxigenase (LOX) activity in tomato plants treated with the EPS of *Lactobacillus plantarum* (EPS Lac), Acibenzolar-S-Methyl (ASM) or distilled water, inoculated (Inoc) or not (NI) with *Xanthomonas gardneri*, 3 (A), 5 (B) and 7 (C) days after the treatments (dat). 5 and 7 dat correspond to 2 and 4 days after inoculation. Bars represent the mean ± standard deviation. Means followed by the same letters do not differ statistically by the Tukey test ($p < 0.05$). (▫) NI and (▪) Inoc

4. Discussion

The use of EPS derived from microorganisms presents a considerable technological application which stands out in various sectors of industry. Nonetheless, their characteristics can range depending upon the microorganism, the fermentation medium composition and growth conditions. The infrared absorption spectrum of a compound is probably its most unique physical property; therefore, the spectrum can be referred as the fingerprint of a molecule (Kacuráková & Wilson, 2001). Polysaccharides, in general, contain a significant number of hydroxyl groups, whose broad absorption band resides around 3.000 cm^{-1}.

Previous studies have shown the similarity of different EPS extracted from *Lactobacillus* species. In general, the infrared spectra show functional characteristics *e.g.*, the OH group whose bands ranges between 3.700 and 3.000 cm^{-1}, and a methyl group at around 2.924 cm^{-1}. Polysaccharides also have high absorption at 1.200-1.000 cm^{-1} (corresponding to the presence of carbohydrates). The position and the intensity of the bands are specific to each polysaccharide, allowing its possible identification (Wang et al., 2008; Ai et al., 2008; Notararigo et al., 2013). Therefore, the position and intensity of the bands of the studied EPS (3.425, 2.931, 1.048, and 1.020 cm^{-1}) allowed its identification as a polysaccharide rich in carbohydrates with OH and CH$_3$ groups. This result reinforces the ability of *Lactobacillus* to maintain the pattern of EPS production, with little variation between species of the genus and bacterial culture conditions. *L. plantarum*, in particular, has a great capacity to adapt to the different cultivation conditions, being able to ferment many different carbohydrates, and high tolerance to low pH. This particularity makes the microorganism easily cultivated under controlled conditions by bioprocesses. Another advantage biotechnological use of *L. plantarum* stems from the abundant production of EPS naturally (Notararigo et al., 2013; Seo et al., 2015).

In this study, *L. plantarum* EPS reduced the severity of tomato bacterial spot when applied at 1.5 and 3.0 mg mL^{-1}. As far as time interval is concerned, the EPS were efficient only when applied at 3 days before the inoculation with *X. gardneri*, indicating a direct effect against the pathogen. Although, *in vitro* assays did not detect an antimicrobial effect of the EPS, nonetheless their bactericidal potential already been demonstrated in other studies against human bacteria (Li et al., 2014; Roselló et al., 2013). Therefore, it is suggested that the EPS could have acted against the pathogen through the activation of resistance mechanisms in tomato plants, even with the plant protection occurring in a short interval between EPS application and the inoculation of the plants (3 days), and the lacking of bacterial spot control when applied 7 days before the treatment.

The possible use of natural compounds, such as EPS of *L. plantarum*, for the control of diseases in plants, meets the growing demand for healthier foods. The demand for and sale of organic food commodities have increased approximately twenty-fold in recent years (APEDA, 2012). Due to this demand, the conventional disease control measures in organic production need to be combined or replaced with effective natural compounds to increased resistance in plants against pathogens.

The EPS capacity to control the bacterial spot of tomato may be related to the plant's ability in recognize molecules from microorganisms (MAMP's). This recognition is known to occur by the receptors located on the cell membrane surface or inside the cell, which will trigger a signal transduction cascade, leading to the activation of defense mechanisms. This type of defense allows plants to respond quickly and efficiently to several pathogens (Zhang & Zhou, 2010, Shang et al., 2007).

The eliciting effects of bacterial EPS in different pathosystems were previously documented. In barley plants, Antoniazzi et al. (2008) observed a reduction of 75% in the disease severity caused by *Bipolaris sorokiniana*. This value was very similar to those observed to the fungicide application. In that study, the control of the disease was attributed to the increase of 1,3-glucanase activities and the increase of phenolic compound levels in plants treated with EPS.

In coffee plants, the application of EPS extracted from *Xanthomonas* spp. (xanthan gum) and commercial xanthan gum were able to induce plant resistance against *Hemileia vastatrix*, reducing the disease severity by 92% (Guzzo et al., 1996). These effects were also observed in studies with tomato plants against *X. gardneri* (Luiz et al., 2015) and in studies with EPS extracted from *X. campestris*, which induced resistance in wheat plants against *Bipolaris bicolor*, *Bipolaris sorokiniana*, and *Drechslera tritici-repentis* (Bach, 1997; Bach et al., 2003). In the present study, the activation of defense mechanisms have been evidenced by the changes in absorbance rates of leaf extract, concentration of phenolics, such as ascorbic and ellagic acid, and also in PAL and GR activities of plants treated with EPS or ASM.

In general, the changes observed in the range of 285-325 nm are related to the class of phenolic compounds, indicating a change in the defense mechanisms in treated plants, especially after the inoculation with the pathogen. Although no changes were observed in the total phenolic content using the reagent Folin-Ciocalteu, probably due the low method sensibility, with the use of a more sensitive technique (HPLC), differences in the concentration of phenolics were found, such as ascorbic and ellagic acid. Phenolic compounds have a great importance in the defense of plants, once after elicitation, several phenolic compounds are oxidized and transformed into antimicrobial compounds, such as quinones (Vaughn & Duke, 1984). In this sense, the presence of ascorbic acid and ellagic acid ensures a reduction in the oxidative damage within the cell, once they act as non-enzymatic antioxidants. Meanwhile, the biosynthesis and regulation of these compounds are directly involved with the activity of phenylalanine ammonia lyase (PAL).

Increased PAL activity is often associated with an increased phenylpropanoid concentration, which would lead to an association with phenolic compound accumulation, since phenylalanine and phenols are both produced by shikimic acid pathway. This enzyme acts in the conversion of L-phenylalanine to trans-cinnamic acid, resulting in several phenolic compounds, such as phytoalexins, flavonoids, and lignin, which confers resistance to the cell wall and act as a signal to the plant defense responses against biotic and abiotic stresses (Gerasimova et al., 2005; Latha et al., 2009).

In this study, EPS and ASM increased PAL activity in inoculated plants. However, plants treated with EPS showed this increase later (4 days after challenge or 7 dat), compared to those plants treated with ASM, in which the increase in activity PAL was observed 2 days after the inoculation (5 dat). Similar trend was observed for the activity of glutathione reductase (GR), where plants treated with the commercial inducer (ASM) showed a faster increase of the enzymatic activity in response to the pathogen when compared to EPS. This may be related to the difference observed in the disease control level provided by the products: ASM activated FAL and GR earlier and more intensely controlled the disease compared to EPS (around 90% against 72%).

GR is an important part of the antioxidant system. In general, the whole set that involves glutathione molecules in plants is related to the tolerance increase to oxidative stress (Sharma, 2012; Sharma & Dubey, 2007). Ge et al. (2013) for example, reported that in melon cultivars resistant to *Colletotrichum lagenarium*, enzymatic antioxidants such as glutathione reductase and ascorbate peroxidase, and non-enzymatic molecules such as glutathione (reduced form) and ascorbic acid, are widely expressed during the defense of the plant. These compounds, along with PR proteins are essential for the defense of melon seedlings, allowing an efficient protection against the pathogen infection. Besides, the glutathione and ascorbic acid are directly involved in the ascorbate-glutathione cycle reactions, and are crucial for the preservation of many metabolic processes of the cell (Drazkiewicz et al., 2003).

The application of EPS in tomato plants also triggered an increase the LOX activity, unlike the commercial inducer. LOX is part of another metabolic pathway, the pathway of jasmonic acid (JA). JA is a phytohormone directly involved in the plant defense responses against stress, and used as a stress indicator. Choudhary (2011) and Ferraz et al. (2014) reported that the application of biotic inducers able to stimulate LOX activity, contributed to the resistance of tomato plants against *Fusarium* sp., and also in soybean plants against *Macrophomina phaseolina*. In both cases, the disease control was attributed to the activation of JA pathway. In this sense, the control of tomato bacterial spot provided by the application of EPS could be the result of the activation of more than one metabolic pathway, unlike the ASM, compound similar to salicylic acid, which primarily triggers those metabolites from the phenylpropanoid pathway (Oostendorp et al., 2001).

It is well known that jasmonic and salicylic acids are usually referred as antagonistic hormones acting against different kinds of pathogens *i.e.*, the plant starts to increase the levels of salicylic acid (against biotrophic pathogens) or jasmonate (against necrotrophic pathogens), promoting specific defense responses, focused on the type of pathogen detected. Though, in some cases, these hormones can occur and act synergistically, *e.g.*, when present in low concentrations in healthy tissues, performing nonspecific immune response (Fu & Dong, 2013). For example, Davidsson et al. (2013) described that the simultaneous activation of salicylate and jasmonate pathways appear to be crucial for the attenuation of virulence of *Pectobacterium* spp. In our study, considering that *Xanthomonas* is a hemibiotrophic pathogen, initially forming an association with living cells of the host, thereafter killing the plant cell to use the nutrients from this process (Chan & Goodwin, 1999), the use of an inducer capable of stimulate both pathways, as we observed for EPS, can be an attractive way for control this bacterial disease.

Also, the obtained results suggest that EPS is able to induce resistance on tomato plants by pre-conditioning them against the pathogen, because plants previously treated with the biopolymer had increased defense mechanisms only after bacterial challenge (priming effect) (Cools & Ishii, 2002; Conrath et al., 2002). In a different way, we observed that the application of ASM increased the content of phenolic acids in tomato plants before the arrival of the pathogen, which could generate costs for the plant at the end of the process. Some studies indicated that ASM, depending of the dose, number of applications, plant growth stage and pathogen pressure, may cause a reduction in the yield, and a delaying on fruit maturity, indicating the efforts of the cells to increase proteins related to defense rather than those related to cell proliferation (Suzuki et al., 2006; Walters et al., 2013).

5 Conclusions

The findings indicate that the EPS of *L. plantarum* were able to decrease the severity of tomato bacterial spot by pre-conditioning the plant, and by increasing the compounds and enzymes related to the plant defense from different pathways. The results provide a basis for further studies in a way that the *Lactobacillus* EPS can be used in agriculture in a few years as an option for the disease control, especially in organic farming.

Acknowledgements

The authors acknowledge the Coordination of Development of High Education Personnel (CAPES) for awarding fellowships, Sakata for the bacterial isolate, and to Dr. Alice M. Quezado-Duval, EMBRAPA, for the identification of the pathogen.

References

Abbasi, P. A., Khabbaz, S. E., Weselowski, B., & Zhang, L. (2015). Occurrence of copper-resistant strains and a shift in *Xanthomonas* spp. causing tomato bacterial spot in Ontario. *Can J Microbiol., 61*(10), 753-61. https://dx.doi/10.1139/cjm-2015-0228

Ai, L., Zhang, H., Guo, B., Chen, W., Wu, Z., & Wu, Y. (2008). Preparation, partial characterization and bioactivity of exopolysaccharides from *Lactobacillus casei* LC2W. *Carbohyd Polym., 74*, 353-357. https://dx.doi/10.1016/j.carbpol.2008.03.004

Antoniazzi, N., Deschamps, C., & Bach, E. E. (2008). Effect of xanthan gum and allicin as elicitors against *Bipolaris sorokiniana* on barley in field experiments. *J Plant Dis Protec., 115*, 104-107 https://dx.doi/10.1007/BF03356248

APEDA (Agricultural and Processed Food Products Export Development Authority). (2012-2014). *Issue of Phytosanitary Certificates Mandatory for Export of Horticultural produts to EU*. Ministry of Commerce and Industry, Govenment of India, New Delhi. Retrieved from http//www.apeda.gov.in/apedawebsite/Announcements/Ministry_of_Agriculture.html

Araújo, E. R., Ferreira, M. A. S. V., & Quezado-Duval, A. M. (2013). Specific primers for *Xanthomonas vesicatoria*, a tomato bacterial spot causal agent. *Eur J Plant Pathol., 137*, 5-9. https://dx.doi/10.1007/s10658-013-0225-4

Axelrod, B. C., Cheesbrough, T. M., & Laasko, S. L. (1981). Lipoxygenase from soybean. *Method Enzymol., 71*, 441-451.

Bach, E. E. (1997). *Morphological distinction and isozyme of Bipolaris spp. and Drechslera tritici-repentis in wheat: Biochemical aspects in the interactions and resistance induction* (Dissertation, University of Sao Paulo).

Bach, E. E., Barros, B. C., & Kimati, H. (2003). Induced resistance against *Bipolaris bicolor*, *Bipolaris sorokiniana* and *Drechslera tritici-repentis* in wheat leaves by xantham gum and heat-inactivate conidia suspension. *J Phytopathol., 151*, 411-418. https://dx.doi/10.1046/j.1439-0434.2003.00742.x

Baque, M. A., Mho, S. H., Lee, E. J., Zhong, J. J., & Paek, K. Y. (2012). Production of biomass and useful compounds from adventitious roots of high-value added medicinal plants using bioreactor. *Biotechnol Adv., 30*, 1255-1267. https://dx.doi/10.1016/j.biotechadv.2011.11.004

Camelini, C. M., Gomes, A., Cardozo, F. T. G. S., Simões, C. M. O., Rossi, M. J., Giachini, A. J., … De Mendonça, M. M. (2013). Production of polysaccharide from *Agaricus subrufescens* Peck on solid-state fermentation. *Appl Microbiol Biot., 97*, 123-133. https://dx.doi/10.1007/s00253-012-4281-z

Carlberg, C., & Mannervik, B. (1985). Glutathione reductase. *Method Enzymol., 113*, 488-495.

Chan, J. W. Y. F., & Goodwin, P. H. (1999). The molecular genetics of virulence of *Xanthomonas campestris*. *Biotechnol Adv., 17*, 489-508. https://dx.doi/10.1016/S0734-9750(99)00025-7

Choudhary, D. K. (2011). Plant growth-promotion (PGP) activities and molecular characterization of rhizobacterial strains isolated from soybean (*Glycine max* L. Merril) plants against charcoal rot pathogen, *Macrophomina phaseolina. Biotechnol Lett., 33*, 2287-2295. https://dx.doi/10.1007/s10529-011-0699-0

Conrath, U., Pieterse, C. M. J., & Mauch-Mani, B. (2002). Priming in plant-pathogen interactions. *Trends Plant Sci., 7*, 210-216. https://dx.doi/10.1016/S1360-1385(02)02244-6

Cools, H. J., & Ishii, H. (2002). Pre-treatment of cucumber plants with acibenzolar-S-methyl systemically primes a phenylalanine ammonia lyase gene (PAL1) for enhanced expression upon attack with a pathogenic fungus. *Physiol Mol Plant P., 61*, 273-282. https://dx.doi/10.1006/pmpp.2003.0439

Coqueiro, D. S., & Di Piero, R. M. (2011). Antibiotic activity against *Xanthomonas gardneri* and protection of tomato plants by chitosan. *J Plant Pathol., 93*, 337-344. https://dx.doi/10.4454/jpp.v93i2.1188

Davidsson, P. R., Kariola, T., Niemi, O., & Palva, E. T. (2013). Pathogenicity of and plant immunity to soft rot pectobacteria. *Front Plant Sci., 4*, 1-13. https://dx.doi/10.3389/fpls.2013.00191

Dimitri, C., & Oberholtzer, L. (2009). *Marketing US Organic Foods: Recent Trends from Farms to Consumers*. ERS, Economic Research Service/USDA.

Drazkiewicz, M., Skórzyńska-Polit, E., & Krupa, Z. (2003). Response of the ascorbate-glutathione cycle to excess copper in *Arabidopsis thaliana* (L.). *Plant Sci., 164*, 195-202. https://dx.doi/10.1016/S0168-9452(02)00383-7

Ebrahim, S., Usha, K., & Singh, B. (2011). Pathogenesis related (PR) proteins in plant defense mechanism. *Science against microbial pathogens*: *Communicating current research and technological advances* (pp. 1043-1054).

FAO (Food and Agriculture Organization of the United Nations). (2014). *FAOSTAT*. Retrieved January 10, 2014, from http://faostat.fao.org/site/567/default.aspx#ancor

Ferraz, H. G. M., Resende, S., Silveira, P. R., Andrade, C. C. L., Milagres, E. A., Oliveira, J. R., & Rodrigues, F A. (2014). Rhizobacteria induces resistance against *Fusarium* wilt of tomato by increasing the activity of defense enzymes. *Bragantia, 73*, 274-283. https://dx.doi/10.1590/1678-4499.0124

Ferreira, D. F. (2003). *SISVAR, 4.3* (Build 45). Federal University of Lavras, Minas Gerais, Brazil.

Feussner, I., & Wasternack, C. (2002). The lipoxygenase pathway. *Annu Rev Plant Biol., 53*, 275-297. https://dx.doi/10.1146/annurev.arplant.53.100301.135248

Fu, Z. Q., & Dong, X. (2013). Systemic acquired resistance: turning local infection into global defense. *Annu Rev Plant Biol., 64*, 839-863. https://dx.doi/10.1146/annurev-arplant-042811-105606

Gao, Q. M., Zhu, S., Kachroo, P., & Kachroo, A. (2015). Signal regulators of systemic acquired resistance. *Front Plant Sci., l*, 1-12. https://dx.doi/10.3389/fpls.2015.00228

Ge, Y., Bi, Y., & Guest, D. I. (2013). Physiological and molecular plant pathology defence responses in leaves of resistant and susceptible melon (*Cucumis melo* L.) cultivars infected with *Colletotrichum lagenarium*. *Physiol Mol Plant P., 81*, 13-21. https://dx.doi/10.1016/j.pmpp.2012.09.002

Gerasimova, N. G., Pridvorova, S. M., & Ozeretskovskaya, O. L. (2005). Role of l-phelylalanine ammonia lyase in the induced resistence and susceptibily of potato plants. *Appl. Biochem. Micro., 41*, 103-105 https://dx.doi/10.1007/s10438-005-0019-3

Guzzo, S. D., Back, E. A., Martins, E. M., & Moraes, W. B. (1993). Crude expolysaccarides (EPS) from *Xanthomonas campestris* pv. *maniohtis*, *Xanthomonas campestris* pv. *campestris* and commercial xanthan gumas inducers of protection in coffee plants against *Hemileia vastatrix*. *J Phytopathol., 139*, 119-128.

Huang, T. K., & Mcdonald, K. A. (2009). Bioreactor engineering for recombinant protein production in plant cell suspension cultures. *Biochem Eng J., 45*, 168-184. https://dx.doi/10.1016/j.bej.2009.02.008

Jones, J. B., Lacy, G. H., Bouzar, H., Stall, R. E., & Schaad, N. W. (2004). Reclassification of the Xanthomonads associated with bacterial spot disease of tomato and pepper. *Syst Appl Microbiol., 27*, 755-762. https://dx.doi/10.1078/0723202042369884

Kacuráková, M., & Wilson, R. H. (2001). Developments in mid-infrared FT-IR spectroscopy of selected carbohydrates. *Carbohyd Polym., 44*, 291-303. https://dx.doi/10.1016/S0144-8617(00)00151-X

Kofalvi, S. A., & Nassuth, A. (1995). Influence of wheat streak mosaic virus infection on phenylpropanoid metabolism and the accumulation of phenolics and lignin in wheat. *Physiol Mol Plant P., 47*, 365-377. https://dx.doi/10.1006/pmpp.1995.1065

Latha, P., Anand, T., Rappathi, N., Prakasam, V., & Samiyappan, R. (2009). Antimicrobial activity of plant extracts and induction of systemic resistance in tomato plants by mixtures of PGPR strains and Zimmu leaf extract against *Alternaria solani*. *Biol Control., 50*, 85-93. https://dx.doi/10.1016/j.biocontrol.2009.03.002

Laws, A., Gu, Y., & Marshall, V. (2001). Biosynthesis, characterisation, and design of bacterial exopolysaccharides from lactic acid bacteria. *Biotechnol Adv., 19*, 597-625. https://dx.doi/10.1016/S0734-9750(01)00084-2

Leemhuis, H., Pijning, T., Dobruchowskaa, J. M., Van Leeuwena, S. S., Kralj, S., & Dijkstrab, B. W. (2013). Three-dimensional structures, reactions, mechanism, α-glucan analysis and their implications in biotechnology and food applications. *J Biotechnol., 163*, 250-272. https://dx.doi/10.1007/s00253-012-3943-1

Li, S., Huang, R., Shah, N. P., Tao, X., Xiong, Y., & Wei, H. (2014). Antioxidant and antibacterial activities of exopolysaccharides from *Bifidobacterium bifidum* WBIN03 and *Lactobacillus plantarum* R315. *J Dairy Sci., 97*, 7334-43. https://dx.doi/10.3168/jds.2014-7912

Luiz, C., Felipini, R. B., Costa, M. E. B., & Di Piero, R. M. (2012). Polysaccharides from aloe barbadensis reduce the severity of bacterial spot and activate disease-related proteins in tomato. *J Plant Pathol., 94*, 387-393. https://dx.doi/10.4454/JPP.FA.2012.046

Luiz, C., Rocha Neto, A. C., & Di Piero, R. M. (2015). Resistance to *Xanthomonas gardneri* in tomato leaves induced bypolysaccharides from plant or microbial origin. *J Plant Pathol., 97*, 119-127. https://dx.doi/10.4454/JPP.V97I1.029

Mahapata, S., & Banerjee, D. (2013). Fungal Exopolysaccharide: Production, Composition and Applications. *Microbiology Insights, 6*, 1-16. https://dx.doi/10.4137/MBI.S10957

Mccue, P., Zheng, Z., Pinkham, J. L., & Shetty, K. (2000). A model for enhanced pea seedling vigour following low pH and salicylic acid treatments. *Biochemistry, 35*, 603-613. https://dx.doi/10.1016/S0032-9592 (99)00111-9

Mello, S. C. M., Takatsu, A., & Lopes, C. A. (1997). Diagrammatic scale for assessment of tomato bacterial spot. *Braz Phytopathol., 22*, 447-448.

Notararigo, S., Nacher-Vazquez, M., Ibarburu, I., Werning, M. L., De Palencia, P. F., Duenas, M. T., … Prieto, A. (2013). Comparative analysis of production and purification of homo- and hetero-polysaccharides produced by lactic acid bacteria. *Carbohyd Polym., 93*, 57-64. https://dx.doi/10.1016/j.carbpol.2012.05.016

Obradovic, A., Jones, J. B., Momol, M. T., Balogh, B., & Olson, S. M. (2004). Management of tomato bacterial spot in the field by foliar application of bacteriophages and SAR inducers. *Plant Dis., 88*, 736-740. https://dx.doi/10.1094/PDIS.2004.88.7.736

Oostendorp, M., Kunz, W., Dietrich, B., & Staub, T. (2001). Induced disease resistance in plants by chemicals. *Eur J Plant Pathol., 107*, 19-28. https://dx.doi/10.1023/A:1008760518772

Potnis, N., Timilsina, S., Strayer, A., Shantharaj, D., Barak, J. D., Paret, M. L., … Jones, J. B. (2015). Bacterial spot of tomato and pepper: Diverse *Xanthomonas* species with a wide variety of virulence factors posing a worldwide challenge. *Mol. Plant Pathol., 16*, 907-920. https://dx.doi/10.1111/mpp.12244

Quezado-Duval, A. M., Leite Junior, R. P., Lopes, C. A., Lima, M. F., & Camargo, L. E. A. (2005). Diversity of *Xanthomonas* spp. associated with bacterial spot of processing tomatoes in Brazil. *Acta Hortic., 695*, 101-108. https://dx.doi/10.17660/ActaHortic.2005.695.11

Raiola, A., Rigano, M. M., Calafiore, R., Frusciante, L., & Barone, A. (2014). Enhancing the human-promoting effects of tomato fruit for bofortified food. *Mediators Inflamm*, 1-16. https://dx.doi/10.1155/2014/139873

Roselló, G., Bonaterra, A., Francés, J., Montesinos, L., Badosa, E., & Montesinos, E. (2013). Biological control of fire blight of apple and pear with antagonistic *Lactobacillus plantarum*. *Eur J Plant Pathol., 137*, 621-633. https://dx.doi/10.1007/s10658-013-0275-7

Salazar, S. M., Grellet, C. F., Chalfoun, N. R., Castagnaro, A. P., & Díaz Ricci, J. C. (2013). Avirulent strain of *Colletotrichum* induces a systemic resistance in strawberry. *Eur J Plant Pathol., 135*, 877-888. https://dx.doi/10.1007/s10658-012-0134-y

Seo, B., Bajpai, V. K., Rather, I. A., & Park, Y. (2015). Partially purified exopolysaccharide from *Lactobacillus plantarum* YML009 with total phenolic content, antioxidant and free radical scavenging efficacy. *Indian J Pharm Educ Res., 49*, 282-292. https://dx.doi/10.5530/ijper.49.4.6

Shan, L., He, P., & Sheen, J. (2007). Endless Hide-and-Seek: Dynamic co-evolution in plant-bacterium warfare. *J. Integr. Plant Biol., 49*, 105-111. https://dx.doi/10.1111/j.1744-7909.2006.00409.x

Sharma, P., & Dubey, R. S. (2005). Drought induces oxidative stress and enhances the activities of antioxidant enzymes in growing rice seedlings. *Plant Growth Regul., 46*, 209-221 https://dx.doi/10.1007/s10725-005-0002-2

Sharma, P., Jha, A. B., Dubey, R. S., & Pessarakli, M. (2012). Reactive oxygen species, oxidative damage, and antioxidative defense mechanism in plants under stressful conditions. *J. Bot.*, 1-26. https://dx.doi/10.1155/2012/217037

Silva, C. M. M. S., & Fay, E. F. (2006). *Environmental impact of the fungicide metalaxyl* (p. 96). Jaguariuna: Embrapa Environment.

Suzuki, K., Nishiuchi, T., Nakayama, Y., Ito, M., & Shinshi, H. (2006). Elicitor-induced down-regulation of cell cycle-related genes in tobacco cells. *Plant Cell Environ., 29*, 183-191. https://dx.doi/10.1111/j.1365-3040.2005.01411.x

Tian, S., Wan, Y., Qin, G., & Xu, Y. (2006). Induction of defense responses against *Alternaria* rot by different elicitors in harvested pear fruit. *App. Microbiol. Biot., 70*, 729-734. https://dx.doi/10.1007/s00253-005-0125-4

Vaughn, K. C., & Duke, S. O. (1984). Function of polyphenol oxidase in higher plants. *Physiol. Plantarum., 60*, 106-112. https://dx.doi/10.1111/j.1399-3054.1984.tb04258.x

Walters, D. R., Ratsep, J., & Havis, N. D. (2013). Controlling crop diseases using induced resistance: Challenges for the future. *J. Exp. Bot., 64*, 1263-1280. https://dx.doi/10.1093/jxb/ert026

Wang, Y., Ahmed, Z., Feng, W., Li, C., & Song, S. (2008). Physicochemical properties of exopolysaccharide produced by *Lactobacillus kefiranofaciens* ZW3 isolated from Tibet kefir. *Int. J. Biol. Macromol., 43*, 283-288. https://dx.doi/10.1016/j.ijbiomac.2008.06.011

Zhang, J., & Zhou, J. M. (2010) Plant immunity triggered by microbial molecular signatures. *Mol Plant., 5,* 783-793. https://dx.doi/10.1093/mp/ssq035

Abbreviations

EPS, exopolysaccharides; PAL, phenylalanine ammonia-lyase; GR, glutathione reductase; LOX, lipoxygenase; GSH, reduced glutathione; GSSG, oxidized glutathione; JA, jasmonic acid; MAMPs, microbe-associated molecular patterns; ASM, Acibenzolar-S-Methyl; HPLC, High performance liquid chromatography; FTIR, Fourier transform infrared spectroscopy; OD, optical density; CFU, colony-forming units; DAT, day after treatmens; DAI, day after inoculation.

19

Auxins Regulations of Branched Spike Development and Expression of *TFL*, a *LEAFY*-Like Gene in Branched Spike Wheat (*Triticum aestivum*)

Wang Yue[1,2], Sun Fulai[3], Gao Qingrong[1,2], Zhang Yanxia[3], Wang Nan[3] & Zhang Weidong[1,2]

[1] State Key Laboratory of Crop Sciences, Shandong Agricultural University, Taian City, Shandong Province, China

[2] Genetic and Breeding Department, Agronomy College, Shandong Agricultural University, Taian City, Shandong Province, China

[3] Bureau of Agriculture, Binzhou City, Shandong Province, China

Correspondence: Sun Fulai, Bureau of Agriculture, Binzhou City, 256600, Shandong Province, China. E-mail: zgbzsfl@163.com

Zhang Weidong, Genetic and Breeding Department, Agronomy College, Shandong Agricultural University, Daizong Street 61, Taian City, 271018, Shandong Province, China.
E-mail: zhangwd@sdau.edu.cn

The research is financed by Key Program for Natural Sciences Fund in Shandong Province in China (ZR2013CZ001).

Abstract

Branched spike wheat is a hexaploid germplasm with branched rachis on its main rachises, and the crucial period for branched rachises occurrence and development is just after the two ridges stage of shoot apex. Natural [indole-3-acetic acid (IAA), indole-3butyric acid (IBA)] and synthetic [(1-naphthaleneacetic acid (NAA), 2,4-Dichlorophenoxyacetic acid (2,4-D)] auxins were applied at this period to investigate the spike traits, seedling growth and photosynthesis related characters and expression of a putative homologue of the *LEAFY* in branched spike wheat. The four types of experienced auxins induced similar effects on these foresaid characters, although the impact extents were different among the auxins treatments. More branched rachis, spikelets, fertile florets and longer branched rachis were obtained in plants with IAA and IBA at 0.1 mM or NAA and 2,4-D at 1.0mM than those plants with no auxin treated. Auxin treatments also increased fresh and dry mass, photosynthetic pigment and parameters. *TFL*, a *LEAFY*-like gene was cloned in branched spike wheat and *TFL* mRNA expression was quantified using real-time reverse transcriptase-PCR. Application of the auxins accelerated the rise in *TFL* expression during the periods of branched rachises occurrence and extension. The data supports the hypothesis that auxins play a central role in the regulation branched spike development and *TFL* might correlate with the development of branched rachises in branched spike wheat.

Keywords: auxins, branched rachis, *LEAFY*, photosynthetic parameters, *TFL*, *Triticum aestivum*

1. Introduction

Branched spike wheat is a special hexaploid wheat germplasm which has branches (branched rachises) on its main rachis (inflorescence axis) and bears an overabundance of spikelets and grains on a spike. The grain number in a branched spike can reach 70-130 while that in a single normal spike generally is 35-70. Branched spike wheat has great application potentials in wheat production areas with high speed of grain filling or long term of grain filling period.

A number of studies have been conducted on field performances of the branched spike wheat. The supernumerary spikelet character in this type of spike was affected by light, temperature, and nutrients conditions (Koric, 1975; Peanell & Halloran, 1983), and environmental factors might play a minor effect on the character expression during spike differentiation stage at the same eco-region (Sun et al., 2000).

Traditionally, plant hormones and synthetic plant growth regulators are used as valuable research tools to elucidated physiological responses of plant or probe biochemical control mechanisms. In many bioassays, it has been shown that auxins play an important role in plant growth and development (Cooke et al., 2002). Indole-3-acetic acid (IAA) and indole-3-butyric acid (IBA) are naturally-occurring, plant hormone of the auxin class. IAA is the most common in the auxins, and has been the subject of extensive studies by plant physiologists (Simen & Petrášek, 2011). IBA is a plant hormone in the auxin family and is an ingredient in many commercial horticultural plant rooting products (Ludwig-Müller, 2000). 1-naphthaleneacetic acid (NAA) and 2,4-Dichlorophenoxyacetic acid (2,4-D) are synthetic plant hormone in the auxin family and also used for plant tissue culture. NAA is a rooting agent and used for the vegetative propagation of plants from stem and leaf cutting (Naduvilpurakkal et al., 2014). 2,4-D is one of the most widely used herbicides in the world. 2,4-D and as a synthetic auxin it is often used in laboratories for plant research and as a supplement in plant cell culture media such as MS medium (Andrew & Reade, 2010).

Our previous studies indicated that as the young spike developed, endogenous IAA contents in branched spike wheat were much higher than those in normal spike lines. Meanwhile, the relative expression of *TalAR3*, an auxin synthesis related gene, in branched spike wheat line was also significantly higher than those in normal spike lines. Higher IAA content might involve in the formation and growth of branched rachis in branched spike wheat (Data published lately).

LEAFY, or *FLORICOULA* (*FLO*)/*LEAFY* (*LFY*) genes were characterized initially as floral meristem genes in *Antirrhinum* and *Arabidopsis* (Coen et al., 1990; Weigel et al., 1992), which act as genetic switches during the choice of floral versus shoot fates (Yanofsky, 1995). Expression of *FLO/LFY* has been observed in young vegetative shoot apices and leaf primordia and might involve in the inflorescence architecture which distinct controlling the initiation of floral meristems in many plant species, including *Oryza sativa* (Kyozuka et al., 1998) and *Z. mays* (Bomblies et al., 2003).

The objective of the present study was to compare the effect of natural and synthetic auxins on the branched spike traits, growth and photosynthesis related characters, and the expression pattern of *TFL* in the shoot apex of plant was also determined in responses to auxins. The obtained results may be important for elucidation of the plant hormone role in branched spike development in wheat.

2. Materials and Methods

2.1 Plant Materials, Growth and Treatments

Fen 33, a released branched spike wheat cultivar in HuangHuan wheat zone in China, was grown in Agronomy Experimental Station of Shandong Agriculture University, Taian City, Shandong Province, China, from 2012 to 2013 growing season. The experiment plot was about 8.0 m^2 and the wheat plants were thinly sowed at 6 cm plant spacing and 22 cm row spacing on 8th October 2012. Cultivation management was similar with common practice at local place.

Sowed wheat lines turned green at mid February in 2013. The young shoots were tripped out and observed under microscopy to determine the development stages of shoot apexes. Auxins dissolved in 50% ethanol were applied at experimental concentrations, 0.1 mM IBA, 0.1 mM IAA, 1.0 mM NAA and 1.0 mM 2,4-D (Sigma-Aldrich Co., USA), respectively. An equal amount of 50% ethanol was added to the control. Auxins treatments were performed three times on 3, 6 and 9 days after the two ridges stage of shoot apex. Plants were foliar sprayed until considerable run-off on the leaf surface occurred at early morning. The experiments were carried out with complete random block design with four replicates.

2.2 Determination of Pike Characters

After wheat maturation at June, twenty spikes of main shoots in each replicate plot were randomly taken out to investigate the number and the length of longest branched rachis, the number of spikelet and fertile floret. Numbers of branched rachis and spikelet were counted. The length of branched rachies was measured by using a meter scale. Fertile floret was represented as the grain number in a spike which was counted at the same time.

2.3 Determination of Growth Traits and Photosynthesis Related Characters

After 30 d of first auxins treatment, the plants were removed from the field along with the soils and were dipped in a bucket filled with water. The plants were moved smoothly to remove the adhering soils particles were weighted for fresh mass, and then placed in an oven run at 80 °C for 24 h. These dried plants were weighed to record the plant dry mass.

Chlorophylls a, b and carotenoids contents (mg g^{-1} FW) were detected according to the methods given by Arnon (1949). Clean leaf materials were homogenized with 80% acetone and centrifuged; the optical density of the acetone extract was measured at 663, 645 and 470 nm using a UV-160A UV Visible Recording Spectrometer, Shimadzu, Japan.

The measured photosynthetic parameters included photosynthesis rate (Photo), intercellular concentration of carbon oxide (Ci), stomatal conductance (Cond) and transpiration rate (Tr). These parameters were detested simultaneously in an open-type leaf pathway by LI-6400 portable photosynthetic apparatus (LI-COR Company, USA). The detesting condition: T = 20±1 °C, air carbon concentration = 380±5 μl (Lu and Gao 2003).

2.4 Isolation of LFY/FLO Homologue TFL in Branched Spike Wheat

The strategy of direct sequencing of a reverse transcription PCR product was used to obtain the putative homologue of *LFY/FLO*, using degenerate primers and cDNA from shoot apical meristems of branched spike wheat plants expected to express a *LFY/FLO* homologue. Total RNA was isolated from young shoot apex tissues using TRIzol reagent (Invitrogen) and RT-PCR was performed using the degenerate forward primer 5′CGC(G)GAGCTC(G)GACGACATGA3′ and reverse primer 5′GCACTGCTCGTAG(C)AGA(G)TGGA3′ to obtain the middle region of *TFL*. The following thermocycling conditions were employed: initial denaturation at 94 °C for 3 min; 30 cycles of 94 °C for 30 s, 59 °C for 45 s, and 72 °C for 1 min; final extension at 72 °C for 10 min. The amplified products were separated on a 1.5% agarose gel, visualised and photographed.

The amplified cDNA fragment was independently cloned at least twice into pGM-T Easy vector (Promega) and sequenced. After sequencing, the sequence was submitted to GenBank to identify similar genes. To obtain the full length cDNA sequence of *TFL*, a set of specific primers were designed for the 5′ and 3′ RACEs (Rapid Amplification of cDNA Ends) by using the SMARTTM RACE cDNA Amplification Kit (Clontech, Palo Alto, CA). These primers were:

5′Race: PF: 5′TTTCTTGGCCTTTCTCGCCA3′;

3′Race: PR: 5′AAGAAGAACGGGCTGGACTA3′;

The amplified sequences were cloned and sequenced. The full length cDNA containing ORF for *TFL* was obtained by PCR with forward primer (5′ATGGATCCATGGATCGCCACGACGCCT3′) and reverse primer (5′TTCTAGCCTGTTGCCCGGAGGATCC3′). The primers contained the generated BamHI recognition site (GGATCC) to facilitate the cloning of the cDNAs.

2.5 Phylogenetic Analysis

Phylogenetic comparisons of amino acid sequences of different *LFY/FLO* homologues were obtained from GenBank (http://www.ncbi.nlm.nih.gov/) and aligned with ClustalW mega (http://www.ebi.ac.uk/Tools/msa/clustalo/). Phylogenetic trees based on the complete sequences were generated using MEGA4 and constructed by the neighbor joining (NJ) method. Bootstrap values were derived from 1000 replicate runs.

2.6 Real-Time RT-PCR

For experiments on apices total RNA was extracted from the apical part of the plants. The samples were taken before the beginning of the auxins treatment and then every 10 d until the 30th d after the beginning of the auxins first treatment. At first sampling the apical part of the plants included apex and a little of non-removable young shoots, and afterwards, only included the young spike. RNA was isolated TRIzol from Invitrogen following manufacturer's protocol. Each RNA sample was treated with RNase-Free. DNase (Promega) following manufacturer's protocol in an effort to remove any residual genomic DNA (gDNA). DNase treated RNA was subjected to reverse transcriptase reactions using oligo-dT primer and PrimeScriptTM Reverse Transcriptase (Takara) according to manufacturer's protocol. The gene-specific primers used in RT-PCR were 5′AGAACGACTGCGACGACGA3′ and 5′CCGCATCTTGGGCTTGTTGA3′. As a control, the cDNA sequence of the actin gene was amplified by using the two primers 5′CACGGCATCGTAAGCAACTG3′ and 5′TCCTTCGTAAATGGGCACGGT3′. The following thermocycling conditions were employed: initial denaturation at 94 °C for 3 min; 30 cycles of 94 °C for 30 s, 57 °C for 45 s, and 72 °C for 1 min; final extension at 72 °C for 7 min. The amplified products were separated on a 1.5% agarose gel, visualised and photographed. The obtained CT values were analyzed by averaging three independently calculated normalized expression values for each sample. PCR products were sequenced to ensure amplification products for the purpose of gene fragment. Expression values are given as the mean of the normalized expression values of the triplicates, calculated according to $2^{-\Delta\Delta CT}$ method (Livak & Schmittgen, 2001).

2.7 Statistical Analysis

Branched spike wheat with auxins treatments were arranged in a completely randomized design with four treatments. Analysis of variance was performed using the SPSS software package. Analysis of variance (ANOVA) was performed on the data to determine the least significant difference (LSD) among treatment at P < 0.05 and Duncan's multiple range tests were applied for comparing the means.

3. Results

The two ridges stage of shoot apex in branched spike wheat in the field ended at about February 20th, in the spring of 2013. The shoot apex entered the glume differentiation stage and the stage lasted approximately 11 days. Afterwards, the floret differentiation stage and stamen and pistil differentiation stage lasted about 12 days and 7 days respectively. The plants entered anther connective stage at about March 22rd. There were about 30 days from the end of two ridges stage to pistil and stamen differentiation stage in the spike lines.

3.1 Effect of Auxins on Branched Spike Characters

Auxins altered notably the characters of branched rachis. The experiment showed that IAA and IBA at 0.1 mM, and PAA and NAA at 1.0 mM were effective in inducing inerratic branched rachis (Figure 1). IBA at 0.1 mM induced the highest increase in the number of branched rachis by 82.6%, 0.1 mM IAA by 78.3%, 1.0 mM NAA by 46.7% and 1.0 mM 2,4-D by 57.6% in comparison with the plants with no auxin treatment. The highest length of branched rachis (19.7mm) was observed in plants treated with IBA. The length of the longest branched rachis were 18.9 mm, 18.4 mm and 14.5 for IAA, NAA and 2,4-D, respectively. For the number of spikelet in a whole spike, IBA and IAA reached 74.9 and 73.6, respectively. The number of spikelet in whole spike treated with 1.0 mM NAA and 2,4-D reached 57.2 and 53.4, respectively. The fertile floret per spike in wheat treated with 0.1 mM IBA, 0.1mM IAA, 1.0 mM NAA and 1.0 mM 2, 4-D reached 90.6, 87.6, 84.9 and 75.4 respectively.

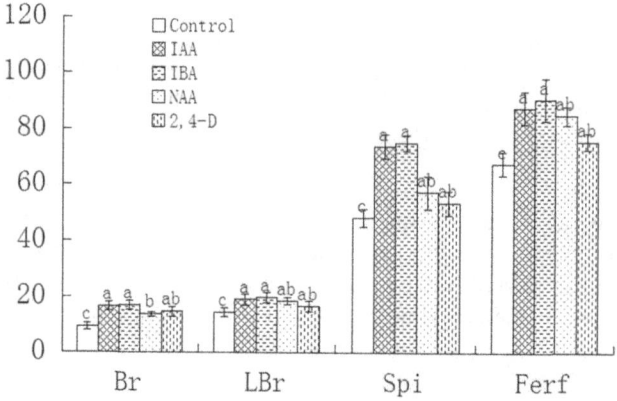

Figure 1. The effect of auxins (IAA, IBA, NAA, 2,4-D) on the number of branched rachies (Br), the length of the longest branched rachies (LBr) (mm), spikelet number (Spi) and fertile floret number (Ferf) in a branched wheat inflorescence after harvest

Note. Data are the means of four independent experiments ± SD. Treatment with at least one letter the same are not significantly different according to Duncan's test.

3.2 Growth Traits and Pigment Contents

Branched wheat treated with IBA at 0.1 mM was characterized by the highest (23.0%) increase in plant fresh after 30 d of first auxin treatment (Table 1). Other auxins showed weaker biological activity in fresh weight with a 14.5% increase in fresh weight level measured in response to 0.1 mM IAA, 4.9% and a 3.3% in the case of 1.0 mM NAA and 1.0 mM 2,4-D, respectively, in comparison with the control. The dry mass of wheat treated with auxins increased by 2.9-24.2% than the control, and the increase order in these auxins was IBA > IAA > NAA > 2,4-D.

Table 1. The effect of auxins (IAA, IBA, NAA, 2,4-D)on growth traits, pigment contents, and photosynthesis related characters in branched spike wheat after 30 days of first treatment.Data are the means of four independent experiments ± SD

	Control	IAA	IBA	NAA	2,4-D
Plant fresh (g)	10.98±1.03	12.57±1.36	13.5±0.98	11.52±0.087	11.34±0.79
Dry mass (g)	2.07±0.31	2.43±0.36	2.57±0.298	2.31±0.234	2.13±0.198
Chloro a (mg·g^{-1} fresh weight)	1.365±0.045	1.47±0.069	1.489±0.036	1.419±0.047	1.398±0.033
Chloro b (mg·g^{-1} fresh weight)	0.357±0.039	0.472±0.052	0.532±0.047	0.387±0.034	0.361±0.059
Carotenoids (mg·g^{-1} fresh weight)	0.462±0.015	0.483±0.021	0.503±0.014	0.493±0.028	0.481±0.015
Photosynthesis rate (μmol CO_2 m^{-2}·s^{-1})	32.7±4.26	34.21±3.69	37.06±2.87	38.29±4.35	34.2±3.96
Intercellular concentration of carbon oxide (μmol CO^2·mol^{-1})	240.6±28.9	298.6±23.6	327.5±28.7	272.3±15.6	268.1±32.1
Stamatal conductance (mmol H_2O m^{-2}·s^{-1})	0.32±0.049	0.35±0.053	0.39±0.035	0.37±0.028	0.34±0.041
Transpiration rate (mmol H_2O m^{-2}·s^{-1})	8.7±0.93	9.31±1.32	9.52±1.21	9.28±1.42	8.86±1.09

IBA applied at 0.1 mM had the most stimulatory effect on chlorophyll a, chlorophyll b and carotenoid accumulation after 30 d of first treatment. Other auxins were characterized by lower stimulatory effects on photosynthetic pigment levels in branched spike wheat. IAA at a concentration of 0.1 mM stimulated chlorophyll a accumulation by 7.69%, chlorophyll b by 32.2% and carotenoid by 4.54% after 30 d of first auxin treatment. The 2.41-3.95% increase in chlorophyll a level and 1.12-8.40% increase in chlorophyll b were noted in response to 1.0 mM NAA and 1.0 m 2,4-D. These auxins also stimulated the carotenoid content by 4.11-8.87% after 30 d of first treatment.

3.3 Photosynthesis Rate, Intercellular Concentration of Carbon Oxide, Stamatal Conductance and Transpiration Rate Determination

Wheat in the presence of 0.1 mM IBA had 13.3%, 36.1%, 21.9% and 9.42% more photosynthesis rate, intercellular concentration of carbon oxide, stamatal conductance and transpiration rate, respectively, than the control after 30 d of first treatment (Table 1). A significant increase in the photosynthetic parameters (4.62% in the case of photosynthesis rate, 24.1% in the case of intercellular concentration of carbon oxide, 9.36% in the case of stamatal conductance and 7.01% in the case of transpiration rate) was also observed with 0.1 mM IAA application. In addition, exposure of branched wheat to 1.0 mM NAA or 2,4-D caused a weaker, but statistically significant increase in the photosynthesis rate (4.58-17.1%), intercellular concentration of carbon oxide (11.4-13.7%), stamatal conductance (6.25-15.6%)and transpiration rate (1.83-6.67%) after 30 d of first treatment.

3.4 Expression of TFL during Spike Development

The effect of auxins on the expression of the *TFL* gene was examined (Genbank accession number: KP408434). The coding region of *TFL* is 1,185 bp, and encodes a putative protein of 394 amino acids, which is 59.2% identical to *WFL* (GenBank accession No: AB231888), a *LFY/FLO* homologous gene from the species of common wheat (Shitsukawa et al., 2006). Comparison of an amino acid sequence alignment containing *TFL* and other *LFY/FLO* proteins showed the presence of several conserved regions. The putative amino acid sequence of *TFL* showed 64.9% homology with *LtLFY* of *Lolium temulentum* (AF321273), 59.9% with *RFL* of *Oryza Sativa* (AB005620), 54.7% with *LEAFY*-like of *B. distachyon* (XM_003580387); 52.9% with *zfl1* of *Zea mays* (AY179881), 53.4% with *LEAFY*-like of *S. italic* (XM_004976618), 52.9% with *LEAFY* like of *O. brachyantha* (XM_006652684), 50.7% with *LEAFY*-like of *S. bicolor* (XM_002446991) (Figures 2 and 3).

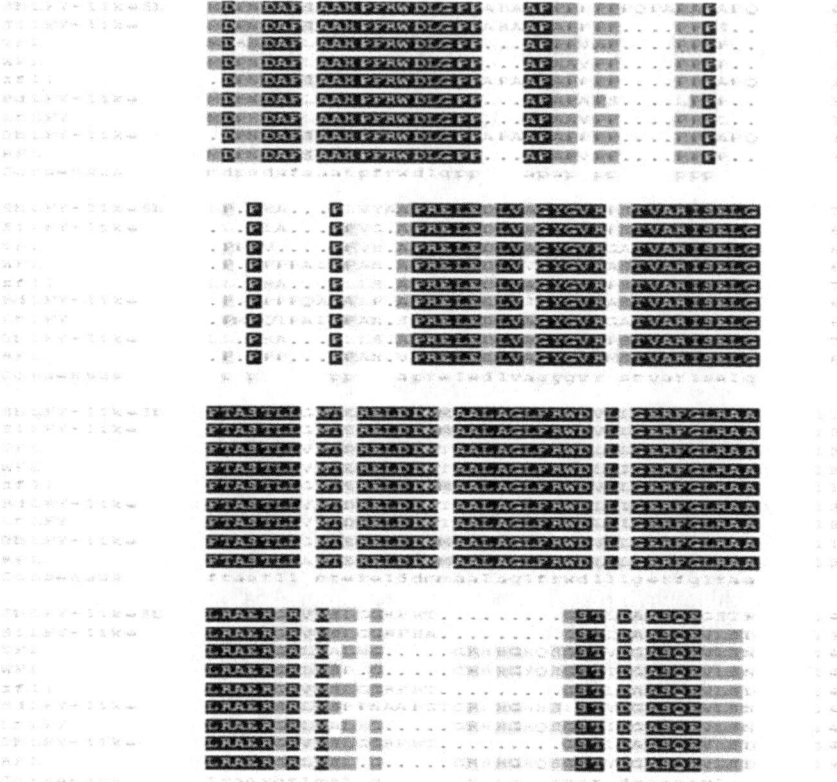

Figure 2. Comparison of amino acid sequences of *TFL* with representative *FLO/LFY*-like proteins

Note. Identical and conserved amino acids are shaded. Sequence segments notoverlapping with the *TFL* fragment were not included. *TFL* (KP408434), branched spike wheat; *WFLa* (AB231888), *Triticum aestivum*; *RFL* (AB005620), *Oryza sativa*; *LtLFY* (AF321273), *Lolium temulentum*; *zfl1* (AY179881), *Zea mays*; *SbLFY*-likeSB (XM_002446991), *Sorghum bicolor*; *ObLFY*-like (XM_006652684), *Oryza brachyantha*; *SiLFY*-like (XM_004976618), *Setaria italica*; *BdLFY*-like (XM_003580387), *Brachypodium distachyon*.

The abundance of transcripts of *TFL* in shoot tips was determined by using Q-PCR while the branched rachis was occurring and growing (Figure 4). In branched spike wheat with no auxin treatment, as spike development, the expression of *TFL* increased slowly at first treatment, reached the highest expression level and then decreased at 30 d. Similar changing patterns appeared under four auxins treatments, but auxins increased notably the expression of *TFL*. Compared with the control at the same time, the expression of *TFL* increased by 5.8-20.5%, 4.6-25.6% and 25-67.8% at 10 d, 20 d and 30 d respectively. The highest increase was IBA at 30d, and the lowest increased was NAA at 20 d.

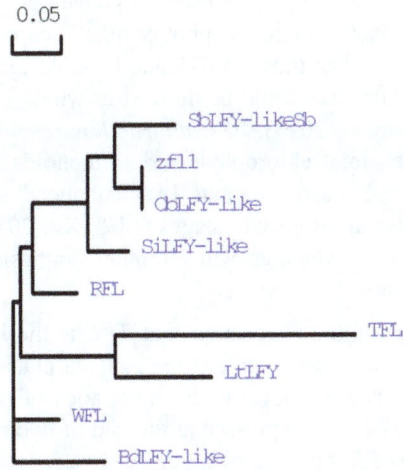

Figure 3. Neighbor-Joining phylogenetic tree of representative *LEAFY* homologues generated with 1,000 bootstrap replicates

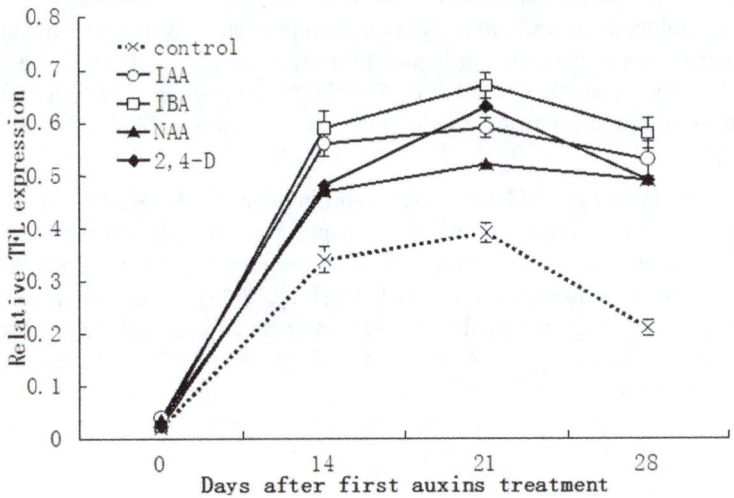

Figure 4. Relative expression of *TFL* in young spikes of branched spike wheat treated with auxins (IAA, IBA, NAA, 2,4-D)

4. Discussion

Our results indicate that auxins play an important role in branched spike architecture. Huang et al. (1990) have proved that adventitious branched rachis produced in some normal spike wheat when high concentration of 2,4-D was applied. This research proved that auxins might promote the occurrence or growth of branched rachis in branched spike wheat. The advantages of branched spike wheat are characterized of branched spike, and the number and length of branched rachises determines the number of spikelet, and then influences the fertile floret, or grain in a spike. Auxins significantly increased these spike parameters. In shoot branching architecture establishment, auxins can influence various stages of bud (shoot branch) development through the regulation of different genes, from a minor role in bud initiation through to a major but indirect effect on bud activity which at least involves different hormonal second messengers (Leyser, 2003). The apical dominance of auxin performed inhibition of lateral bud development, and auxin regulation involved in lateral meristem development and occurrence (Foo et al., 2005; Beveridge, 2006). Auxin also has an important role in promoting lateral root development. Before lateral root primordia breakthrough the root periderm, auxin is required to induce the expression of loading protein *LAX3*, and then triggering the activation of recombinant enzymes in the epidermal cells to promote cell separation and root formation (Swarup et al., 2008). The branches from shoots, roots and inflorescences seemed all required involvement of auxins.

Auxin plays an important role in the process of plant growth and development (Hohm et al., 2013). 0.1 mM IBA were the most effective in promoting plant growth, and photosynthetic capability. The effects of synthetic auxins NAA and 2, 4-D were significantly lower than those of IBA and IAA. A significant increment in growth, pigment content, and activities of antioxidant enzymes could be noticed in wheat young seedlings when the seeds were pretreated with IAA (Agami & Mohamed, 2013). In *Catharanthus roseus*, IBA, IAA and NAA significantly improved plant fresh and dry weights, total chlorophyll and carotenoids content, and net photosynthetic rate (Alam et al., 2012). In *Panax ginseng*, that exogenous IAA enhanced significantly net photosynthetic rate, stomatal conductance, transpiration rate at all growth stages (Li & Xu, 2014). The results in this study suggest there were no negative effects of auxins on wheat growth and photosynthesis when more fertile floret obtained in branched spike wheat by auxins application.

Evidence provided in this study supports the suggestion that *TFL* is the branched spike wheat homologue of *FLO/LFY*. A high degree of similarity was seen between the isolated cDNA molecule and fragments of known *FLO/LFY* homologues. *TFL* expressed at a low level in the shoot apex of control and auxin-treated plants at the start of the experiment, showing that *TFL* was expressed at the end of two ridges stage of wheat, at this time the shoot apex did not enter into the stage of floral organogenesis.

In no auxins and auxins treated plants, the quantities of *TFL* unregulated at the time when spike were undergoing the branched rachises differentiated and elongation. That implicates *TFL* might involved in branched rachis transition and possibly in branched rachis organogenesis.

In flower plants, *FLO/LFY* encodes a plant-specific transcriptional factor (Blazquez et al., 1997). The primary function of *FLO/LFY* homologous genes is to repress the development of vegetative organs and to promote the formation of flower meristem and then also influence blossum time (Mandel et al., 1995; Wada et al., 2002; Molinero-Rosales et al., 1999). The other function of *FLO/LFY* homologous genes is to maintain the activity of flower meristem and to activate the genes that specify the development of different flower organs (Weigel et al., 1992; Carmona et al., 2002).

Previous studied documented that *FLO/LFY* homologs might involve in inflorescence architecture. In maize, *zfl1* and *zfl2* are duplicate *FLO/LFY* homologs in maize. Transposon insertions into the two maize genes led to a disruption of floral organ identity and patterning, as well as to defects in inflorescence architecture and in the vegetative to reproductive phase transition (Bomblies & Doebley, 2005). In normal spike wheat, the expression pattern of *WFL*, a *FLO/LFY* ortholog, indicated that *WFL* is associated with spikelet formation rather than floral meristem identity (Shitsukawa et al., 2006). Auxins induced increasement of the expression of *TFL* over this time, although the extent to which *TFL* levels increased were associated with the kinds of auxins. This indicates that *TFL* may be involved in the formation and branching, and its expression might be induced by auxins.

5. Conclusions

The four types of experienced auxins induced similar effects on various physiological processes, including growth traits, photosynthetic parameters, although the impact extents were different among them. The data supports the hypothesis that auxins play a central role in the regulation branched spike characters and *TFL* might correlate with the development of branched rachises in branched spike wheat.

References

Agami, R. A., & Mohamed, G. F. (2013). Exogenous treatment with indole-3-acetic acid and salicylic acid alleviates cadmium toxicity in wheat seedlings. *Ecotoxicology and Environmental Safety, 94*, 164-171. http://dx.doi.org/10.1016/j.ecoenv.2013.04.013

Alam, M. M., Naeem, M., Idrees, M., Masroor, M. A., & Moinuddin, K. (2012). Augmentation of Photosynthesis, Crop Productivity, Enzyme Activities and Alkaloids Production in Sadabahar (*Catharanthus roseus* L.) through Application of Diverse Plant Growth Regulators. *Crop Sci. Biotech., 15*, 117-129. http://dx.doi.org/10.1007/s12892-011-0005-7

Andrew, H. C., & Reade, J. P. H. (2010). *Herbicides and Plant Physiology* (2th ed.). Wiley-Blackwell.

Arnon, D. I. (1949). Copper enzymes in isolated chloroplast. Polyphenol-oxidase *Beta vulgaris* L. *Plant Physiol, 24*(1), 1-5.

Beveridge, C. A. (2006). Axillary bud outgrowth: sending a message. *Current Opinion in Plant Biology, 9*(1), 35-40. http://dx.doi.org/10.1016/j.pbi.2005.11.006

Blazquez, M. A., Soowal, L. N., Lee, I., & Weigel, D. (1997). *LEAFY* expression and flower initiation in Arabidopsis. *Development, 124*(19), 3835-3844.

Bomblies, K., & Doebley, J. F. (2005). Molecular evolution of *FLORICAULA/LEAFY* orthologs in the *Andropogoneae (Poaceae)*. *Molecular Biology and Evolution, 22*(4), 1082-1094. http://dx.doi.org/10.1093/molbev/msi095

Bomblies, K., Wang, R. L., Ambrose, B. A., Schmidt, R. J., Meeley, R. B., & Doebley, J. (2003). Duplicate *FLORICAULA/LEAFY* homologs *zfl1* and *zfl2* control inflorescence architecture and flower patterning in maize. *Development, 130*, 2385-2395. http://dx.doi.org/10.1242/dev.00457

Carmona, M. J., Cubas, P., & Martinez-Zapater, M. J. (2002). *VFL*, the Grapevine *FLORICAULA/LEAFY* Ortholog, Is Expressed in Meristematic Regions Independently of Their Fate. *Plant Physio, 130*(1), 68-77. http://dx.doi.org/10.1104/pp.002428

Coen, E. S., Romero, J., Doyle, S., Elliot, R., Murphy, G., & Carpenter, R. (1990). Floricaula: A homeotic gene required for flower development in Antirrhinum majus. *Cell, 63*(6), 1311-1322. http://dx.doi.org/10.1016/0092-8674(90)90426-F

Cooke, T. J., Poli, D. B., Sztein, A. E., & Cohen, J. D. (2002). Evolutionary patterns in auxin action. *Plant Mol. Biol., 49*, 319-338. http://dx.doi.org/10.1007/978-94-010-0377-3_5

Foo, E., Buillier, E., Goussot, M., Foucher, F., Rameau, C., & Beveridge, C. A. (2005). The branching gene *RAMOSUS1* mediates interactions among two novel signals and auxin in pea. *The Plant Cell, 17*(2), 464-474. http://dx.doi.org/10.1105/tpc.104.026716

Hohm, T., Preuten, T., & Fankhauser, C. (2013). Phototropism: Translating light into directional growth. *American Journal of Botany, 100*(1), 47-59. http://dx.doi.org/10.3732/ajb.1200299

Huang, G., & Yan, J. (1990). Development genetics of Branched spike in *Triticum aestivum* L. *Southwest China Journal of Agricultural Sciences, 2*, 001.

Koric, S. (1975). Genetic basis for high spike productivity. *Proceedings of the 2nd Intern Winter wheat Conference, Zagreb, Yugoslavia* (pp. 188-144). U.S. Department of Agriculture.

Kyozuka, J., Konishi, S., Nemoto, K., Izawa, T., & Shimamoto, K. (1998). Down-regulation of *RFL*, the *FLO/LFY* homolog of rice, accompanied with panicle branch initiation. *Proceedings of the National Academy of Sciences, 95*(5), 1979-1982. Retrieved from http://www.pnas.org/content/95/5/1979.short

Leyser, O. (2003). Regulation of shoot branching by auxin. *Trends in Plant Science, 8*(11), 541-545. http://dx.doi.org/10.1016/j.tplants.2003.09.008

Li, X., & Xu, K. (2014). Effects of exogenous hormones on leaf photosynthesis of *Panax ginseng*. *Photosyntehtica, 52*(1), 152-156. http://dx.doi.org/10.1007/s11099-014-0005-1

Livak, K. J., & Schmittgen, T. D. (2001). Analysis of relative gene expression data using real-time quantitative PCR and the $2^{-\Delta\Delta C_T}$ method. *Methods, 25*(4), 402-408. http://dx.doi.org/10.1006/meth.2001.1262

Lu, J. Y., Shan, L., & Gao, J. F. (2003). Effects of drought stress on photosynthesis and some physiological characteristics in flag leaf during grain filling of wheat. *Agricultural Research in the Arid Areas, 2*, 017.

Ludwig-Müller, J. (2000). Indole-3-butyric acid in plant growth and development. *Plant Growth Regulation, 32*(2-3), 219-230. http://dx.doi.org/10.1023/A:1010746806891

Mandel, M. A., & Yanofsky, M. F. (1995). A gene triggering flower development in Arabidopsis. *Nature, 377*(6549), 522-524.

Molinero-Rosales, N., Jamilena, M., & Zurit, A. S. (1999). *FALSIFLORA*, the tomato orthologue of *FLORICAULA* and *LEAFY*. Controls flowering time and floral meristem identity. *The Plant Journal, 20*(6), 685-693. http://dx.doi.org/10.1046/j.1365-313X.1999.00641.x

Naduvilpurakkal, B. S., Radhakrishnan, S., Usha, K. A., & Kadavilpparampu, M. (2014). Radical chemistry of glucosamine naphthalene acetic acid and naphthalene acetic acid: A pulse radiolysis study. *Journal of Physical Organic Chemistry, 27*(6), 478-483. http://dx.doi.org/10.1002/poc.3285

Peanell, A. L., & Halloran, G. M. (1983). Inheritance of supernumerary spikelet development in wheat. *Euphytca, 32*, 767-776.

Shitsukawa, N., Takagishi, A., Ikari, C., Takumi, S., & Murai, K. (2006). *WFL*, a wheat *FLORICAULA/LEAFY* ortholog, is associated with spikelet formation as lateral branch of the inflorescence meristem. *Genes & Genetic Systems, 81*(1), 13-20. http://doi.org/10.1266/ggs.81.13

Simon, S., & Petrášek, J. (2011). Why plants need more than one type of auxin. *Plant Science, 180*(3), 454-460. http://dx.doi.org/10.1016/j.plantsci.2010.12.007

Sun, D. F., Zhu, X. D., Wan, Z. B., & Cai, J. (2000). The stability and expression in F_1 generation of supernumerary spikelets in bread wheat. *Journal Huazhong (Central China) Agricultural University, 19*(3), 213-218.

Swarup, K., Benková, E., Swarup, R., Casimiro, I., Péret, B., Yang, Y., & Parry, G. (2008). The auxin influx carrier *LAX3* promotes lateral root emergence. *Nature Cell Biology, 10*(8), 946-954. http://dx.doi.org/10.1038/ncb1754

Wada, M., Cao, Q. F., & Kotada, N. (2002). Apple has two orthologues of *FLORICAULA /LEAFY* involved in flowering. *Plant Molecular Biology, 49*(6), 567-577. http://dx.doi.org/10.1023/A:1015544207121

Weigel, D., Alvarez, J., Smyth, D. R., Yanofsky, M. F., & Meyerowitz, E. M. (1992). *LFY* controls floral meristem identity in Arabidopsis. *Cell, 69*(5), 843-859. http://dx.doi.org/10.1016/0092-8674(92)90295-N

Yanofsky, M. (1995). Floral meristems to floral organs: Genes controlling early events in Arabidopsis flower development. *Annual Review of Plant Biology, 46*(1), 167-188. http://dx.doi.org/10.1146/annurev.pp.46.060195.001123

20

Early Screening of Some Kurdistan Wheat (*Triticum aestivum* L.) Cultivars under Drought Stress

Nariman S. Ahmad[1], Shadia H. S. Kareem[1], Kamil M. Mustafa[1] & Dastan A. Ahmad[1]

[1] Crop Science Department, Faculty of Agricultural Sciences, University of Sulaimani, Kurdistan, Iraq

Correspondence: Nariman S. Ahmad, Faculty of Agricultural Sciences, Sulaimani University, Sulaimani, Kurdistan 46001, Iraq. E-mail: nariman.ahmad@univsul.edu.iq

Abstract

Due to the rapid climatic change drought becomes abiotic constraint globally. A factorial laboratory experiment was designed with CRD to evaluate the effects of kernel priming on wheat cultivars under induced drought stress. Seven common wheat cultivars in Kurdistan (Adana, Maxipak, Sham4, Sham6, Aras, Azadi and Rizgari) were tested under different negative osmotic solutions (0, -0.5, -1 and -1.5 Mpa), using Polyethylene glycol (PEG-6000). Among different cultivars Azadi exhibited better survival at high levels of drought stress for germination and its related traits. It also revealed high performance for shoot growth under the water stress, which was affirmed by the principal component analysis and cluster analysis. The superiority of this cultivar might be refer to exposing of this genotype to natural selection for a long duration under semiarid conditions of the local environment. Rizgari also had better performance mostly for the seedling characteristics, being a suitable cultivar for the late induced drought. The other cultivars had an intermediate response to the induced drought stress. This method could assist the plant breeder for rapid detection of drought tolerant genotypes in a large population with the reduced cost and labor compared to field trials.

Keywords: wheat cultivars, drought stress, polyethylene glycol, germination, cluster analysis

1. Introduction

Drought stress is considered to be one of the major abiotic constraints worldwide that limits the growth and productivity of crop plants (Jain, Mittal, & Gadre, 2013). Plants are varied in their capacity to adjust their metabolism and growth. They can tolerate the particular abiotic stress by establishing a metabolic homeostasis being in less stress for this condition. While the sensitive plants are unable to launch metabolic homeostasis that results in the growth and yield reduction, ending by the plant death (Jogaiah, Govind, & Tran, 2013). Under rain-fed conditions, water shortage is a severe limiting factor for germination and seedling establishment. These stages are extremely important to determine the growing period and the yield of crops (Khakwani et al., 2011). Among different plant species, wheat is one of the major crops worldwide, it accounts for about 20% of the human food supply. Wheat consumption has increased by 5% a year in the developing countries for nearly the last 70 years (Marmar, Baenziger, Dweikat, & El Hussein, 2013). The global production of wheat is significantly affected by the climate change and water scarcity in the grown environment (Al-Ghamdi, 2009; Bano, Ullah, & Nosheen, 2012). Water shortage at germination and seedling stage is among the factors to influence the yield of wheat crop (Noorka & Khaliq, 2007). The study programs of inducing drought tolerance in wheat should address the problem in a multi-disciplinary approach (Marmar et al., 2013). Selection of physiological traits associated with the drought tolerance is essentially enhanced to increase the efficiency of selection (Ciucă, Bănică, David, & Săulescu, 2010). Early screening as a physiological dissection of drought tolerance is one of the approaches to assist plant breeder in rapid detection of suitable genotype to be involved in the next breeding program.

To raise the productivity of wheat crop, it is crucial to identify the genotype that tolerate higher level of drought. This can be obtained by exploring maximum genetic potential from available wheat germplasm (Chachar et al., 2014). Screening for drought tolerance based on the field trials is costly and time-consuming, in addition to the typical condition required to express their effective genes responsible for the studied characteristics. Therefore, preliminary screening methods are commanded for the field criteria (Kim, Yun, H. K. Park, & M. S. Park, 2001). Identifying the drought tolerant wheat genotypes at germination and seedling growth stage under low osmotic potential is practiced as a reliable physiological indicator by the researchers (Chachar et al., 2016). Hence, the investigation of water stress based on the germination of different varieties is a forward step to identify the most tolerant genotype (s) under drought stress. Selection for drought tolerance at germination and the early seedling stage is frequently accomplished using simulated drought induced by chemicals like polyethylene glycol (PEG6000). It imposes water stress under *in vitro* conditions that maintains a uniform water potential throughout an experimental period, whereby a large set of genotypes can be screen accurately (Manoj & Uday, 2007).

The advantage of using Poly Ethylene Glycol (PEG) compared to others osmotic solutions is that due to the high molecular weight (6000-8000) PEG cannot enter the plant cells, instead, the water is withdrawn from the cell and cell wall without affecting or hurting the cell structure (Van den Berg & Zeng, 2006). While other osmotic solutions of low molecular weight could be toxic to plant as they are easily be taken by the plant (Hamza, 2012). Polyethylene glycol molecules are known to be inert, no-ionic, virtually impermeable to cell membranes and can induce uniform water stress without causing direct physiological damage (Kulkarni & Deshpande, 2005). PEG as a drought stress causing factor can reduce water potential, resulting in the growth reduction of germinated seeds and seedling (Zhu, Kang, Tan, & Xu, 2006).

The objective of this study was to evaluate the wheat cultivars for drought resistance at germination and early growth stage, utilizing PEG-6000 as an osmoticum to induce different levels of stress conditions to allow rapid screening for the most tolerant wheat genotypes to water stress.

2. Methods

2.1 Plant Materials

The experimental material was consisted of seven wheat cultivars (Table 1). They represented the commercial cultivars of common wheat in Kurdistan region. They were obtained from Bakrajo research station of Sulaimani and Erbil research station.

Table 1. Names and the sources of seven wheat cultivars used in the study

No.	Name	Pedigree/origin	Source
G_1	Adana	Turkey	
G_2	Maxipak	(Frontana × Kenya 58 – New thatch/Norin 10 Brevor) × Gabo 55. Pakistan. Local for Iraq	
G_3	Sham4	ICARDA	Bakrajo research Station, Sulaimani
G_4	Sham6	PLC "S" – Ruff "S" × Gta "S" – RTTE Cm-12904-1M-3M-1Y-1Y-OSK-OAP. ICARDA	
G_5	Aras	(Sonora 64 × Lerma Rojoo 64) × Sentaclena. Mexico. Local For Kurdistan region	
G_6	Azadi	Local For Kurdistan region	Erbil Research Station
G_7	Rizgari	Local For Kurdistan region	

2.2 Osmotic Stress Experiment

A laboratory experiment was conducted to estimate the drought stress of seven wheat cultivars. The experiment was laid out in completely randomized design (CRD) with two factors: genotypes and water stresses. Four osmotic solutions (including distilled water) were applied during the germination period on the common wheat cultivars with three replicates, using Poly Ethylene Glycol (PEG) of molecular weight 6000. The levels of negative osmotic solutions were prepared for the potentials of 0, -0.5, -1 and -1.5 Mpa (0, -5, -10 and -15 bar) by dissolving separately calculated amounts of PEG 6000 in distilled water (0, 17.0, 21.9, 25.3g PEG 6000/100 ml), respectively, at 30 °C.

Kernels were surface sterilized with ethanol 70% for 15 min. Residual ethanol was removed by thorough washing with sterilized distilled water. Twenty grains from each variety with three replicates were placed in Petri dish (90 mm diameter) for all treatments. Two layers of Whatman filter paper were used and moistened with 10 ml of distilled water and the Petri dishes were placed in a dark incubator for 24 hours for an "imbibition period" at 25 °C. Five ml of designated treatment solution was applied every three days into each petri dish after thorough washing and draining the previous left solution. The Petri dishes were kept under laboratory condition in an incubator (M 7040 R Electro.mag) at 25±2 °C for 12 days.

The germinated seeds were counted daily for the experiment duration, started from the second day after sowing. Seeds were considered germinated when they exhibited radicle extension more than 2 mm. The following characteristics were measured:

➢ Germination percentage (%): Counted after 4, 8 and 12 days with some modifications according to International Seed Testing Association (ISTA, 1993).

$$\text{Germination \%} = (\text{The number of germinated seeds until the i day/total number of seeds}) \times 100 \qquad (1)$$

➢ Mean Daily Germination (MDG): It is an index of daily germination speed determined by the following equation (Ellis & Roberts, 1981):

$$\text{Mean Daily Germination (MDG)} = FGP/d \qquad (2)$$

Where, FGP: final germination percentage (Viability), d: day(s) spent from the first to final germination.

➢ Coefficient of velocity of germination (CVG): This index is the velocity and acceleration of seed germination, gives an indication of the germination rapidity. It increased when the number of germinated seeds increases and the time required for germination decreases. It was calculated by the following equation (Kader & Jutzi, 2004):

$$\text{CVG (\%.d}^{-1}) = 100 \times \Sigma Ni/\Sigma \ (NiTi)) \qquad (3)$$

Where, N is the percentage of germinated seed in day i, and Ti is the sequence of day from sowing seed.

➢ Germination rate index (GRI): Reflects the percentage of germination on each day for the germination period. Higher GRI values indicate higher and faster germination. It was calculated by the following equation (Kader, 2005):

$$\text{GRI (\%.d}^{-1}) = \Sigma \ (Ni/i) \qquad (4)$$

Where, N is the percentage of germinated seed in day i.

➢ Mean germination time (MGT): Lower MGT is the faster of germinated seeds. It was calculated by the following equation   (Kader, 2005):

$$\text{MGT (d)} = \Sigma \ (NiTi)/(\Sigma Ni) \qquad (5)$$

Where, N is the percentage of germinated seed in day i, and Ti is the sequence of days from sowing.

➢ Root length, shoot length and root length/shoot length ratio: The length (mm) and weight (mg plant^{-1}) of seedling root and shoot were measured and recorded at the 12th day after sowing (true leaf initiation stage). Root to shoot length ratio were estimated by dividing root length to shoot length.

2.3 statistical Analysis

Data was subjected to analysis of variance (ANOVA), using XLSTAT 2015.4.01.20780 software and the comparisons of trait' means for both factors and their interactions were made using Duncan's multiple range test at 5% level of probability. To interpret the relationships among studied criteria a biplot derived from principal component analysis (PCA) was conducted based on the rank correlation matrix of the two-way data from both selection criteria (germination and seedling traits) and genotypes. The analysis compared and grouped the wheat cultivars based on the studied characteristics. Cluster analysis based on squared Euclidean distance was also performed to classify the genotypes using the same software.

3. Results and Discussion

The yield of wheat as like as many other crops has been reduced significantly with the effect of drought. Detecting the genotypes those can thrive on limited water resource is critical to promote the wheat production under rain-fed condition (Ahmad, Shabbir, Minhas, & Shah, 2013). Better use of water through the development of crop varieties with less water requirement and more drought tolerant is promising to satisfy the food demand for steadily increasing of the world population (El-Shafey, Hassaneen, Gabr, & El-Sheihy, 2009; Xinqing, Kun, Shi-Kui, Xiao-Xia, & Mu-Yi, 2006). Selection of tolerated wheat genotypes to water scarcity will help the breeding program in early tagging of drought tolerant genotype under stressed regions. Survival ability was investigated for seven common wheat cultivars in Kurdistan for the first time to tolerate chemical dehydration by PEG during the germination and early growth stage. The analysis results indicated significant variance of cultivars and PEG concentrations for most of the traits studied. However, their interactions declared a significant variance in shoot length trait only (Table 2). Significant effect and differential response of wheat varieties at seedling stage to PEG treatment were reported by some other researchers (Bayoumi, Eid, & Metwali, 2008; Dhanda, Sethi, & Behl, 2004).

Plant Genetics

Table 2. Analysis of variance for the studied traits of seven wheat cultivars during *in vitro* drought stress, induced by four concentrations (0 to -15 bars) of polyethylene glycol 6000

Sources of variance	d.f.[a]	Mean square										
		Germination (%)	MDG[b] (d)	CVG[c] (%.d^{-1})	GRI[d] (%.d^{-1})	MGT[e] (d)	RL (cm)	SHL (cm)	RW (mg·plant^{-1})	SHW (mg·plant^{-1})	RL/SHL[f]	RW/SHW[g]
Cultivars	6	3874.44**	24.10**	4.97**	710.63**	3.70**	189.12**	106.04**	181.29**	113.157	1.15**	0.613*
PEG conc.	3	5065.33**	17.05**	0.66	206.01**	0.36	37.37**	282.72**	74.37	86.541	0.66**	0.822*
Cultivars × PEG conc.	18	462.15**	1.00	0.28	22.77	0.22	9.43	7.20**	27.73	73.167	0.11	0.176
Residual	56	177.95	0.80	0.31	19.17	0.23	7.70	3.08	29.30	56.695	0.08	0.101
Minimum		10.00	0.417	8.333	0.833	8.000	4.200	2.140	0.400	4.600	0.341	0.027
Maximum		100.00	8.333	12.500	40.417	12.000	25.200	20.560	46.000	56.000	2.079	2.609
Mean		67.67	3.244	11.748	14.680	8.559	9.571	11.058	17.056	16.827	0.957	1.116
Standard deviation		25.99	1.765	0.809	8.757	0.693	4.719	4.639	6.448	8.089	0.432	0.425

Note. [a]: Degree of freedom; [b]: Mean Daily Germination; [c]: Coefficient of Velocity of Germination; [d]: Germination Rate Index; [e]: Mean Germination Time; [f]: Root length/shoot length ratio; [g]: Root weight/shoot weight ratio.

*, **: Significant at 5% and 1% probability level, respectively.

3.1 Germination Percentage (%)

Drought stress leads to drop the germination percent, due to less availability of free water to the kernel, as the PEG lowers the osmotic potential of the external medium (Datta, Mondal, Banerjee, & Mondal, 2011). The germination percentages of all the seven cultivars were at the highest level for control treatment and started to decrease by increasing the level of water stress using PEG. These findings are in line with the result of Moayedi, Boyce, Barakba, and Ghodsi (2009) who reported that the decline in germination percentage was observed with increasing the osmotic stress up to -0.9 Mpa.

It can be concluded that reducing in germination percentage is associated with either decreasing the water absorption into the seeds or delaying of the germination events (Khan et al., 2013). The decreasing trend in germination was generally the same for all three germination periods, counting for most of the cultivars except the germination after four days for Azadi cultivar (Figure 1). At this stage, the germination was linearly increased with the increased water stress. Short exposing of the kernel (only four days) to the induced condition had less stress effect on the germination process, which might be a reason of the lineared increased germination. While at the eighth and twelves days the germination of this cultivar was more affected by the induced stress due to longer exposing to the induced water stress. Azadi had less germination ration compared to control, but with a little effect compared to other cultivars, as it showed better performance to resist the stressed condition.

Regardless of the water stress, Azadi cultivar had the highest percentage of germination at 12th day. The last stress condition (at -15 bars of osmotic pressure) exhibited better germination performance (98%) for Azadi compared to other cultivars. It has been indicated that when a grain accomplishes a critical level of hydration it will be proceed toward a full germination without stopping (Almaghrabi, 2012). This cultivar considered being drought tolerant as their germination percentage did not reduce significantly with the increased moisture stress during germination, because some plants can develop their biochemical and physiological function to tolerate the water deficient condition (Chachar et al., 2016). However, the physiological changes below the critical level of hydration could be effective and inhibit the germination, attaining a critical level of hydration for this genotype (40%) made the germination to be proceed without cessation (Chachar et al., 2014). Adana cultivars at the second osmotic solution (-5 bar) had higher germination percent compared to control condition after Azadi; then it was linearly reduced to less germination percent compared to control treatment. Among all the cultivars Rizgari was the most affected by water stress, giving the lowest and significant germination percent (Table 3).

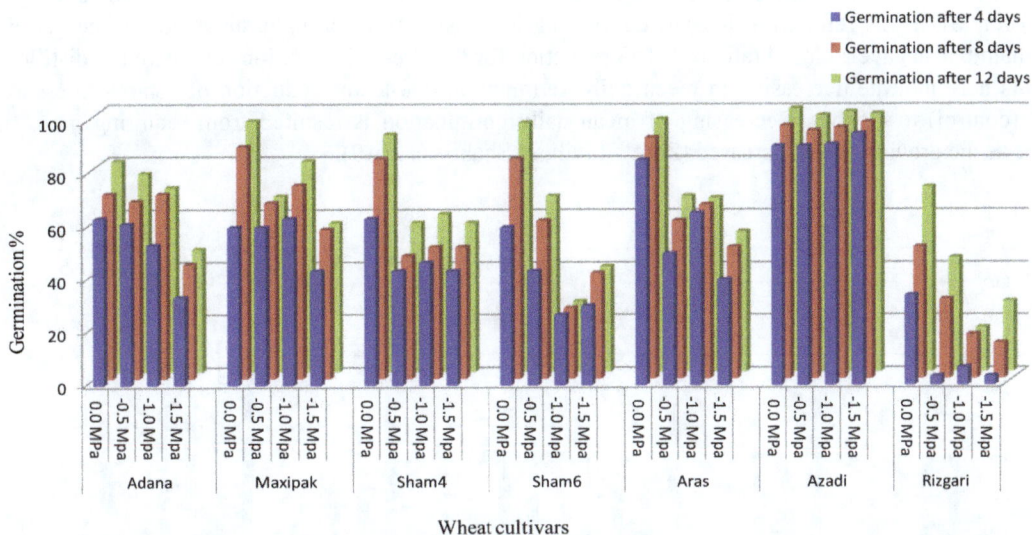

Figure 1. Effect of different levels of osmotic stress induced by PEG 6000 (-MPa) on the germination of seven wheat cultivars at three different times (after 4, 8 and 12 days)

Different respond of the cultivars to drought stress could be resulted from the amount of water absorbed, size of the seeds, and the features of surface coating of the seeds (Mohammadi & Mojaddam, 2014). This certain criteria is important and can be used in selecting drought-resistant cultivars (Qayyum, Razzaq, Ahmad, & Jenks, 2011). Although, some other researchers (Baloch et al., 2012; Sayar, Khemira, Kameli, & Mosbahi, 2008) stated non-significant discrimination among some wheat genotypes for germination. The high variation of the current cultivars responded to different osmotic stresses indicates a wide range of genetic variation among them, being suitable genetic materials in developing new genotypes for the arid condition, as the tolerant genotypes could be screened and tagged at seedling stage before extensive and expensive field trial (Baloch et al., 2012).

Table 3. Effect of cultivars and different drought stress conditions induced by PEG 6000 on the studied characters of seven wheat cultivars

Factor		Germination (%)	MDG[a] (d)	CVG[b] (%.d⁻¹)	GRI[c] (%.d⁻¹)	MGT[d] (d)	RL (cm)	SHL (cm)	RW (mg·plant⁻¹)	SHW (mg·plant⁻¹)	RL/SHL[e]	RW/SHW[f]
Cultivars	Adana	68.000bc	2.917bc	12.001a	13.854bc	8.337b	6.564c	8.909 de	13.233c	13.083b	0.809cd	1.065b
	Maxipak	74.667b	3.299b	11.916a	14.913b	8.402b	6.151c	9.50cde	13.842bc	13.700b	0.778cd	1.064b
	Sham4	65.833bc	2.847bc	11.886a	12.691bc	8.422b	6.769c	8.057 e	14.931bc	17.268ab	1.140b	0.980b
	Sham6	57.000c	2.500c	11.951a	10.781c	8.384b	6.364c	10.719c	18.563b	22.425a	0.647 d	0.971b
	Aras	70.500b	3.333b	12.078a	16.198b	8.292b	13.218ab	13.524b	15.755bc	16.333ab	1.006bc	1.000b
	Azadi	98.500a	6.181a	12.106a	29.826a	8.263b	12.600b	16.533a	18.583b	17.233ab	0.768cd	1.116b
	Rizgari	39.167 d	1.632 d	10.300b	4.497 d	9.810a	15.334a	10.168cd	24.468a	17.746ab	1.551a	1.614a
PEG conc.	0.0 MPa	89.429a	4.563a	11.586b	19.206a	8.649a	11.322a	16.456a	14.638b	18.771a	0.694b	0.824b
	-0.5 Mpa	67.524b	3.056b	11.744ab	13.978b	8.565a	9.656ab	9.872b	16.531ab	14.002a	1.016a	1.188a
	-1.0 Mpa	59.714b	2.778b	11.995a	13.522b	8.371a	9.194b	9.650b	18.238a	17.596a	1.042a	1.185a
	-1.5 Mpa	54.000b	2.579b	11.668ab	12.014b	8.650a	8.114b	8.255c	18.818a	16.938a	1.077a	1.265a

Note. [a]: Mean Daily Germination; [b]: Coefficient of Velocity of Germination; [c]: Germination Rate Index; [d]: Mean Germination Time; [e]: Root length/shoot length ratio; [f]: Root weight/shoot weight ratio.

3.2 Mean Daily Germination

The mean square results showed a significant effect of cultivars and polyethylene glycol on mean daily germination at 1% level. As like as germination percent the highest value of mean daily germination (6.181 days) was for Azadi cultivar (Table 3). Rizgari was the most affected cultivar with the drought stress (Figure 2) giving the minimum MDG value (1.632 days). The highest reduction percent of the induced stress compared to control (58.73%) for MDG was referred to Rizgari confirming its sensitivity to drought stress during the germination period. Sham 6 also gave a closed ratio of MDG reduction for the stressed conditions compared to distilled water. The results here indicate decreasing in mean daily germination due to the reduction of osmotic pressure from 0.00 bar (control) to -15 bar. Decreasing of mean daily germination is resulted from requiring more time to germinate as the drought stress increased (Zare', Tavili, & Shahbazi, 2010).

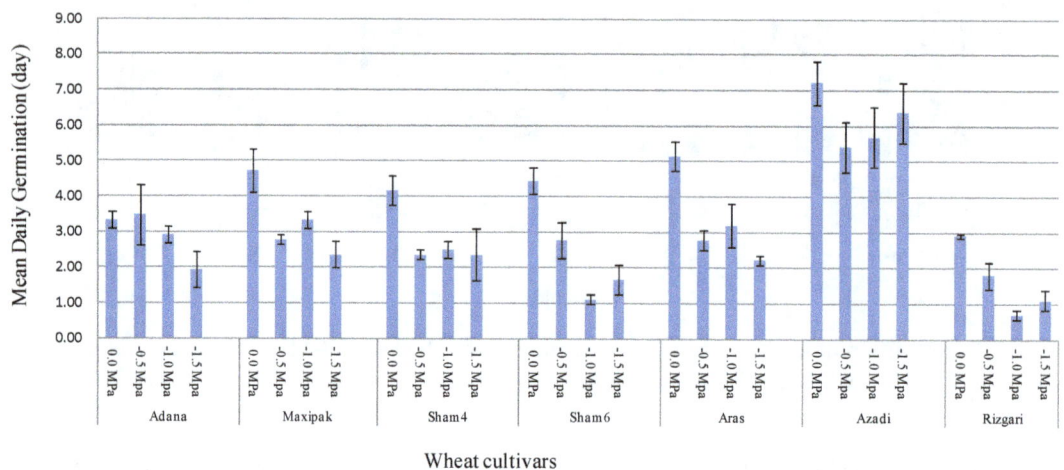

Figure 2. Effect of water stress on the Mean daily germination of seven wheat cultivars counted for 12 days

3.3 Coefficient of Velocity of Germination (CVG)

The mean square value of CVG was significantly different at 1% level among the various cultivars. While non-significant mean square was observed for different levels of drought stress and their interaction with the cultivars. The lowest coefficient of the velocity of germination was observed for control treatment and it was increased with the increased drought stress. This trend was realized in most of the cultivars (Figure 3). Azadi had the highest CVG value of 12.106, followed by Aras cultivar, as they were appeared to be more tolerant to drought stress conditions. Despite the necessity of high energy for biological processes under the stress condition, the cultivars under study were able to continue in their physiological activities to give higher CVG value compared to control treatment. Similar results were obtained for some wheat cultivars induced to water stress by Almaghrabi (2012) using Polyethylene Glycol at different concentration levels. Based on the result obtained for CVG, Azadi performed higher adaptation to drought stress condition compared to other cultivars.

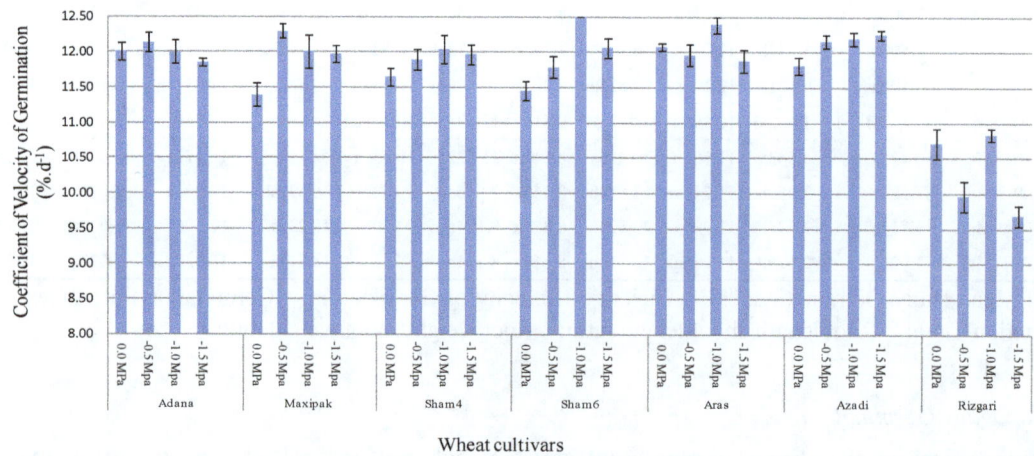

Figure 3. Effect of water stress on the coefficient of velocity of germination of seven wheat cultivars

3.4 Germination Rate Index (GRI)

It has been reported by Gholamin and Khayatnezhad (2010) that water stress in wheat starts with dropping the rate of germination and seedling growth. As like as germination the mean squares of cultivars and PEG concentrations of GRI were highly significant (1%), while the mean square values of their interaction were not significant (Table 2). The highest value of germination rate index was observed for Azadi cultivars (29.826), indicating the high tolerance of the cultivar to water stress (Figure 4). While the lowest value was recorded for Rizgari, being the most sensitive to give the maximum reduction of GRI (66.40%) compared to control. This fact is in accordance with what was obtained by Mollasadeghi, Ghanifathi, Masoumzadeh, & Aghahasanbeyglo (2014) who reported the high susceptibility of germination rate index in some bread wheat cultivars to the variation in osmotic potential.

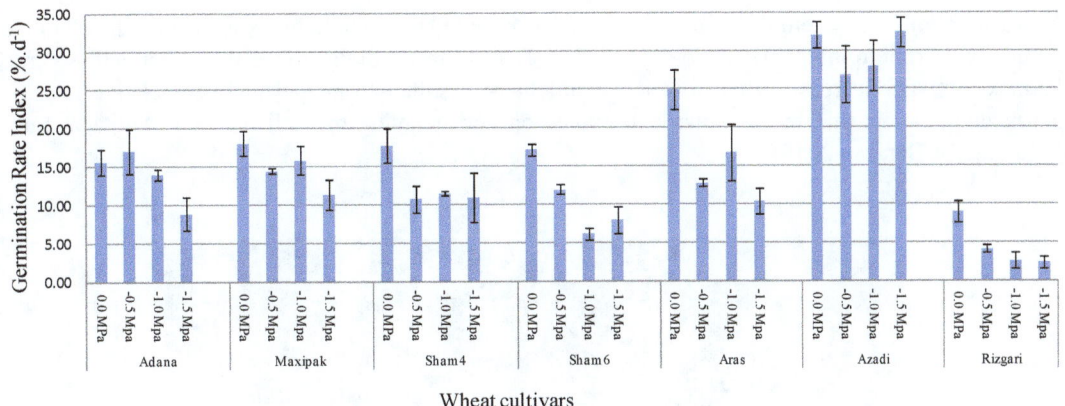

Figure 4. Effect of water stress on the germination rate index of seven wheat cultivars

3.5 Mean Germination Time (MGT)

It is clear from data presented in Table 2 that MGT was significantly affected by different wheat cultivars while no significant differences were observed in term of Polyethylene concentration and their interaction with cultivars. Significant differences was also obtained among different wheat genotypes for MGT at different levels of PEG 6000 and putrescine (Aydin, Pour, Halİloğlu, & Tosun, 2015). The minimum value of mean germination time (8.26 days) was found for Azadi, showing higher tolerant to the increased concentration of PEG when compared to other cultivars. Increased PEG concentration had a negative and significant impact on mean germination time for Rizgari cultivar (Figure 5), giving higher and significant value of 9.810 days. Less water availability under high concentration of PEG 8000 is one of the reasons to increase mean germination time (Iqbal & Ashraf, 2006).

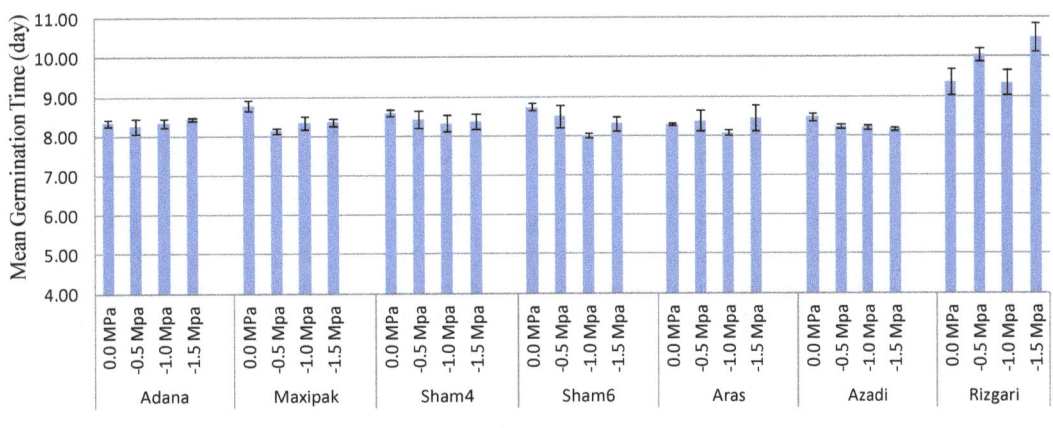

Figure 5. Effect of water stress on the mean germination time of seven wheat cultivars

3.6 Root Length (cm)

Root length in wheat crop considered as the most powerful trait among the seedling growth parameters in drought tolerance selection program (Baloch et al., 2012; Hassan, Mohamed, El-Rawy, & Amein, 2016). The increased concentrations of PEG at seedling growth stage had influenced the root length of wheat seedling. Rizgari cultivars had recorded the maximum root length of 15.33 cm (Table 3) compared to the others, showing the ability to perform root formation at the drought stress. Measuring such growth parameters to drought at the seedling stage may alleviate consequent depression and indicate the existence of potential variability resources of the genotypes to drought resistance (Biesaga-Kościelniak et al., 2014). However Azadi cultivar had less root length compared to Rizgari and Aras, the increase in root length had indicated by 28% compared to control when induced to the different PEG concentrations (Figure 6). It has been indicated that the root architecture has an influence on the agronomic trait and the yield of crop. A deeper and more extended root system allow the seedlings to extract more water in surrounded soil (de Dorlodot et al., 2007).

The development of root system in Azadi cultivar under drought stress could be explained by the stimulation of certain gene (s) of root formation (Placido et al., 2013). The shortest root due to the PEG treatment was recorded for Maxipak cultivar. Decrease in the root length might be resulted from the diminish relative turgidity and protoplasm dehydration, that brings down cell expansion and delaying the cell division (Mujtaba, Summiya, Khan, Mumtaz, & Barakat, 2016).

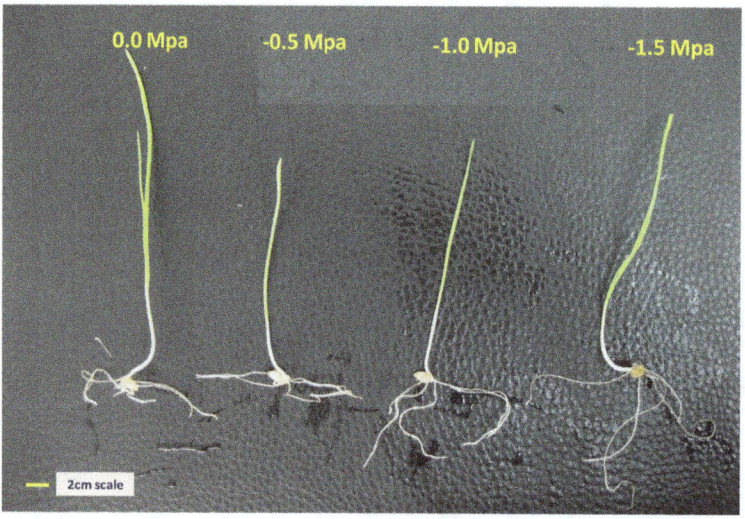

Figure 6. Rooting performance for Azadi cultivar under different stress levels (0.0Mpa, -0.5Mpa, -1.0Mpa and -1.5 Mpa)

3.7 Shoot Length (cm)

Presence of the increased PEG 6000 concentrations during the seedling growth had certainly reduced the shoot development and wheat seedling survival. The drought resistance is qualified by a small reduction of shoot growth under drought stressed condition (Moucheshi, Heidari, & Assad, 2012). The cultivars were differed in the length of seedling shoot under different levels of water stress. Maximum shoot length of 16.5 cm was recorded for Azadi cultivar (Table 3), giving less reduction percent of the induced stresses (23%) compared to the control treatment. While the lowest length was observed for Sham4 (Figure 7), recording the maximum reduction of 65.6% when compared to the control. A significant decrease in root and shoot length of wheat under water stress has been realized by other researchers (Gholamin & Khayatnezhad, 2010). Longer coleoptiles under water stress are one of the advantages of allowing the seedling emergence and establishment (Chachar et al., 2014).

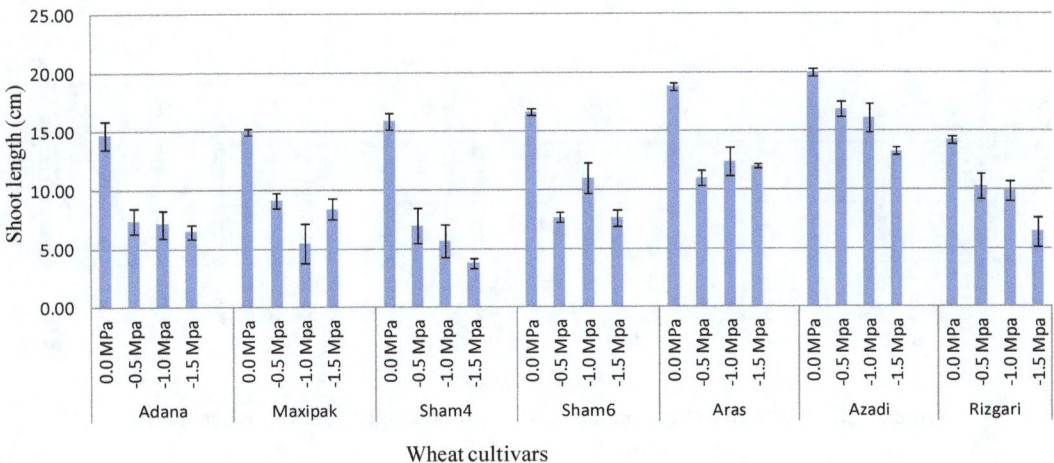

Figure 7. Effect of water stress on the shoot length of seven wheat cultivars

Apart from the root and shoot lengths, root/shoot ratio also being a good indicator to tag the drought tolerant genotype. The present study revealed significant variations for the root/shoot ratio between the cultivars (Figure 8). It has been indicated that drought resistant genotypes had balanced root and shoot growth (Dhanda, Sethi, & Behl, 2004). Less differences in root/shoot ratio for different drought stresses was for Rizgari to give 13.4%, followed by Aras and Azadi. For some of the cultivars, the increased value of root/shoot ratio was observed with the increase in PEG concentration. Physiological activities of the root system are realized to be less sensitive to low water content, while sap transferring to the upper part of seedling required higher water potential (Sani & Boureima, 2015).

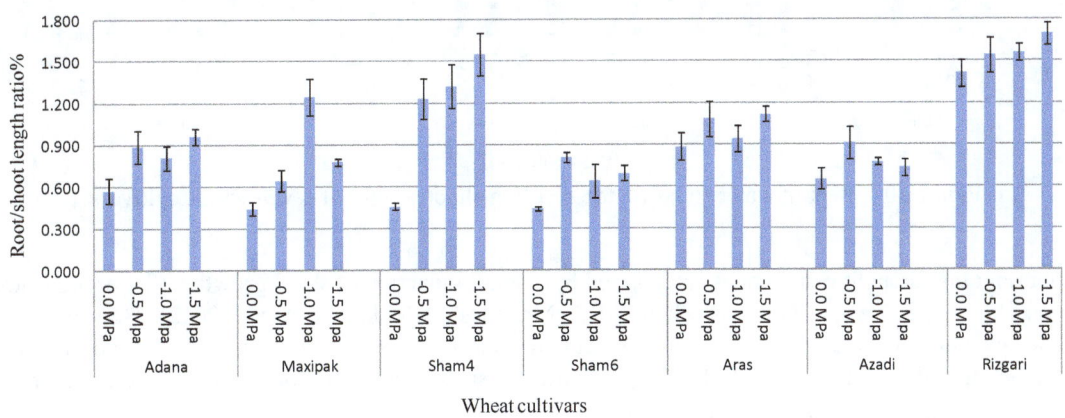

Figure 8. Effect of water stress on the coefficient of root/shoot length ratio of seven wheat cultivars

3.8 Root and Shoot Dry Weight (mg/seedling)

Racing the concentration of PEG had recorded the progressive increasing in root weight at different rates for all cultivars except Aras (Figure 9). Under high PEG treatment (-10 and -15 bar), the maximum root weight was recorded for Rizgari and Sham6 cultivars.

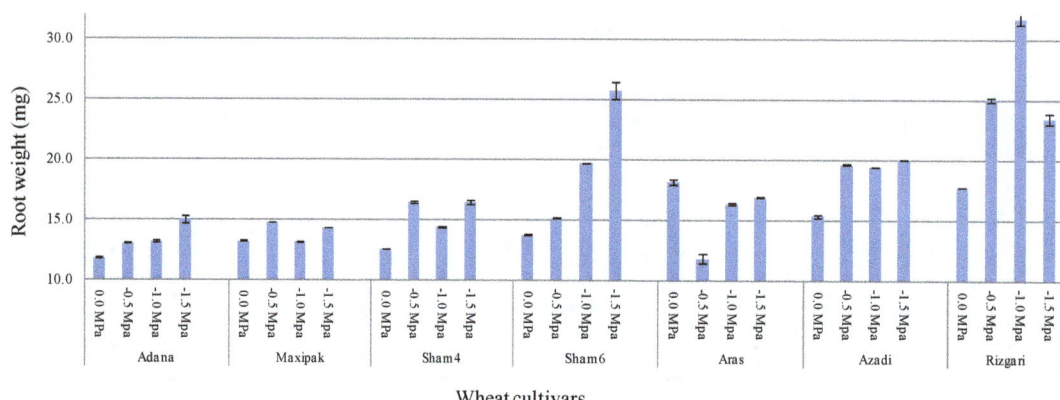

Figure 9. Effect of water stress on the seedling root weight of seven wheat cultivars

PEG prompted a decrease in shoot weight in number of cultivars (Figure 10). The reduction in shoot fresh weight was attributed to less number and smaller leave development with the increased PEG concentration of the growth media. The minimum shoot weight was observed for Adana while the maximum shoot weight was recorded for Sham6, followed by Rizgari (Table 3).

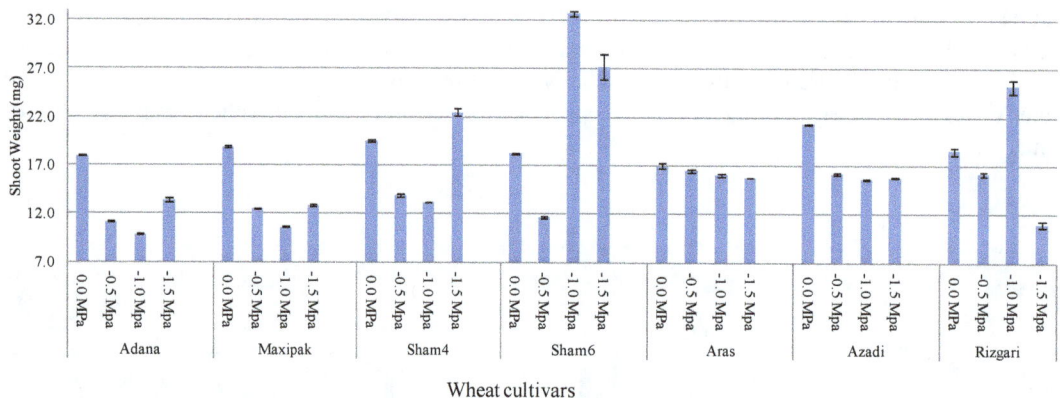

Figure 10. Effect of water stress on the seedling shoot weight of seven wheat cultivars

It is important that drought resistance is categorized by a small reduction of shoot growth under drought stressed condition (Ming, Pei, Naeem, Gong, & Zhou, 2012; Moucheshi, Heidari, & Assad, 2012). However the high root/shoot weight ratio was observed for Rizgari cultivar, the most persistent cultivar was Azadi, being more stable in the ratio of root/shoot weight (23.12%) for different stress conditions compared to control treatment.

3.9 Principal Component Analysis (PCA)

The PCA analysis was performed, as the most frequently used multivariate method, to assess the relationships between all attributes to identify superior cultivar (s) for the water-stressed condition. The relationships among different parameters were graphically displayed in a biplot of PCA1 and PCA2, based on the rank correlation matrix. Biplot diagram revealed that the first and second components justified 54.77% and 28.83% of total variation, respectively, with different characteristics studied (Table 4) and accounted for 83.60% of total variation.

Table 4. Principal component analysis for the studied traits of seven wheat cultivars

Traits	F1	F2
Germination	0.308	0.348
MDG	0.308	0.348
CVG	0.393	-0.112
GRI	0.332	0.310
MGT	-0.392	0.122
RL	-0.181	0.459
SHL	0.161	0.492
RWT	-0.295	0.328
SHWT	-0.079	0.092
RL/SHL	-0.349	0.106
RWT/SHWT	-0.344	0.234
Eigenvalue	6.025	3.172
Variability (%)	54.771	28.833
Cumulative %	54.771	83.605

Three groups were identified considering both components simultaneously (Figure 11). The germination, MDG, GRI and SHL were clustered in group I, while MGT, RL, RWT, SHWT, RL/SHL, RWT/SHWT were associated with group II and CVG with group III.

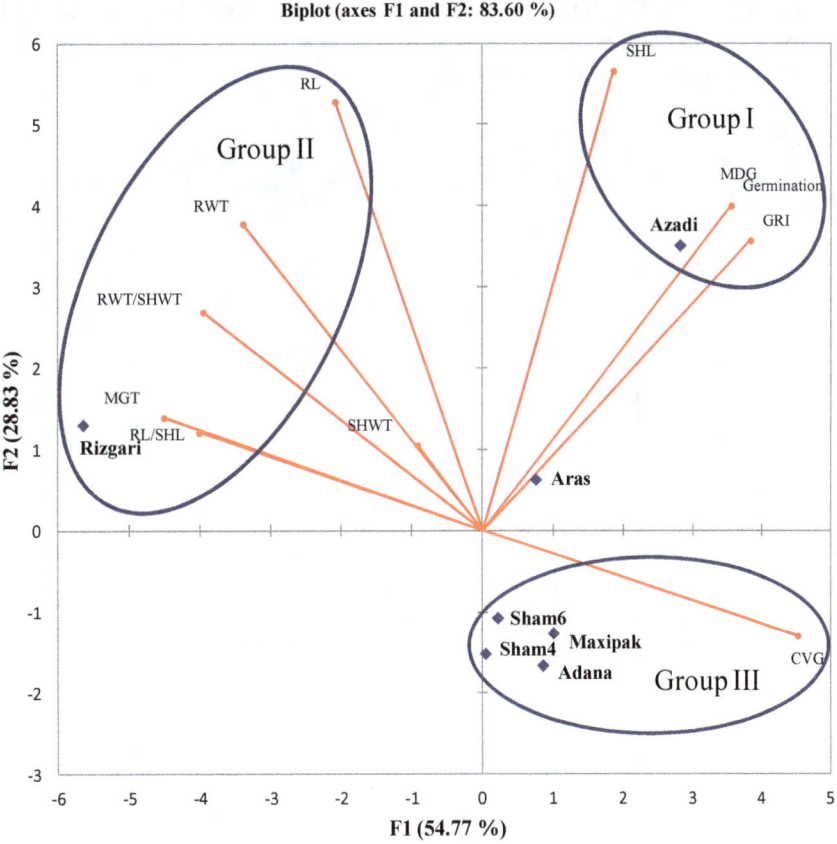

Figure 11. Biplot of principal component analysis of seven wheat cultivars and studied traits. Group I – high water stress resistant cultivars during germination process; Group II – high water stress tolerant for seedling characteristics; Group III – low water stressed cultivars for germination and seedling characteristics

Group I consists of Azadi cultivar, which has a good performance for germination, MDG, GRI and SHL under induced water stress (Figure 11). The cosine of the angle between the vectors of two characteristics approximates the correlation between them. Shoot length found to be in a positive and significant association with germination and MDG and GRI, meaning that prediction of the Shoot length as one of the drought-resistant trait could be explained through germination. Ahmad, Shabbir, Minhas, and Shah (2013) had similar result in identifying a strong positve correlation between Shoot lenght and germination pecent. In the second group (II), Rizgari was superior in MGT and most of the seedling traits, explaining its resistance in seedling traits under induced water stress and all of these traits were in a strong and negative association with CVG. Group III included most of the cultivars studied having a good germination velocity, showing negative and significant association with most of the traits of second group. The cultivars here were variable in their resistance to water-stressed conditions according to the traits studied. Biplot analysis has been used widely by other researchers for screening drought tolerant cultivars of wheat (El-Mohsen, El-Shafi, Gheith, & Suleiman, 2015; Farshadfar, Elyasi, & Aghaee, 2012). However the variation angles of the dataset with Biplot analysis does not precisely translated into correlation coefficients, the angles are informative enough to reflects the importance of the largest contributor to the total variation at each axis of differentiation (Abdi & Williams, 2010).

3.10 Cluster Analysis

As like as PCA, the cluster analysis classified the cultivars for the induced drought, based on the traits studied, into three groups of 1, 1 and 5 cultivars, respectively (Figure 12). The first group contained Azadi with higher germination rate and shoot length, and it was considered as a drought resistant group especially at germination stage. The second group consisted of Rizgari with higher performance for the seedling characteristics when exposed to the drought condition using PEG-6000. While the rest of cultivars were grouped in the third cluster showing their reasonable performance for CVG only with the variable response for other traits studied during the water stress conditions. Biplot results of the current study are in agreement with cluster analysis in identifying the same tolerant genotypes for the induced water stress. Cluster analysis has been utilized to describe the variation and grouping the genotypes based on drought tolerance indices (El-Mohsen, El-Shafi, Gheith, & Suleiman, 2015; Golabadi, Arzani, & Maibody, 2006).

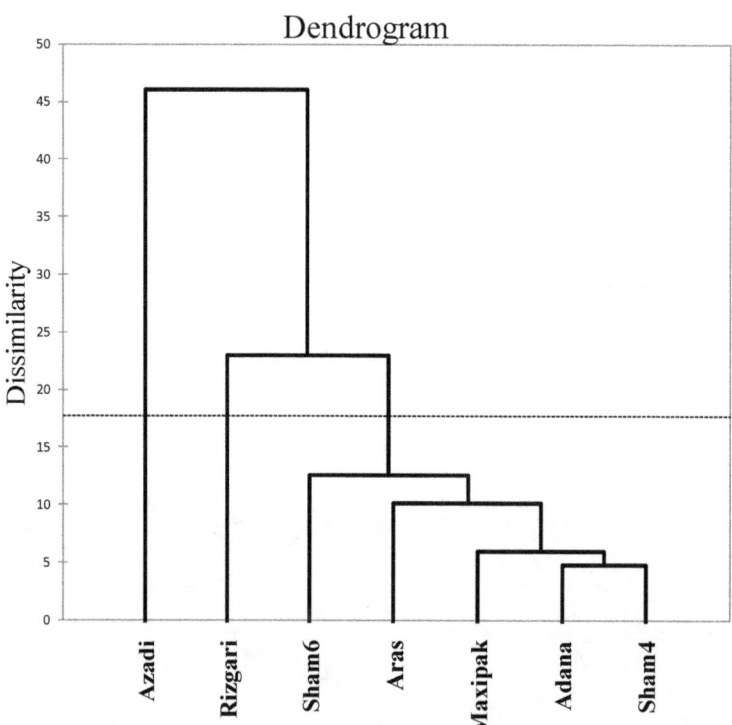

Figure 12. Dendrogram of seven wheat cultivars based on cluster analysis using various germination and seedling characteristics. Group I including Azadi cultivar (drought tolerant during germination), Group II including Rizgari cultivar (drought tolerant at seedling stage) and, Group III including five cultivars of Sham6, Aras, Maxipak, Adana and Sham4 (variable drought resistant)

4. Conclusion

In the current study osmotic stress caused significant effects on the germination and seedling traits of the studied cultivars. Azadi cultivar had the best performance for most of the traits under the drought stress conditions, which indicates the contributor characteristics of this genotype in tolerating drought stress. The introgression of desired allele (s) of stress tolerance from the related wild species into Azadi cultivar under local condition could be the reason of superiority of this cultivar under water stress. Rizgari cultivar had better performance for the seedling characteristics more than germination progress. It shows higher root length and root weight than other cultivars under induced water stress. Hence, Rizgari could tolerate drought stress at the seedling and later stages if passed the germination period at regular water status. This cultivar suits the local condition as most of the drought stress will come after the germination period. The best response of these two cultivars to drought stress condition could assure their suitability to be cultivated in arid and semiarid areas of Kurdistan.

Principle component analysis and Cluster analysis were in support to the analysis of variance to discriminate the genotypes at different level of drought stress through the studied characteristics. All three analysis positioned the genotype into three distinct groups; Azadi was tolerant to water stress throughout most of the germination characteristics, making the first group. Rizgar as the second group was suitable to resist the water stress during the seedling growth. While other cultivars had variable level of resistance to the induces stress for the studied characteristics.

The parameters here are found to be a useful index to discriminate drought tolerant genotypes at early growing stage under induced water-stressed environments. The current study will help the breeder for rapid selection of the tolerant genotype for any breeding program, avoiding extensive field trials.

References

Abdi, H., & Williams, L. J. (2010). Principal component analysis. *Wiley Interdisciplinary Reviews: Computational Statistics, 2*(4), 433-459. https://doi.org/10.1002/wics.101

Ahmad, M., Shabbir, G., Minhas, N., & Shah, M. (2013). Identification of drought tolerant wheat genotypes based on seedling traits. *Sarhad J. Agric., 29*(1), 21-27. Retrieved from https://www.researchgate.net/publication/263002286

Al-Ghamdi, A. A. (2009). Evaluation of oxidative stress tolerance in two wheat (*Triticum aestivum*) cultivars in response to drought. *International Journal of Agriculture and Biology, 11*(1), 7-12. Retrieved from https://www.fspublishers.org/published_papers/76090_..pdf

Almaghrabi, O. A. (2012). Impact of drought stress on germination and seedling growth parameters of some wheat cultivars. *Life Science Journal, 9*(1), 590-598. Retrieved from http://www.lifesciencesite.com/lsj/life0901/087_8160life0901_590_598.pdf

Aydin, M., Pour, A. H., Haliloğlu, K., & Tosun, M. (2015). Effect of Putrescine Application and Drought Stress on Germination of Wheat (*Triticum aestivum* L.). *Journal of the Faculty of Agriculture, 46*(1), 43-55. Retrieved from http://e-dergi.atauni.edu.tr/ataunizfd/article/download/5000186286/5000163975.

Baloch, M. J., Dunwell, J., Khakwani, A. A., Dennett, M., Jatoi, W. A., & Channa, S. A. (2012). Assessment of wheat cultivars for drought tolerance via osmotic stress imposed at early seedling growth stages. *Journal of Agricultural Research, 50*(3), 299-310. Retrieved from http://www.jar.com.pk/curntissu%20(2).php?p=547

Bano, A., Ullah, F., & Nosheen, A. (2012). Role of abscisic acid and drought stress on the activities of antioxidant enzymes in wheat. *Plant Soil Environ., 58*(4), 181-185. Retrieved from http://www.agriculturejournals.cz/publicFiles/63008.pdf

Bayoumi, T., Eid, M. H., & Metwali, E. (2008). Application of physiological and biochemical indices as a screening technique for drought tolerance in wheat genotypes. *African Journal of Biotechnology, 7*(14), 2341-2352. Retrieved from http://www.academicjournals.org/AJB

Biesaga-Kościelniak, J., Ostrowska, A., Filek, M., Dziurka, M., Waligórski, P., Mirek, M., & Kościelniak, J. (2014). Evaluation of spring wheat (20 varieties) adaptation to soil drought during seedlings growth stage. *Agriculture, 4*(2), 96-112. https://doi.org/10.3390/agriculture4020096

Chachar, M., Chachar, N., Chachar, S., Chachar, Q., Mujtaba, S., & Yousafzai, A. (2014). In-vitro screening technique for drought tolerance of wheat (*Triticum aestivum* L.) genotypes at early seedling stage. *Journal of Agricultural Technology, 10*(6), 1439-1450. Retrieved from http://www.ijat-aatsea.com/pdf/v10_n6_14_november/7_IJAT_10(6)_2014_M.H._Chachar-Crop%20Physiology.pdf

Chachar, Z., Chachar, N., Chachar, Q., Mujtaba, S., Chachar, G., & Chachar, S. (2016). Identification of Drought Tolerant Wheat Genotypes Under Water Deficit Conditions. *International Journal of Research GRANTHAALAYAH, 4*(2), 206-2014. Retrieved from http://granthaalayah.com/Articles/Vol4Iss2/23_ IJRG16_A02_84.pdf

Ciucă, M., Bănică, C., David, M., & Săulescu, N. N. (2010). SSR markers associated with the capacity for osmotic adjustment in wheat (*Triticum aestivum* L.). *Romanian Agricultural Research, 27*(1), 1-5. Retrieved from https://www.researchgate.net/publication/268261853

Datta, J., Mondal, T., Banerjee, A., & Mondal, N. (2011). Assessment of drought tolerance of selected wheat cultivars under laboratory condition. *Journal of Agricultural Science and Technology, 7*, 383-393. Retrieved from http://www.ijat-aatsea.com/pdf/April_v7_n2_11/16%20IJAT2010_28FT.pdf

de Dorlodot, S., Forster, B., Pagès, L., Price, A., Tuberosa, R., & Draye, X. (2007). Root system architecture: Opportunities and constraints for genetic improvement of crops. *Trends in Plant Science, 12*(10), 474-481. https://doi.org/10.1016/j.tplants.2007.08.012

Dhanda, S., Sethi, G., & Behl, R. (2004). Indices of drought tolerance in wheat genotypes at early stages of plant growth. *Journal of Agronomy and Crop Science, 190*(1), 6-12. https://doi.org/10.1111/j.1439-037X. 2004.00592.x

El-Mohsen, A. A. A., El-Shafi, M. A., Gheith, E., & Suleiman, H. (2015). Using Different Statistical Procedures for Evaluating Drought Tolerance Indices of Bread Wheat Genotypes. *Advance in Agriculture and Biology, 4*(1), 19-30.

El-Shafey, N. M., Hassaneen, R. A., Gabr, M. M., & El-Sheihy, O. (2009). Pre-exposure to gamma rays alleviates the harmful effect of drought on the embryo-derived rice calli. *Australian Journal of Crop Science, 3*(5), 268. Retrieved from http://www.cropj.com/Nadia_3_5_2009_268_277.pdf

Ellis, R., & Roberts, E. (1981). The quantification of ageing and survival in orthodox seeds. *Seed Science and Technology (Netherlands), 9*(2), 373-409. Retrieved from http://agris.fao.org/agris-search/search.do? recordID=XE8182678

Farshadfar, E., Elyasi, P., & Aghaee, M. (2012). *In vitro* selection for drought tolerance in common wheat (*Triticum aestivum* L) genotypes by mature embryo culture. *American Journal Sciences Research, 48*, 102-115.

Gholamin, R., & Khayatnezhad, M. (2010). Effects of Polyethylene Glycol and NaCl Stress on Two Cultivars of Wheat *Triticum durum* at Germination and Early Seeding Stages. *American-Eurasian Journal of Agricultural and Environmental Sciences, 9*, 86-90. Retrieved from http://www.idosi.org/aejaes/ jaes9(1)/14.pdf

Golabadi, M., Arzani, A., & Maibody, S. M. (2006). Assessment of drought tolerance in segregating populations in durum wheat. *African Journal of Agricultural Research, 1*(5), 162-171. Retrieved from http://citeseerx.ist.psu.edu/viewdoc/download?doi=10.1.1.128.5995&rep=rep1&type=pdf

Hamza, J. H. (2012). Seed Priming of Bread Wheat to Improve Germination Under Drought Stress. *Iraqi Journal of Agricultural Sciences, 43*(2), 100-107. Retrieved from http://repository.uobaghdad.edu.iq/ ArticleShow.aspx?ID=3655

Hassan, M. I., Mohamed, E. A., El-Rawy, M. A., & Amein, K. A. (2016). Evaluating interspecific wheat hybrids based on heat and drought stress tolerance. *Journal of Crop Science and Biotechnology, 19*(1), 85-98. https://doi.org/10.1007/s12892-015-0085-x

Iqbal, N., & Ashraf, M. Y. (2006). Does seed treatment with glycinebetaine improve germination rate and seedling growth of sunflower (*Helianthus annuus* L.) under osmotic stress. *Pakistan Journal of Botany, 38*(5), 1641-1648. Retrieved from http://www.pakbs.org/pjbot/PDFs/38(5)/PJB38(5)1641.pdf

ISTA. (1993). International Rules for Seed Testing. *Seed Science and Technology, 21*, 1-288. Retrieved from http://trove.nla.gov.au/version/13460205

Jain, M., Mittal, M., & Gadre, R. (2013). Effect of PEG-6000 imposed water deficit on chlorophyll metabolism in maize leaves. *Journal of Stress Physiology & Biochemistry, 9*(3), 262-271. Retrieved from http://www.jspb.ru/issues/2013/N3/JSPB_2013_3_262-271.pdf

Jogaiah, S., Govind, S. R., & Tran, L.-S. P. (2013). Systems biology-based approaches toward understanding drought tolerance in food crops. *Critical Reviews in Biotechnology, 33*(1), 23-39. https://doi.org/10.3109/07388551.2012.659174

Kader, M. (2005). A comparison of seed germination calculation formulae and the associated interpretation of resulting data. *Journal and Proceeding of the Royal Society of New South Wales, 138*, 65-75. Retrieved from http://royalsoc.org.au/images/pdf/journal/138_Kader.pdf

Kader, M., & Jutzi, S. (2004). Effects of thermal and salt treatments during imbibition on germination and seedling growth of sorghum at 42/19 °C. *Journal of Agronomy and Crop Science, 190*(1), 35-38. https://doi.org/10.1046/j.0931-2250.2003.00071.x

Khakwani, A. A., Dennett, M. D., & Munir, M. (2011). Drought tolerance screening of wheat varieties by inducing water stress conditions. *Songklanakarin J. Sci. Technol, 33*(2), 135-142. Retrieved from http://www.thaiscience.info/journals/Article/SONG/10761800.pdf

Khan, M., Shabbir, G., Akram, Z., Shah, M., Ansar, M., Cheema, N., & Iqbal, M. (2013). Character association studies of seedling traits in different wheat genotypes under moisture stress conditions. *SABRAO Journal of Breeding and Genetics, 45*(3), 458-467. Retrieved from http://www.sabrao.org/journals/vol45_3_dec2014/SABRAO%20J%20Breed%20Genet%2045(3)%20458-467%20Khan.pdf

Kim, Y.-J., Yun, S.-J., Park, H.-K., & Park, M.-S. (2001). A Simple Method of Seedling Screening for Drought Tolerance in Soybean. *Korean Journal of Crop Science, 46*(4), 284-288. Retrieved from http://ocean.kisti.re.kr/downfile/volume/kscs/JMHHBK/2001/v46n4/JMHHBK_2001_v46n4_284.pdf

Kulkarni, M., & Deshpande, U. (2005). *In vitro* screening of tomato genotypes for drought resistance using polyethylene glycol (PEG). *Vegetable Science, 32*(1), 11-14. Retrieved from http://www.ajol.info/index.php/ajb/article/viewFile/56885/45294

Manoj, K., & Uday, D. (2007). In vitro screening of tomato genotypes for drought resistance using polyethylene glycol. *African Journal of Biotechnology, 6*(6), 691-696. Retrieved from https://www.researchgate.net/publication/27797690

Marmar, A., Baenziger, S., Dweikat, I., & El Hussein, A. A. (2013). Preliminary screening for water stress tolerance and genetic diversity in wheat (*Triticum aestivum* L.) cultivars from Sudan. *Journal of Genetic Engineering and Biotechnology, 11*(2), 87-94. https://doi.org/10.1016/j.jgeb.2013.08.004

Ming, D., Pei, Z., Naeem, M., Gong, H., & Zhou, W. (2012). Silicon alleviates PEG-induced water-deficit stress in upland rice seedlings by enhancing osmotic adjustment. *Journal of Agronomy and Crop Science, 198*(1), 14-26. https://doi.org/10.1111/j.1439-037X.2011.00486.x

Moayedi, A. A., Boyce, A. N., Barakba, S. S., & Ghodsi, M. (2009). The Effects of Different Levels of Osmotic Stress on Germination and Seedling Growth in Promising Durum Wheat Genotypes. *Middle Eastern and Russian Journal of Plant Science and Biotechnology, 3*(1), 10-14. Retrieved from http://www.globalsciencebooks.info/Online/GSBOnline/images/0906/MERJPSB_3(SI1)/MERJPSB_3(SI1)10-14o.pdf

Mohammadi, N., & Mojaddam, M. (2014). The Effect of Water Deficit Stress on Germination Components of Grain Sorghum Cultivars. *Indian Journal of Fundamental and Applied Life Sciences, 4*(4), 289-291. Retrieved from http://www.cibtech.org/J-LIFE-SCIENCES/PUBLICATIONS/2014/Vol-4-No-4/JLS-043-046-MANI-EFFECT-CULTIVARS.pdf

Mollasadeghi, V., Ghanifathi, T., Masoumzadeh, B., & Aghahasanbeyglo, A. A. (2014). Bread wheat tolerance against drought at early growth stages and grain filling period. *Applied mathematics in Engineering, Management and Technology, 2*(2), 50-59. Retrieved from http://amiemt-journal.com/test/vol2-2/7.pdf

Moucheshi, A., Heidari, B., & Assad, M. (2012). Alleviation of drought stress effects on wheat using arbuscular mycorrhizal symbiosis. *International Journal of AgriScience, 2*(1), 35-47. Retrieved from http://shirazu.ac.ir/dro/sites/dro/files/droPaG3Heydari-10.pdf

Mujtaba, S. M., Summiya, F., Khan, M. A., Mumtaz, A., & Barakat, K. (2016). Physiological Studies on Six Wheat (*Triticum aestivum* L.) Genotypes for Drought Stress Tolerance at Seedling Stage. *Agricultural Research & Technology: Open Access Journal, 1*(2), 1-6.

Noorka, I. R., & Khaliq, I. (2007). An efficient technique for screening wheat (*Triticum aestivum* L.) germplasm for drought tolerance. *Pakistan Journal of Botany, 39*(5), 1539-1546. Retrieved from http://www.pakbs.org/pjbot/PDFs/39(5)/PJB39(5)1539.pdf

Placido, D. F., Campbell, M. T., Folsom, J. J., Cui, X., Kruger, G. R., Baenziger, P. S., & Walia, H. (2013). Introgression of novel traits from a wild wheat relative improves drought adaptation in wheat. *Plant Physiology, 161*(4), 1806-1819. https://doi.org/10.1104/pp.113.214262

Qayyum, A., Razzaq, A., Ahmad, M., & Jenks, M. A. (2011). Water stress causes differential effects on germination indices, total soluble sugar and proline content in wheat (*Triticum aestivum* L.) genotypes. *African Journal of Biotechnology, 10*(64), 14038-14045. https://doi.org/10.5897/AJB11.2220

Sani, D. O., & Boureima, M. M. (2015). Effect of polyethylene glycol (PEG) 6000 on germination and seedling growth of pearl millet [*Pennisetum glaucum* (L.) R. Br.] and LD50 for *in vitro* screening for drought tolerance. *African Journal of Biotechnology, 13*(37), 3742-3747.

Sayar, R., Khemira, H., Kameli, A., & Mosbahi, M. (2008). Physiological tests as predictive appreciation for drought tolerance in durum wheat (*Triticum durum* Desf.). *Agronomy Research, 6*(1), 79-90. Retrieved from http://agronomy.emu.ee/vol061/p6108.pdf

Van den Berg, L., & Zeng, Y. (2006). Response of South African indigenous grass species to drought stress induced by polyethylene glycol (PEG) 6000. *South African Journal of Botany, 72*(2), 284-286. https://doi.org/10.1016/j.sajb.2005.07.006

Xinqing, S., Kun, W., Shi-Kui, D., Xiao-Xia, H., & Mu-Yi, K. (2006). Regionalisation of suitable herbages for grassland reconstruction in agro-pastoral transition zone of northern China. *New Zealand Journal of Agricultural Research, 49*(1), 73-84. https://doi.org/10.1080/00288233.2006.9513696

Zare', S., Tavili, A., & Shahbazi, R. (2010). The effect of different levels of salicylic acid on the improvement of germination components in berry plant under salinity and drought stress. *Journal of Range and Watershed, Iranian Journal of Natural Resources, 3*(1), 29-39.

Zhu, J., Kang, H., Tan, H., & Xu, M. (2006). Effects of drought stresses induced by polyethylene glycol on germination of *Pinus sylvestris* var. mongolica seeds from natural and plantation forests on sandy land. *Journal of Forest Research, 11*(5), 319-328. https://doi.org/10.1007/s10310-006-0214-y

21

Antioxidant Activity and Phytochemical Content of Fresh and Freeze-Dried *Lepisanthes fruticosa* Fruits at Different Maturity Stages

Mirfat Ahmad Hasan Salahuddin[1], Zaulia Othman[2], Joanna Cho Lee Ying[2], Erny Sabrina Mohd Noor[1] & Salma Idris[3]

[1] Agrobiodiversity and Environment Research Centre, MARDI Headquarters, Selangor, Malaysia

[2] Horticultural Research Centre, MARDI Headquarters, Selangor, Malaysia

[3] Seed and Genebank Centre, MARDI Headquarters, Selangor, Malaysia

Correspondence: Mirfat Ahmad Hasan Salahuddin, Agrobiodiversity and Environment Research Centre, Malaysian Agricultural Research and Development Institute (MARDI) Headquarters, Persiaran MARDI-UPM, 43400 Serdang, Selangor, Malaysia. E-mail: mirfat@mardi.gov.my

Abstract

Antioxidant and phytochemical compounds of fruits can vary widely depending on many factors such as processing and maturity stage as one of the major contributors. Therefore, this study investigated the antioxidant activity and phytochemical attributes of fresh and freeze-dried *Lepisanthes fruticosa* fruit extracts at eight different maturity stages. The freeze-dried extracts were obtained by lyophilisation using a freeze-dryer. In general, antioxidant activity and phytochemical contents of both fresh (FLF) and freeze-dried (FDLF) extracts showed a decrease with fruit maturation. Among the eight maturity stages developed for *L. fruticosa,* the lower maturity (unripe) stages exhibited the strongest potential, with stage 1 being the most notable. The FDLF fruit extracts were found to be significantly ($P < 0.05$) stronger radical scavenger than FLF extracts at all maturity stages tested. The IC_{50} values of FDLF for the eight maturity stages were more effective, with stage 1 showing the lowest IC_{50} (1.57 mg/ml). Total phenolic content of FDLF was also significantly ($P < 0.05$) higher than FLF at all eight stages tested, with the highest also being shown at stage 1 (15848.96 ± 401.82 mg/100 g). On the contrary, FLF extracts displayed significantly ($P < 0.05$) higher total flavonoid content than FDLF at almost all stages except for 2, 3 and 6. The highest content was shown in stage 1 with 37.35 ± 0.77 mg/100 g. These findings showed that antioxidant activity and phytochemical content of *L. fruticosa* fruits were significantly affected by processing and maturity. The obtained results are important for the promotion of use of *L. fruticosa* fruit extracts as a natural antioxidant in functional food production in the future.

Keywords: *Lepisanthes fruticosa*, freeze-drying, radical scavenging activity, total phenolic, total flavonoid

1. Introduction

Lepisanthes fruticosa (Roxb) Leenh or locally known as ceri Terengganu is a non-seasonal underutilised fruit species that produces fruits throughout the year (Mirfat & Salma, 2015). It belongs to the family Sapindaceae and can be found in South East Asia which comprise Malaysia, Myanmar, Indo-China, Thailand, Philippines and Indonesia (Lim, 2013). The species is found growing naturally in the forests and only occasionally cultivated. In Malaysia, this species is widely distributed in Johor and the East Coast of Peninsular Malaysia (Mirfat & Salma, 2015). A number of species from the genus *Lepisanthes* are widely used in traditional and folk medicine systems in different parts of the world (Kuspradini, Ritmaleni, & Mitsunaga, 2012). Based on the ethnobotanical studies, *L. fruticosa* is usually consumed as food source and also used in traditional medicine by rural folks. The seed is eaten roasted and the root is used in a compound poultice to relieve itching and to lower temperature during fever (Mirfat & Salma, 2015). Wetwitayaklung, Charoenteeraboon, Limmatvapirat, and Phaechamud (2012) have reported that *L. fruticosa* root has antipyretic properties and the ripe fruit has antidiarrhoea properties.

Recently, interest in *L. fruticosa* fruits has arisen due to its strong antioxidant capacity as compared to a numbers of underutilised fruits and commercial fruits (Mirfat & Salma, 2015). Our earlier study also revealed that *L. fruticosa* ripe fruits showed the highest free radical scavenging activity and had a great source of total phenolic contents among many other fruits tested (Mirfat & Salma, 2015). Phenolics are one of the major phytochemicals that play vital roles in the inhibition of free radicals to prevent or slow the oxidative damage in our body (Mirfat, Salma, Razali, & Umi Kalsum, 2015). However, it is interesting to note that antioxidant and phytochemical compounds of fruits can vary widely depending on many factors such as processing and maturity stage as one of the major contributors.

To the best of our knowledge, antioxidant and phytochemical content of *L. fruticosa* fruits at different maturity stages have never been compared and evaluated. Maturity is one of the factors which can strongly influence the nutritional value as well as physiological properties such as developmental of colours and texture of fruits (Mohd. Zainudin, Abdul Hamid, Anwar, Osman, & Saari, 2014). Previous reports also stated that changes in antioxidant properties were found to be varied different fruits such as olive, orange and tomato (Mohd. Zainudin et al., 2014). Changes of the antioxidant and phytochemical compounds during maturation hold great significance from both dietary and nutritional point of view. There is a great importance to identify the harvest maturity with enhanced levels of antioxidant activity and phytochemical compounds targeting increased functional properties (Fu, He, Zhao, Yang, & Mao, 2009).

In order to meet different requirements of consumers and preserve fresh plant materials, fruits could also be processed. One of the widely used processing method of fruits is freeze-drying. It was suggested that dried fruits showed higher antioxidant activity and polyphenolic content than fresh fruits due to their low moisture content with increased shelf life (Vijaya Kumar Reddy, Sreeramulu, & Raghunath, 2010).

Therefore, this study was carried out with the objective of evaluating the antioxidant activity and phytochemical content of fresh and freeze dried *L. fruticosa* fruits at different maturity stages in order to come up with the best harvesting maturity. The fruit processing can be explored as a viable method for processing fruits retaining the maximum amount of these antioxidants.

2. Methods

2.1 Plant Materials

Lepisanthes fruticosa fruits were sampled from MARDI fruit genebank in Serdang, Selangor. The flowers of *L. fruticosa* were tagged initially after fully bloom and let to grow until full maturity. The fruits were harvested manually every week after tagging (Figure 1) to evaluate the characteristics, colour and shape of fruits. After evaluating three cycles of the fruit development, colour indices were developed.

Figure 1. Different colour indices of *L. fruticosa* during fruit development

2.2 Sample Preparation and Extraction

Fruit samples were washed with running tap water before being weighed for the edible portion parts. For the freeze-dried samples, the samples were transferred into freeze dryer bottle in and kept at -80 °C freezer overnight. Then, the frozen samples were lyophilised at -85 °C, pressure 250 mtorr for 3 days using a bench-top freeze dryer (Virtis, USA). The freeze-dried samples were finely ground and kept in an airtight container prior to extraction. The fresh and the freeze-dried samples were then extracted using methanol (1:10) and shaken for approximately 1 hour. The extracts were then centrifuged for 10 minutes at 10,000 rpm. The residue was separated from the supernatant and the procedure was repeated twice. The two resulting supernatants were mixed together to obtain the crude extracts which were stored at -80 °C prior to analysis.

2.3 Determination of Antioxidant Activity

Scavenging activity of the fruit extracts on 2,2-diphenyl-1-picrylhydrazyl (DPPH) radicals was assayed according to Molyneux (2004) with some modifications. Various concentrations of the crude extracts in methanol were prepared to give a final volume of 7 µl and were mixed with 280 µl of methanolic solution containing DPPH (Sigma, USA) radicals resulting in a final concentration of 0.06 mM. The mixture was vigorously shaken and left to stand for 30 min. in the dark. The absorbance was measured at 517nm and ascorbic acid (Sigma, USA) was used as the positive control. The results were expressed as IC_{50} value (mg/ml), which is the inhibitory concentration at which DPPH radicals were scavenged by 50%. All procedures were carefully carried out with minimum exposure of light.

2.4 Determination of Total Phenolic Content

Total phenolic content of the extracts was estimated by a colorimetric assay as described by Singleton and Rossi (1965) with some modifications. Briefly, 50 µl of the crude extracts were mixed with 100 µl of Folin Ciocalteau's phenol reagent (Merck, Germany). After 3 mins, 100 µl of 10% sodium carbonate (Na_2CO_3) (Sigma Aldrich, USA) was added to the reaction mixture and allowed to stand in the dark for 60 mins. The absorbance was measured at 725 nm and the total phenolic content was obtained from a calibration curve using gallic acid (0-10 µg/ml) as a standard reference. Estimation of the phenolic content was carried out in triplicate. The results were mean values ± standard deviations and expressed as mg gallic acid per 100 g samples. All procedures were carefully carried out with minimum exposure of light.

2.5 Determination of Total Flavonoid Content

Measurement of flavonoid concentration of the extracts was based on the method described by D. Kim, Chun, Y. Kim, Moon, and Lee (2003) with some modifications. An aliquot of 100 µl of fruit extract was diluted with 400 µl of distilled water. Afterwards, 30 µl of 5% sodium nitrite ($NaNO_2$), (Sigma, USA) solution was added and allowed to react for 5 min. Following this, 20 µl of 10% aluminium chloride ($AlCl_3·6H_2O$) (Sigma, USA) was added and left to stand for 5 min. Finally, 200 µl of sodium hydroxide (NaOH) (Sigma, USA) was added and the mixture was well-mixed with a vortex. All samples were analyzed in triplicate and the absorbance was measured immediately at 510 nm. Rutin (Sigma, USA) was used to calculate the standard curve and the results were expressed as mg rutin per 100 g samples.

2.6 Statistical Analysis

The results of this study are means and standard deviations of three measurements. Data were further analysed using SAS9.2 (32) statistical software and procedures.

3. Results

Maturity study showed that one cycle of L. fruticosa fruit development (from flowering to fully red skin colour) was about 5 weeks. Eight maturity indices were developed according to the progressive fruit development as indicated by the colour changes. Figure 2 shows all the maturity indices developed for L. fruticosa fruits based on the following criteria; 1 = fully green (2 weeks after flower bloom (FB)), 2 = green with trace of light red (3 weeks after FB), 3 = more green than light red in whole fruit (3-4 weeks after FB), 4 = more light red than green (4 weeks after FB), 5 = light red with trace of green (4-5 weeks after FB), 6 = light red with trace of bright red (5 weeks after FB), 7 = bright red with trace of light red (> 5 weeks after FB), 8 = fully bright red (>5 weeks after FB).

Results of the antioxidant activity of fresh and freeze dried L. fruticosa at eight different maturity stages are presented in Figure 3. From the results, the free radical scavenging activities of both L. fruticosa fruit extracts were observed to be increased with fruit maturation, indicating the decrease of antioxidant activity. The activity of freeze-dried L. fruticosa (FDLF) was significantly different (P < 0.05) as compared to fresh (FLF) at all maturity stages tested. The lowest IC_{50} value was observed at stage 1 with a significant difference (P < 0.05) between FDLF (1.57 ± 0.89 mg/ml) and FLF (3.04 ± 1.05 mg/ml). The IC_{50} values of FDLF for the eight maturity stages ranged from 1.57-6.51 mg/ml. Lee et al. (2007) reported that IC_{50} values lower than 10 mg/ml are indicative of the effective antioxidant activity. However, there was no significance difference (P < 0.05) shown among FDLF extracts when they compared to each other. Fresh extracts (FLF) was observed to show significance difference (P < 0.05) only at stage 5, 6, 7, 8.

Figure 2. *Lepisanthes fruticosa* fruits at eight different maturity stages

Note. 1 = fully green, 2 = green with trace of light red, 3 = more green than light red in whole fruit, 4 = more light red than green, 5 = light red with trace of green, 6 = light red with trace of bright red, 7 = bright red with trace of light red, 8 = fully bright red.

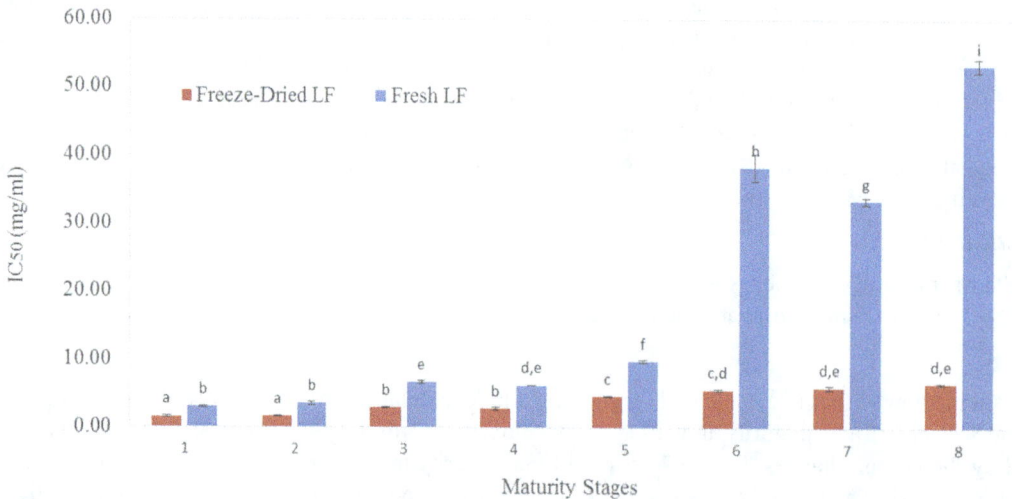

Figure 3. Antioxidant activity (IC_{50}) of freeze-dried and fresh *L. fruticosa* fruit extracts at eight maturity stages

Note. Values marked with different letters are significantly different at $P < 0.05$. The higher the IC_{50} values indicate the stronger ability to act as radical scavengers. Data expressed as means ± standard deviations (n = 3).

Figure 4 shows the total phenolic content (TPC) of FDLF and FLF extracts at different maturity stages. Similar to antioxidant activity, total TPC of FDLF were significantly ($P < 0.05$) higher than FLF at all eight stages tested. Total phenolic content of FDLF extracts ranged from 953.59-158848.96 mg/100 g, with the highest being shown at stage 1 (15848.96 ± 401.82 mg/100 g). The trend was also observed to be decreasing with fruit maturation. Of all the maturity stages, FDLF showed no significance difference ($P < 0.05$) at stage 7 to 8. Meanwhile, there was no significance difference observed for fresh at all maturity stages tested.

On the other hand, results of total flavonoid content (TFC) were on the contrary. From Figure 5, TFC of FLF extracts were significantly higher ($P < 0.05$) than FDLF at almost all stages except for 2, 3 and 6. The highest TFC was observed at stage 1 with FLF and FDLF showed 37.35 ± 0.77 mg/100 g and FD 32.53 ± 1.32 mg/100 g, respectively. FLF was found to be significantly varied ($P < 0.05$) only at stage 1 and 6, whereas FDLF showed significance at stage 1, 2 and 5. Similar to that shown by that TPC (Figure 4), TFC was also decreased as the fruits ripened at most of the maturity stages tested.

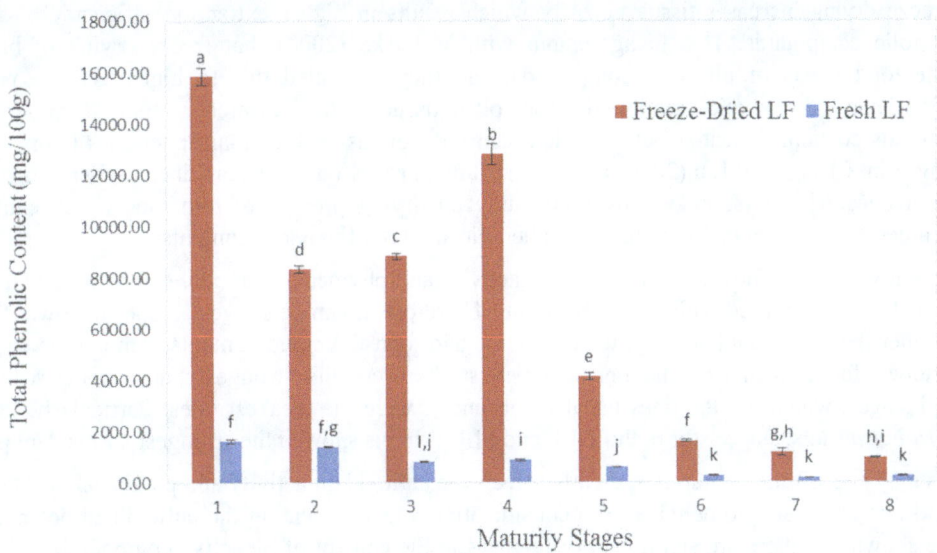

Figure 4. Total phenolic content of *L. fruticosa* fruit extracts at eight maturity stages; LF; *Lepisanthes fruticosa*

Note. Values marked with different letters are significantly different at P < 0.05. Data expressed as means ± standard deviations (n = 3).

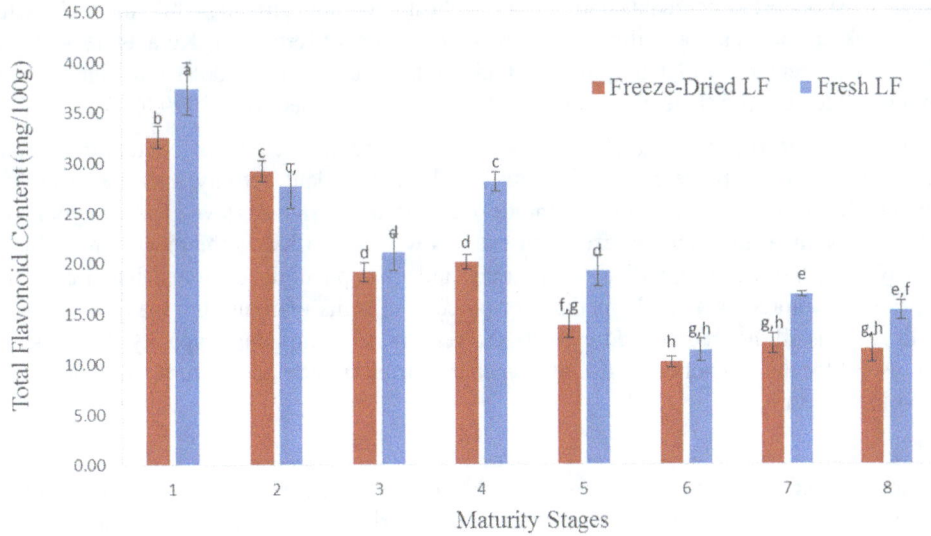

Figure 5. Total flavonoid content of *L. fruticosa* fruit extracts at eight maturity stages; LF; *Lepisanthes fruticosa*

Note. Values marked with different letters are significantly different at P < 0.05. Data expressed as means ± standard deviations (n = 3).

4. Discussion

Phenolics and flavonoids are the main phytochemicals responsible for the antioxidant capacity of fruits. They are important for human health because of their activities as radical scavengers (Hertog, Hollman, & Van de Putte, 1993). The findings have revealed that freeze drying could enhance the antioxidant activity and total phenolic content in *L. frutiosa* fruits, which is against the notion that processed food has lower antioxidant properties than the fresh ones. The FDLF extracts were found to be significantly ($P < 0.05$) stronger radical scavenger than FLF at all maturity stages tested with all IC_{50} values of lower than 10 mg/ml. This is in accordance to Lee et al. (2007) who reported that IC_{50} values lower than 10 mg/ml are indicative of the effective antioxidant activity. Freeze drying process has been popularly applied in laboratory practice and food industry to preserve fresh plant material. Freeze drying increases tissue porosity which results in higher extraction efficiency of antioxidants such as phenolic compounds. This in agreement with Materska (2004) that freeze drying of blackcurrants increased the total levels of phenolic compounds, and they attributed this to higher sample porosity and extraction efficiency of phenolic compounds from plant tissue. Other researcher reported that freeze drying protects materials containing heat-sensitive antioxidant compounds such as plant phenolics (Norshahida et al., 2011). Chang, Lin, Chang, and Liu (2006) also reported that processed fruits contained higher nutritional values than the fresh ones. This is supported by their study on drying process of tomatoes which enhanced their nutritional values by increasing parts of the total phenolic and total flavonoid contents.

The present study has also shown that antioxidant activity and phytochemical contents of both FDLF and FLF extracts seemed to be decreased with fruit maturation. Therefore, it can be suggested that the lower the maturity index, the higher the antioxidant activity, total phenolic and total flavonoid contents. Among the eight maturity stages developed for *L. fruticosa*, the unripe green stages were the strongest antioxidants, while the red, well-matured stages, which are the ones usually consumed, were the weakest ones. Particularly, the youngest stage of *L. fruticosa* fruits, stage 1 for both FDLF and FLF extracts showed the strongest antioxidant potential.

The findings may suggest that maturation greatly affects the antioxidant activity and phytochemical contents of *L. fruticosa*, and this factor should be taken into consideration when evaluating the antioxidant potential of fruits. Studies have shown that there are significant differences in the content of bioactive compounds and antioxidant capacity of tropical fruits at various stages of ripening. Considering quantitative changes of total bioactive compounds, unripe fruits have been reported to have highest level of bioactivities, which decreased at the semi-mature stage and remained unchanged at the commercial harvest maturity (Mphahlele, Stander, Fawole, & Opara, 2014). This is in accordance to Arancibia-Avila et al. (2008) that ripe durian had the highest antioxidant capacity and bioactive compounds. Total polyphenols and flavonoids in the ripe durian were higher than in mature and overripe. Other studies also showed that the unripe yellow fruits of the Greek Service Tree (*Sorbus domestica*) were the strongest antioxidants and had the highest phenolic contents, while the well matured brown fruits were the weakest ones and gave the lowest phenolic contents (Termentzi, Kefalas, & Kokkalou, 2009). This is also the in agreement with Huseyin, Mustafa, Sedat, and Cemal (2008) that during maturation of cranberry, from green to dark red stage, the total phenolic content decreased from 7990 to 4745 mg GAE/kg FW.

In conclusion, the study clearly showed that antioxidant activity and phytochemical content of *L. fruticosa* fruits were significantly affected by processing and maturity. The antioxidant activity and the naturally occurring antioxidant phtytochemicals, phenols and flavonoids reached their highest levels at the youngest stage as indicated by the lowest maturity stage. Freeze-dried extracts were identified as the most remarkable antioxidant with significantly ($P < 0.05$) higher radical scavenging and total phenolic contents than fresh fruit extracts. Freeze drying can be explored as a viable method for processing fruits retaining the maximum amount of these antioxidant properties as all the freeze-dried extracts showed greater antioxidant capacity than fresh ones. These results may be useful for the promotion of use of *L. fruticosa* fruit extracts as a natural antioxidant in functional food production in the future.

References

Arancibia-Avila, P., Toledo, F., Yong, S. P., Soon, T. J., Seong, G. K., Buk, G. H., … Gorinstein, S. (2008). Antioxidant properties of durian fruit as influenced by ripening. *Food and Science Technology*, 1-8. http://dx.doi.org/10.1016/j.lwt.2007.12.001

Chang, C. H., Lin, H. Y., Chang, C. Y., & Liu, Y. C. (2006). Comparisons on the antioxidant properties of fresh, freeze-dried and hot-air-dried tomatoes. *Journal of Food Engineering*, *77*, 478-485. http://dx.doi.org/10.1016/j.jfoodeng.2005.06.061

Fu, M., He, Z., Zhao, Y., Yang, J., & Mao, L. (2009). Antioxidant properties and involved compounds of daylily flowers in relation to maturity. *Food Chemistry*, *114*, 1192-1197. http://dx.doi.org/10.1016/ j.foodchem.2008.10.072

Hertog, M. G. L., Hollman, P. C. H., & Van de Putte, B. (1993). Content of potentially anti-carcinogenic flavonoids of tea infusions, wines and fruit juices. *Journal of Agricultural and Food Chemistry, 41*, 1242-1246. http://dx.doi.org/10.1021/jf00032a015

Huseyin, C., Mustafa, O., Sedat, S., & Cemal, K. (2008). Phytochemical accumulation and antioxidant capacity at four maturity stages of cranberry fruit. *Scientia Horticulturae, 117*, 345-348. http://dx.doi.org/10.1016/j.scienta.2008.05.005

Kim, D., Chun, O. Kim, Y., Moon, H., & Lee, C. (2003). Quantification of phenolics and their antioxidant capacity in fresh plums. *Journal of Agricultural and Food Chemistry, 51*, 6509-6515. http://dx.doi.org/10.1021/jf0343074

Kuspradini, H., Ritmaleni, D. S., & Mitsunaga, T. (2012). Phytochemical and comparative study of anti-microbial activity of *Lepisanthes amoena* leaves extract. *Journal of Biology, Agriculture and Healthcare, 2*(11), 80-86.

Lee, Y. L., Tsung, Y. M., & Mau, J. L. (2007). Antioxidant properties of various extracts from *Hypsizigus marmoreus. Food Chemistry, 104*, 1-9. http://dx.doi.org/10.1016/j.foodchem.2006.10.063

Lim, T. K. (2013). *Edible Medicinal and Non-Medicinal Plants* (Vol. 6, Fruits). Netherlands: Springer. http://dx.doi.org/10.1007/978-94-007-5628-1

Materska, M. (2014). Bioactive phenolics of fresh and freeze-dried sweet and semi-spicy pepper fruits (*Capsicum annuum* L.). *Journal of Functional Foods, 7*, 269-277. http://dx.doi.org/10.1016/j.jff.2014.02.002

Mirfat, A. H. S., & Salma, I. (September 2, 2015). Ceri Terengganu: The future antioxidant superstar. *MARDI Scientia, 6*, 6.

Mirfat, A. H. S., Salma, I., Razali, M., & Umi Kalsum, H. Z. (2015). Antioxidant and nutritional values of selected underutilised *Mangifera* species in Malaysia. *Indian Journal of Plant Genetic Resources, 28*(1), 72-79. http://dx.doi.org/10.5958/0976-1926.2015.00010.8

Mohd. Zainudin, M. A., Abdul Hamid, A., Anwar, F., Osman, A., & Saari, N. (2014). Variation of bioactive compounds and antioxidant activities of carambola (*Averrhoa carambola* L.) at different ripening stages. *Scientia Horticulturae, 172*, 325-331. http://dx.doi.org/10.1016/j.scienta.2014.04.007

Molyneux, P. (2004). The use of the stable free radical diphenylpicryl-hydrazyl (DPPH) for estimating antioxidant activity. *Songklanakarin Journal of Science and Technology, 26*(2), 211-219.

Mphahlele, R. M., Stander, M. A., Fawole, O. A., & Opara, U. L. (2014). Effect of fruit maturity and growing location on the postharvest contents of flavonoids, phenolic acids, vitamin C and antioxidant activity of *Pomegranate juice* (cv. Wonderful). *Scientia Horticulturae, 179*, 36-45. http://dx.doi.org/10.1016/j.scienta.2014.09.007

Norshahida, M. S., Azizah, A. H., Azizah, O., Nazamid, S., Farooq, A., Mohd. Sabri, P. D., & Muhammad Redzuan, H. (2011). Effect of freeze drying on the antioxidant compounds and antioxidant activity of selected tropical fruits. *International Journal of Molecular Sciences, 12*(7), 4678-4692. http://dx.doi.org/10.3390/ijms12074678

Singleton, V. L., & Rossi, J. A. (1965). Colorimetry of total phenolics with phosphomolybdic-phosphotungstic acid reagents. *American Journal of Enology and Viticulture, 16*, 144-158.

Termentzi, A., Kefalas, P., & Kokkalou, E. (2009). Antioxidant activities of various extracts and fractions of *Sorbus domestica* fruits at different maturity stages. *Food Chemistry, 98*, 599-608. http://dx.doi.org/10.1016/j.foodchem.2005.06.025

Vijaya Kumar Reddy, C., Sreeramulu, D., & Raghunath, M. (2010). Antioxidant activity of fresh and dry fruits commonly consumed in India. *Food Research International, 43*, 285-288. http://dx.doi.org/10.1016/j.foodres.2009.10.006

Wetwitayaklung, P., Charoenteeraboon, J., Limmatvapirat, C., & Phaechamud, T. (2012). Antioxidant activities of some Thai and exotic fruits cultivated in Thailand. *Research Journal of Pharmaceutical, Biological and Chemical Sciences, 3*(1), 12-21.

22

Morphological Characteristics, Nutritional Quality, and Bioactive Constituents in Fruits of Two Avocado (*Persea americana*) Varieties from Hainan Province, China

Yu Ge[1,†], Xiongyuan Si[2,†], Jianqiu Cao[3], Zhaoxi Zhou[1], Wenlin Wang[4] & Weihong Ma[1]

[1] Haikou Experimental Station, Chinese Academy of Tropical Agricultural Sciences, Hainan, China

[2] Biotechnology Center, Anhui Agricultural University, Anhui, China

[3] College of Horticulture and Landscape Architecture, Hainan University, Hainan, China

[4] Guangxi South Subtropical Agricultural Science Research Institute, Guangxi, China

Correspondence: Yu Ge, Haikou Experimental Station, Chinese Academy of Tropical Agricultural Sciences, Hainan, China. E-mail: geyu@catas.cn

Weihong Ma, Haikou Experimental Station, Chinese Academy of Tropical Agricultural Sciences, Hainan, China. E mail: zjwhma@163.com

[†] *These authors contributed equally to this work.*

Abstract

We studied the morphological characteristics, nutritional quality, and the bioactive compounds in fruits of two avocado accessions, RN-7 and RN-8, produced in Hainan province, China. Edible and non-edible parts of the fruit (pulp and seed) were compared to evaluate their possible contribution to improve the sustainability of the food and pharmaceutical industries. The basic characteristics evaluated were moisture, ash, total lipid, fatty acid composition, soluble sugars, titratable acid, soluble protein, and minerals. We also measured the concentrations of six types of bioactive compounds; total phenolics, flavonoids, tannin, ascorbic acid, tocopheryl acetate, and carotenoids. Our analyses of the nutritional compositions demonstrated that the pulp of the RN-7 and RN-8 proved to be rich in moisture, total lipid, and soluble protein. The seed, in turn, had higher soluble sugar, titratable acidity, sodium, potassium, calcium, iron, copper, and zinc contents. Other nutritional compositions (ash, magnesium, and manganese) had little differences between the pulp and seed of avocado fruit. With regard to the contents of bioactive compounds, the seed was superior to the pulp in the contents of total phenolics, flavonoids, and tannin. Regarding the concentrations of ascorbic acid, tocopheryl acetate, and total carotenoids, the highest values were found in the pulp. The results of fatty acid compositions displayed that the palmitic, palmitoleic, stearic, oleic, and linoleic acid contents of the pulp were higher than those of the seed, while myristic and arachic acid had higher contents in the seed.

Keywords: chemical compositions, pulp, seed

1. Introduction

Avocado (*Persea americana* Mill.) belongs to the botanical family Lauraceae, and originated in Mexico, or possibly Central or South America (Dreher & Davenport, 2013). Avocado fruit is rich in lipids, proteins, minerals, vitamins, and other nutritients and active ingredients (Dreher & Davenport, 2013; Galvão et al., 2014). The lipid content can comprise 15-30% of the fresh weight of the fruit depending on the cultivar, season, and growing conditions (Meyer & Terry, 2008). It is remarkable is that the lipids in avodado fruit contain ~60% monounsaturated fatty acids and ~13% essential fatty acids such as linoleic and linolenic acid, which are beneficial to human cardiovascular health (Villa-Rodríguez et al., 2011; Giraldo & Moreno-Piraján, 2012; Dreher & Davenport, 2013; Donetti & Terry, 2014; Pedreschi et al., 2016). In contrast to lipid content, the sugar content of avocado fruit is relatively low (Meyer & Terry, 2008, 2010). Hence, avocado fruit is generally recommended for people suffering from diabetes because it is a high-energy food. Natural antioxidants, in particular flavonoids and other groups of polyphenols, have potential uses in the pharmaceutical and food industries because of their many benefits such as reducing the risk of inflammatory diseases and preventing lipid oxidation (Chen et al., 2014). Avocado fruit contains more phenolic compounds than other kinds of tropical and subtropical fruits (Kosinska et al., 2012; Vinha et al., 2013; Chen et al., 2014). Recent studies have also demonstrated that avocado contains other bioactive compounds that are equally beneficial to human heath, such as vitamins and phytochemicals (vitamin C, vitamin E, carotene, etc.) and minerals (phosphorus, sodium, potassium, calcium, and magnesium) (Dreher & Davenport, 2013; Vinha et al., 2013).

It is widely recognized that the avocado was initially introduced to China in 1918 (Papademetriou, 2000). Hundreds of avocado varieties were also introduced to China from the USA, Israel, and other countries in the late 1950s (He, 2012; Zhang et al., 2015). Traditional selective breeding and hybridization were widely used and continue to this day at the Chinese Academy of Tropical Agricultural Sciences, resulting in the introduction of more than a dozen high-quality avocado varieties (Zhang et al., 2015). At present, several Chinese superior avocado varieties are widely grown in various regions of Hainan province.

The avocado cultivars 'Fuerte' and 'Hass' are the most commercially valuable varieties and account for up to two-thirds of the avocado production around the world. Hence, most studies of avocado quality characteristics use these two cultivars (Ashton et al., 2006; Meyer & Terry, 2008, 2010; Hurtado-Fernandez et al., 2011, 2014, 2015; Rodríguez-Carpena et al., 2011; Villa-Rodríguez et al., 2011; Reddy et al., 2012; Donetti & Terry, 2014; Ferreyra et al., 2016; Pedreschi et al., 2016; Rohman et al., 2016). However, no similar studies on Chinese native avocado varieties have been published to date. Thus, the objectives of this study were to determine the morphological characteristics, nutritional quality, and the compositions of bioactive compounds in the fruits of two avocado varieties, RN-7 and RN-8. These two varieties have been recommended by the Chinese Academy of Tropical Agricultural Sciences, located in the province of Hainan, China. The non-edible seeds were also investigated in order to evaluate their potential use as cheap waste production for the food, pharmaceutical, and dermocosmetic industries.

2. Materials and Methods

2.1 Plant Material, Reagents, and Sample Preparation

Fruits of the two avocado accessions, RN-7 and RN-8, used in the present study were obtained from the garden of the avocado germplasm resource affiliated with the Chinese Academy of Tropical Agricultural Sciences in Danzhou city, Hainan province, China (North latitude: 19°11′, East longitude: 108°50′). Eighteen mature fruits of each accession were collected and selected for their firmness and absence of mechanical damage and visible decay. The fruits were immediately transported in standard polystyrene foam boxes that are used for export packaging and held at 5-6 °C until they reached the laboratory. The pulp and seeds were separated from the fruits, homogenized using a kitchen blender, and stored at 4 °C until analysis, which was conducted within one week.

2.2 Morphological Characteristics and Physicochemical Assays

Fruit skin color was determined from eight points on the equatorial area of each fruit per accession using a SPAD-502 Plus colorimeter. Values were obtained in CIELAB scale (L*, a*, b*), and Hue angle and chroma values were calculated. Length, width, and weight were measured for each fruit and its seed. The measurements were performed on nine fruits of each accession.

2.3 Quantification of Nutritional Compositions

2.3.1 Moisture Assay

Fresh avocado pulp and seed samples (5 g) were homogenized separately using a high-speed homogenizer and placed in an air dry oven (GZX-9146 MBE, Shanghai, China) at 105 °C for 6 h. Dry weights of avocado pulp and seed were measured, and moisture contents (g) were calculated from the differences between fresh and dry weight. The results are expressed as g/100 g on a fresh weight basis. The measurements were performed in three replications per accession.

2.3.2 Ash Assay

A quartz crucible was placed in a muffle furnace at 550 °C for 0.5 h, removed when the temperature had dropped below 200 °C, and weighed after reaching room temperature, followed by repeated burning to attain a constant weight. The quartz crucible containing fresh avocado pulp or seed (5 g) was weighed, the samples were fully carbonized until smoke-free, placed in the muffle furnace at 550 °C for 4 h, and removed when the temperature was < 200 °C. The quartz crucible containing the ash was weighed again after cooling to room temperature. The results are displayed as g/100 g on a fresh weight basis. The experiments were performed in triplicate for each accession.

2.3.3 Oil Content Assay

Oil content was evaluated by the method of Villa-Rodríguez et al. (2011) with slight modifications. The avocado pulp and seed were dried and ground to a powder, and the dry powders (5 g) were transferred to a filter paper cylinder after addition of absolute ether at 50 °C. The ratio of material to ether was 1:20. The filtered solutions were extracted until no more oil was present using a Soxhlet extractor. The extracts were then evaporated on a rotary evaporator and weighed. The results are expressed as g/100 g on a fresh weight basis. The experiments were performed in triplicate for each avocado accession.

2.3.4 Soluble Sugar Assay

The total sugar content was determined using the colorimetric anthrone method described by Meyer and Terry (2008) with some modifications. Fresh avocado pulp and seed samples (1 g) were homogenized using a high-speed homogenizer with 5 ml of ethanol (80%) and transferred to 10 mL test tubes. The sample solutions were placed in a boiling water-bath for 10 min after decolorization with activated carbon. After cooling, the solutions were filtered and transferred to 25 mL triangular flasks, and the volumes were adjusted to 25 mL with ethanol (80%). Filtrate samples (1 mL) were mixed with 5 mL anthrone reagent, shaken gently, and then placed in a boiling water-bath for 10 min. After cooling, the absorbance was measured at 620 nm using a spectrophometer (1 mL distilled water plus 5 mL anthrone reagent was used as the blank control). The total sugar content was calculated based on a calibration curve for glucose (R^2 = 0.997). The soluble sugar content was expressed as g/100 g fresh weight of sample. All measurements were performed in triplicate for each accession.

2.3.5 Titratable Acidity Assay

Fresh avocado pulp and seed (5 g) samples were homogenized using a high-speed homogenizer (Heidolph, Diax 900, Germany). The samples were transferred to 50 mL triangular flasks, mixed with 30 mL of distilled water, and incubated in a water-bath at 80 °C for 90 min with constant stirring. After cooling, the solutions were transferred to 50 mL centrifuge tubes, centrifuged at 8000 rpm for 10 minutes. The supernatants were transferred to 50 ml calibrated flasks and the volumes were adjusted to 50 mL with distilled water. Titratable acidity was determined by titrating 15 ml of avocado aqueous extracts with 0.01 M NaOH, using phenolphthalein (1%) as indicator. Distilled water was the blank control. Results were expressed as grams of tartaric acid per 100 g of sample, according to the methodology described by Vinha et al. (2013). The experiments were carried out in triplicate for each accession.

2.3.6 Soluble Protein Assay

Analyses of soluble protein contents were carried out using the Coomassie Blue staining method of Bradford (1976) with some modifications. Fresh avocado pulp and seed samples (1 g) were homogenized using a high-speed homogenizer in 5 ml of distilled water and transferred to 25 mL triangular flasks. The volume was then adjusted to 25 mL with distilled water. After filtration, 0.1 mL samples of the filtrates were transferred to test tubes, 5 mL of the Coomassie Brilliant Blue G-250 reagent was added to each, and they were mixed completely. After incubation for 2 min at room temperature, the absorbance was measured at 595 nm using a Shimadzu UV-1800 spectrophotometer. Distilled water was the blank control. The soluble protein was calculated using 0-100 µg/mL and 1000 µg/mL calibration curves based on bovine serum albumin (R^2 = 0.997). The soluble protein content was expressed as g/100 g fresh weight of sample. All measurements were performed in triplicate for each accession.

2.3.7 Fatty Acid Composition Assay

Analyses of the fatty acid profiles were performed using the method of Villa-Rodríguez et al. (2011) with some modifications. The oils extracted from avocado pulp and seed (40 µL) were saponified at 80 °C for 30 min after addition of 5 mL NaOH-MeOH (0.2 mol/L). After cooling, the solutions were mixed with 2.5 mL BF$_3$-MeOH (14%) and incubated at 80 °C for 30 min to produce methyl esters of the fatty acids. Following this, 2 mL of saturated NaCl and 4 mL n-hexane were added, and the resulting solutions were refluxed for 15 min. The upper layers were then removed, filtered through 0.22 µm membranes, and used for fatty acid GC-MS analyses.

The analyses were performed on an Agilent7890B-7000B GC-MS equipped with a DB-5MS (60 m × 0.25 mm i.d., 0.25 µm film thickness) column using helium (1.2 mL/min) as the carrier gas. The oven temperature was programmed as follows: initial temperature of 100 °C held for 3 min, increased to 180 °C at 3 °C/min, held for 1 min, increased to 220 °C at 1 °C/min, held for 1 min, and finally increased to 280 °C at 5 °C/min and held for 5 min. Injector and detector temperatures were 250 °C and 230 °C, respectively. The mass spectrometer was operated in the electron impact mode at 70 eV in the scan range of 35-400 m/z. The fatty acid methyl esters

(FAMEs) were identified by comparing peaks retention times to those of commercial standards and by comparing the respective ion chromatograms with those reported in the NIST 2011 library. The FAMEs were quantified against methyl nonadecanoate that was added as an internal standard. Quantifications were evaluated from calibration curves of the respective FAMEs ($R^2 \geq 0.995$). FAMEs were expressed as mg/100 g on a fresh weight basis. The experiments were carried out in triplicate for both avocado accessions.

2.3.8 Mineral Elements Assay

Avocado pulp and seeds were dried and ground to a powder. The dry powders (0.5 g) were transferred to 50 mL beakers containing 5 mL concentrated nitric acid and digested at a temperature of 140 °C for 1 h on an electric hot plate. Determination of Na, Mg, K, Ca, Mn, Fe, Cu, and Zn in the previously mineralised samples was conducted in an AAnalyst 400 atomic absorption spectrometer (Perkin Elmer Ltd., Shanghai, China). Concentrated nitric acid was used as the blank control. Quantification was obtained from a calibration curve of certified reference materials ($R^2 \geq 0.995$). Mineral elements were expressed as mg/100 g on a fresh weight basis. The experiments were performed in triplicate for both accessions.

2.4 Quantification of Bioactive Compounds

2.4.1 Total Polyphenolic Assay

Total phenolic content was determined based on the method of Villa-Rodríguez et al. (2011) with slight modifications. Dry powder residues of avocado pulp and seed (0.2 g, after lipid removal) were dissolved in 5 mL ethanol (50%), subjected to ultrasonic extraction for 15 min, and centrifuged at 8000 rpm for 10 minutes. The 0.1 mL extracts were transferred to centrifuge tubes, mixed with 3 mL distilled water, 0.25 mL Folin and Ciocalteau phenol reagent (1 N), and 0.75 mL Na_2CO_3 (20%), and then diluted with distilled water to a volume of 5 mL. After incubation for 30 min in the dark at room temperature, the absorbance was read at 760 nm using a Shimadzu UV-1800 spectrophotometer. Distilled water was used as the blank control. Total phenolic compounds were calculated using a calibration curve of gallic acid ($R^2 = 0.997$) and displayed as gallic acid equivalents (mg GAE/100 g on a fresh weight basis). The experiments were performed in triplicate for the two accessions.

2.4.2 Tannin Assay

Fresh avocado pulp and seed samples (5 g) were homogenized separately in 80 mL distilled water using a high-speed homogenizer (Heidolph, Diax 900, Germany), placed in a boiling water-bath for 30 min, and then adjusted to a volume of 100 mL with distilled water. Subsamples (5 mL) of the solutions were transferred to centrifuge tubes and centrifuged at 8000 rpm for 5 minutes. For the assays, 1 mL aliquots of the extracts were combined with 1 mL $Na_2WO_4{\cdot}2H_2O$ and Na_2MoO_4 mixture solution, 3 mL Na_2CO_3, and 5 mL distilled water, and incubated for 2 hour in the dark at room temperature to allow for color development. The 0 mg/L sample was used as the blank control. Tannin content was measured spectrophotometrically on a Shimadzu UV-1800 spectrophotometer at 765 nm. The values on the calibration curve were referred to total polyphenolic assay and displayed as gallic acid equivalents. The results are displayed as mg GAE/100 g on a fresh weight basis. The experiments were performed in triplicate.

2.4.3 Total Flavonoid Assay

Total flavonoid concentrations were determined by the method of Villa-Rodríguez et al. (2011) with some modifications. As in the polyphenol assay, dry powder residues of avocado pulp and seed (0.2 g, after lipid removal) were dissolved in 5 mL ethanol (50%), extracted ultrasonically for 15 min, and centrifuged at 8000 rpm for 10 minutes. Aliquots of the extracts (0.1 mL) were transferred to centrifuge tubes, mixed with 2 mL distilled water and 0.15 mL $NaNO_2$ (5%), and incubated for 6 min. After 0.15 mL of $AlCl_3{\cdot}6H_2O$ (10%) was added, the extracts were allowed to stand for 1 min, and 2 mL of NaOH (1 M) were then added. The volume was adjusted to 10 mL with distilled water. For the seed samples, 0.5 mL aliquots of the extracts were diluted with distilled water to 6 mL, and the absorbance was determined at 510 nm with a Shimadzu UV-1800 spectrophotometer. Distilled water was used as the blank control. Total flavonoid concentrations were calculated using a calibration curve of rutin ($R^2 = 0.995$) and expressed as rutin equivalents (mg RE/100 g on a fresh weight basis). The experiments were performed in triplicate.

2.4.4 Ascorbic Acid Assay

Ascorbic acid content was determined using the modified 2,6-dichlorophenolindophenol method (Franck et al., 2003). Samples of fresh avocado pulp and seed (5 g) were homogenized separately using a high-speed homogenizer (Heidolph, Diax 900, Germany) in 5 ml of oxalic acid (2%). The volume was adjusted to 100 mL with distilled water. The sample solutions were transferred to centrifuge tubes, centrifuged at 8000 rpm for 10 minutes, and 10 mL aliquots of the extracts were titrated with 2,6-dichlorophenolindophenol. Oxalic acid (2%)

was used as the blank control. Ascorbic acid was expressed as mg/100 g on a fresh weight basis. The experiments were performed in triplicate.

2.4.5 Tocopheryl Acetate Assay

Samples of fresh avocado pulp and seed (5 g) were homogenized using a high-speed homogenizer as described above for the moisture, acidity, and soluble protein assays. After addition of 50 mL ethanol (95%), 10 mL ethyl ether, and 20 mL NaOH solution (50%), the solutions were transferred to 250 mL saponification flasks and saponified by refluxing in a boiling water bath for 30 min. The saponification reactions were transferred to 250 mL separatory funnels and mixed with 40 mL distilled water and 50 mL ethyl ether. After shaking vigorously for 1 min, the supernatants were transferred to 250 mL volumetric flasks, and the volumes were adjusted to 250 mL with ethyl ether. Samples (5 mL) of the extracts were heated at 90 °C in a water-bath until the flask contents were reduced almost to dryness. The end-products were dissolved in methanol to a final volume of 10 mL. The absorbance at 284 nm was measured using a Shimadzu UV-1800 spectrophotometer. Methanol was used as the blank control. Tocopheryl acetate was expressed as mg/100 g on a fresh weight basis. The experiments were performed in triplicate.

2.4.6 Total Carotenoid Assay

Fresh avocado pulp and seed samples (5 g) were homogenized using a high-speed homogenizer as described in the previous section. The volume was adjusted to 100 mL with petroleum ether. The sample solutions were filtered through sodium sulphate, transferred to 100 ml volumetric flasks, and then diluted to 100 ml with petroleum ether. After incubation for 24 h at room temperature, total carotenoid content was measured spectrophotometrically at 445 nm using a Shimadzu UV-1800 spectrophotometer. Petroleum ether was the blank control. The results are presented as β-carotene equivalents (mg/100 g on a fresh weight basis). The experiments were carried out in triplicate.

2.5 Statistical Analyses

The data were analyzed using SPSS version 20.0 software (SPSS Inc., Chicago, IL, USA). The results were presented as the mean ± standard deviation of three or nine measurements.

3. Results and Discussion

3.1 Morphological Characteristics and Physicochemical Analyses

Table 1 shows the morphological parameters of the fruits of two avocado accessions. The fruits of RN-7 are ovate, large in size (505.56±39.09 g), and are larger than those of many of the widely cultivated avocado varieties such as 'Hass', 'Fuerte', 'Gwen', and 'Lamb Hass', etc. (Gómez-López et al., 1999, 2002; Schaffer et al., 2012). While the fruits of RN-8 were pyriform and of medium size (336.67±26.46 g). The seeds of RN-7 and RN-8 were all of medium size (50.90±9.08 and 69.73±4.26 g) and nearly spherical in shape. Others have reported the seed weight of more than thirty avocado cultivars from around the world from 26.54 g for Duke to 115.82 g for Ceniap 2 (Gómez-López et al., 1999, 2002; Rodríguez-Carpena et al., 2011; Galvão et al., 2014). This indicates that these two Chinese native avocado accessions had the medium-sized seeds in comparison with foreign avocado varieties. Lightness and chroma of RN-7 avocado fruit were greater than those of RN-8; however, the hue angle of RN-7 avocado fruit was the same as RN-8, which suggest that the degree of saturation and color intensity of RN-7 avocado fruit are higher than those of RN-8 avocado fruit, but peel color of RN-7 and RN-8 were very similar.

Table 1. Morphological characteristics and physicochemical parameters (mean value ± standard deviation, $n = 9$) of fruits of two Chinese avocado accessions

Morphological characteristic	RN-7	RN-8
Fruit weight (g)	505.56±39.09	336.67±26.46
Fruit length (cm)	12.30±0.50	16.88±1.30
Fruit diameter (cm)	8.92±0.38	6.91±0.31
Seed weight (g)	69.73±4.26	50.90±9.08
Seed length (cm)	4.49±0.24	5.90±0.60
Seed diameter (cm)	4.71±0.18	4.23±0.22
L*	45.27±2.35	38.35±2.36
Hue (°)	1.09±0.02	1.09±0.04
Chroma (%)	40.37±3.02	31.01±4.66

Note. L* = Luminance; Hue (°) = Hue angle.

3.2 Nutritional Compositions Analyses

The avocado pulp has a higher water and total lipid content than does the seed (Table 2), which is in agreement with previous studies (Rodríguez-Carpena et al., 2011; Vinha et al., 2013; Galvão et al., 2014). The total lipid contents in the pulp of RN-7 and RN-8 fruits were all ≤ 8%, placing them in the group of varieties with low oil content, which are inferior to the widely grown 'Hass' cultivar (Gómez-López, 1999, 2002; Rodríguez-Carpena et al., 2011; Villa-Rodríguez et al., 2011; Dreher and Davenport, 2013). The pulp of RN-7 and RN-8 contains more soluble protein (0.42 g/100 g), more than twice that measured in the seed in the present study (Table 2), although previous studies found that soluble protein levels in the pulp are lower than levels in the peel and seed (Rodríguez-Carpena et al., 2011; Vinha et al., 2013; Galvão et al., 2014). Comparable levels of ash differed between the pulp and the seed depending on the variety (Rodríguez-Carpena et al., 2011; Vinha et al., 2013; Galvão et al., 2014). Similarly, we found that comparisons of the ash content between the pulp and seed were the exact opposite for RN-7 and RN-8 (Table 2). The soluble sugar content and titratable acidity of the pulp were higher than in the seed, which agreed with the results of a previous study (Vinha et al., 2013). In addition, recent research has shown that soluble sugars may be the precursors of lipid synthesis in avocado fruit (Kilaru et al., 2015). This was supported in the present study, where we found that the higher the lipid content, the lower the soluble sugar levels in the pulp and seed of avocado fruit (Table 2). The contents of six mineral elements (sodium, potassium, calcium, iron, copper, and zinc) were higher in the seed than in the pulp, but two other mineral elements (magnesium and manganese) showed very small differences between the pulp and seed (Table 2).

Table 2. Nutritional compositions (mean value ± standard deviation, proximates for g/100 g FW and minerals for mg/100 g FW, *n* = 3) of pulp and seed of fruits of two Chinese avocado accessions

Nutritional composition	RN-7		RN-8	
	Pulp	Seed	Pulp	Seed
Proximates				
Moisture	82.85±0.17	69.61±0.20	83.59±0.32	69.71±0.38
Ash	0.52±0.00	0.64±0.01	0.74±0.01	0.63±0.01
Total lipid	7.33±0.15	1.40±0.04	6.53±0.14	3.18±0.17
Soluble sugar	0.56±0.02	1.78±0.03	0.72±0.05	2.43±0.03
Titratable acidity	1.78±0.03	2.57±0.08	2.63±0.03	2.87±0.03
Soluble protein	0.42±0.03	0.19±0.02	0.42±0.02	0.16±0.01
Minerals				
Sodium	0.52±0.04	1.54±0.01	0.47±0.04	1.11±0.09
Magnesium	1.40±0.01	1.41±0.06	1.73±0.02	1.83±0.03
Potassium	247.01±10.58	336.45±25.24	240.24±1.24	310.95±0.64
Calcium	10.87±0.52	18.51±0.76	9.01±0.20	14.32±0.47
Manganese	0.03±0.00	0.05±0.00	0.03±0.00	0.03±0.00
Iron	0.91±0.03	1.18±0.09	0.80±0.11	1.48±0.05
Copper	0.18±0.00	0.35±0.00	0.16±0.00	0.33±0.03
Zinc	0.05±0.00	0.07±0.00	0.06±0.00	0.10±0.00

3.3 Fatty Acid Profiles Analyses

The fatty acid compositions of the pulp and seed oils are presented in Table 3. Eight fatty acids were detected in the pulp and seeds of fruits of avocado accessions RN-7 and RN-8. The palmitic (C16:0), palmitoleic (C16:1), stearic (18:0), oleic (C18:1), and linoleic (18:2) acid contents of the pulp were higher than those in the seed. Myristic (C14:0) and arachic (C20:0) acid levels were higher in the seed, while the linolenic acid (C18:3) levels differed considerably between the two accessions. These results differed slightly from those reported previously, since Galvão et al. (2014) suggested that levels of oleic acid (C18:1) in the pulp and linoleic (18:2) in the seed were higher among the cultivars 'Fortuna', 'Collinson', and 'Barker'. Nevertheless, other fatty acids such as myristic (C14:0), palmitic (C16:0), palmitoleic (C16:1), stearic (18:0), and arachic (C20:0) acids showed different comparable levels between the pulp and seed among these three cultivars. The contents of palmitic (C16:0), oleic acid (C18:1), and linoleic (18:2) acids in the pulp of RN-7 and RN-8 avocado fruits all exceeded 1000 mg/100 g fresh weight, and represented the majority of the total fatty acids quantified. These results agree with those reported previously by other authors for avocado cultivars such as 'Hass' and 'Fuerte', etc. (Ozdemir & Topuz, 2004; Meyer & Terry, 2008, 2010; Villa-Rodríguez et al., 2011; Dreher & Davenport, 2013; Donetti & Terry, 2014; Galvão et al., 2014; Ferreyra et al., 2016; Pedreschi et al., 2016; Rohman et al., 2016). In all of them, more than 63% of total fatty acids (TFA) of the pulp and seed of avocado were unsaturated, the remaining were saturated (37%) (Table 3). Total unsaturated fatty acids (ΣUFA), total saturated fatty acids (ΣSFA), and TFA were all much higher in the pulp of two accessions than in the seed (Table 3). The ratios of ΣUFA/ΣSFA were larger than 1.0 for the pulp and seed of two accessions, especially in the pulp of RN-8, which indicated that avocado could serve as a food supplement in the diet to decrease the level of cholesterol and fats, preventing the risk of cardiovascular disease (Richard et al., 2008).

Table 3. Fatty acid compositions (mean value ± standard deviation, mg/100 g FW, $n = 3$) of the pulp and seed oils obtained from fruits of two Chinese avocado accessions

Fatty acids	RN-7		RN-8	
	Pulp	Seed	Pulp	Seed
Saturated fatty acids (SFA)				
Myristic acid (C14:0)	15.46±0.44	28.01±1.00	12.90±0.16	25.92±1.17
Palmitic acid (C16:0)	2431.81±97.27	288.07±29.22	1727.92±6.73	359.00±24.52
Stearic acid (18:0)	83.71±4.23	44.92±1.81	73.20±2.77	46.31±2.00
Arachic acid (C20:0)	25.38±0.37	40.59±0.11	24.02±0.24	40.62±0.15
Mono-unsaturated fatty acids (MUFA)				
Palmitoleic acid (C16:1)	526.63±26.46	53.99±4.30	393.11±6.36	48.32±5.52
Oleic acid (C18:1)	2090.11±169.83	266.08±14.39	1999.57±181.68	174.51±10.41
Poly-unsaturated fatty acids (PUFA)				
Linoleic acid (18:2)	1615.41±98.16	420.23±6.78	1481.21±15.94	626.85±82.75
Linolenic acid (C18:3)	83.20±5.49	74.37±4.70	44.46±3.54	61.30±5.49
ΣSFA	2556.36±38.73	401.59±12.19	1838.04±9.9	471.85±12.27
ΣUFA	4315.35±84.65	814.67±27.32	3918.35±48.61	910.98±26.05
TFA	6871.71±123.38	1216.26±39.51	5756.39±58.51	1382.83±38.32
ΣUFA/ΣSFA	1.69±0.02	2.03±0.01	2.13±0.02	1.93±0.01

Note. ΣSFA = total unsaturated fatty acids; ΣUFA = total unsaturated fatty acids; TFA = total fatty acids.

3.4 Bioactive Compounds Analyses

Remarkably, the total phenolic, flavonoid, and tannin contents of the seed were 10- to 40-fold greater than in the pulp (Table 4). Previous studies also found that total phenolic and flavonoid contents of the seed far exceeded those of the pulp, and these compounds possess strong *in vitro* antioxidant activity and antimicrobial potential (Hidalgo et al., 2010; Rodríguez-Carpena et al., 2011; Kosinska et al., 2012; Vinha et al., 2013). Therefore, the avocado seed, as a byproduct, could be an interesting and inexpensive raw material for a functional food ingredient or an antioxidant additive (Rodríguez-Carpena et al., 2011; Kosinska et al., 2012). Vinha et al. (2013) suggested that the ascorbic acid and total carotenoid contents of the seed are superior to those in the pulp. However, considering the contents of ascorbic acid and total carotenoids, we found that the highest values were in the pulp (Table 4). Tocopheryl acetate was not detected in the seed of either of the two accessions; however, tocopheryl acetate was present in almost the same amounts in the pulp and seed (Vinha et al., 2013).

Table 4. Bioactive constituents (mean value ± standard deviation, mg/100 g FW, $n = 3$) present in the pulp and seed of fruits of two Chinese avocado accessions

Bioactive constituents	RN-7		RN-8	
	Pulp	Seed	Pulp	Seed
Total phenolics	44.39±2.81	685.58±13.45	109.39±5.29	798.52±54.04
Flavonoids	43.85±6.37	1636.25±50.88	-	936.60±56.91
Tannin	0.09±0.01	2.02±0.04	0.05±0.01	2.45±0.09
Ascorbic acid	18.57±0.31	10.03±0.00	17.15±1.28	12.02±0.20
Tocopheryl acetate	1.28±0.03	-	2.32±0.11	-
Total carotenoids	2.14±0.18	2.02±0.04	2.02±0.02	1.98±0.05

4. Conclusions

This study is the first to characterize and evaluate the morphological characteristics and the contents of nutrients and bioactive constituents in fruits of two avocado cultivars grown in Hainan province, China. The pulp and seeds were thoroughly compared in fruits from cultivars RN-7 and RN-8. For the main nutrients, the highest lipid and soluble protein contents were found in the pulp, but total phenolics, flavonoids, and tannin concentrations in the seed were 10- to 40-fold higher than in the pulp. The pulp and seed of both avocado accessions were found to contain a variety of fatty acids with carbon chain lengths of C14, C16, C18, and C20 and varying degrees of unsaturation. Relatively high levels of palmitic, oleic, and linoleic acids were present in the pulp.

References

Ashton, O. B. O., Wong, M., & McGhie, T. K. (2006). Pigments in avocado tissue and oil. *Journal of Agricultural Food Chemistry, 54*, 10151-10158. http://dx.doi.org/10.1021/jf061809j

Bradford, M. M. (1976). A rapid and sensitive method for the quantitation of microgram quantities of protein utilizing the principle of protein-dye binding. *Analytical Biochemistry, 72*, 248-254. http://dx.doi.org/10.1016/0003-2697(76)90527-3

Chen, G. L., Chen, S. G., Zhao, Y. Y., Luo, C. X., Li, J., & Gao, Y. Q. (2014). Total phenolic contents of 33 fruits and their antioxidant capacities before and after *in vitro* digestion. *Industrial Crops and Products, 57*, 150-157. http://dx.doi.org/10.1016/j.indcrop.2014.03.018

Donetti, M., & Terry, L. A. (2014). Biochemical markers defining growing area and ripening stage of imported avocado fruit cv. Hass. *Journal of Food Composition and Analysis, 34*, 90-98. http://dx.doi.org/10.1016/j.jfca.2013.11.011

Dreher, M. L., & Davenport, A. J. (2013). Hass avocado composition and potential health effects. *Critical Reviews in Food Science and Nutrition, 53*, 738-750. http://dx.doi.org/10.1080/10408398.2011.556759

Ferreyra, R., Selles, G., & Saavedra, J. (2016). Identification of pre-harvest factors that affect fatty acid profiles of avocado fruit (*Persea americana* Mill) cv. 'Hass' at harvest. *South African Journal of Botany, 104*, 15-20. http://dx.doi.org/10.1016/j.sajb.2015.10.006

Franck, C., Baetens, M., Lammertyn, J., Scheerlinck, N., Davey, M. W., & Nicola, B. M. (2003). Ascorbic acid mapping to study core breakdown development in "Conference" pears. *Postharvest Biology and Technology, 30*, 133-142. http://dx.doi.org/10.1016/S0925-5214(03)00108-X

Galvão, M. D. S., Narain, N., & Nigam, N. (2014). Influence of different cultivars on oil quality and chemical characteristics of avocado fruit. *Food Science and Technology, Campinas, 34*, 539-546. http://dx.doi.org/10.1590/1678-457x.6388

Giraldoa, L., & Moreno-Piraján, J. C. (2012). Lipase supported on mesoporous materials as a catalyst in the synthesis of biodiesel from *Persea americana* mill. oil. *Journal of Molecular Catalysis B: Enzymatic, 77*, 32-38. http://dx.doi.org/10.1016/j.molcatb.2012.01.001

Gómez-López, V. M. (1999). Characterization of avocado (*Persea americana* Mill.) varieties of low oil content. *Journal of Agricultural Food Chemistry, 47*, 2707-2710. http://dx.doi.org/10.1021/jf981206a

Gómez-López, V. M. (2002). Fruit characterization of high oil content avocado varieties. *Scientia Agricola, 59*, 403-406. http://dx.doi.org/10.1590/S0103-90162002000200030

He, G. X. (2012). The situation and developing prospect of avocado research and production in Guangxi. *Journal of Guangxi Vocational and Technical College, 5*, 1-6.

Hidalgo, M., Sánchez-Moreno, C., & Pascual-Teresa, S. (2010). Flavonoid-flavonoid interaction and its effect on their antioxidant activity. *Food Chemistry, 121*, 691-696. http://dx.doi.org/10.1016/j.foodchem.2009.12.097

Hurtado-Fernandez, E., Carrasco-Pancorbo, A., & Fernandez-Gutierrez, A. (2011). Profiling LC-DAD-ESI-TOF MS method for the determination of phenolic metabolites from avocado (*Persea americana*). *Journal of Agricultural Food Chemistry, 59*, 2255-2267. http://dx.doi.org/10.1021/jf104276a

Hurtado-Fernandez, E., Pacchiarotta, T., Mayboroda, O. A., Fernandez-Gutierrez, A., & Carrasco-Pancorbo, A. (2014). Quantitative characterization of important metabolites of avocado fruit by gas chromatography coupled to different detectors (APCI-TOF MS and FID). *Food Research International, 62*, 801-811. http://dx.doi.org/10.1016/j.foodres.2014.04.038

Hurtado-Fernandez, E., Pacchiarotta, T., Mayboroda, O. A., Fernandez-Gutierrez, A., & Carrasco-Pancorbo, A. (2015). Metabolomic analysis of avocado fruits by GC-APCI-TOF MS: effects of ripening degrees and fruit varieties. *Analytical and Bioanalytical Chemistry, 407*, 547-555. http://dx.doi.org/10.1007/s00216-014-8283-9

Kilaru, A., Cao, X., Dabbs, P. B., Sung, H. J., Rahman, M. M., Thrower, N., ... Ohlrogge, J. B. (2015). Oil biosynthesis in a basal angiosperm: transcriptome analysis of *Persea Americana* mesocarp. *BMC Plant Biology, 15*, 203. http://dx.doi.org/10.1186/s12870-015-0586-2

Kosinska, A., Karamac, M., Estrella, I., Hernandez, T., Bartolome, B., & Dykes, G. A. (2012). Phenolic compound profiles and antioxidant capacity of *Persea americana* Mill. peels and seeds of two varieties. *Journal of Agricultural Food Chemistry, 60*, 4613-4619. http://dx.doi.org/10.1021/jf300090p

Meyer, M. D., & Terry, L. A. (2008). Development of a rapid method for the sequential extraction and subsequent quantification of fatty acids and sugars from avocado mesocarp tissue. *Journal of Agricultural Food Chemistry, 56*, 7439-7445. http://dx.doi.org/10.1021/jf8011322

Meyer, M. D., & Terry, L. A. (2010). Fatty acid and sugar composition of avocado, cv. Hass, in response to treatment with an ethylene scavenger or 1-methylcyclopropene to extend storage life. *Food Chemistry, 121*, 1203-1210. http://dx.doi.org/10.1016/j.foodchem.2010.02.005

Ozdemir, F., & Topuz, A. (2004). Changes in dry matter, oil content and fatty acids composition of avocado during harvesting time and post-harvesting ripening period. *Food Chemistry, 86*, 79-83. http://dx.doi.org/10.1016/j.foodchem.2003.08.012

Papademetriou, M. K. (2013). *Avocado Production in Asia and the Pacific*. Bangkok, BK: FAO Publisher.

Pedreschi, R., Hollak, S., Harkema, H., Otma, E., Robledo, P., Somhorst, D., ... Defilippi, B. G. (2016). Impact of postharvest ripening strategies on 'Hass' avocado fatty acid profiles. *South African Journal of Botany, 103*, 32-35. http://dx.doi.org/10.1016/j.sajb.2015.09.012

Reddy, M., Moodley, R., & Jonnalagadda, S. B. (2012). Fatty acid profile and elemental content of avocado (*Persea americana* Mill.) oil effect of extraction methods. *Journal of Environmental Science and Health, Part B, 47*, 529-537. http://dx.doi.org/10.1080/03601234.2012.665669

Richard, D., Kefi, K., Barbe, U., Bausero, P., & Visioli, F. (2008). Polyunsaturated fatty acids as antioxidants. *Pharmacological Research, 57*, 451-455. http://dx.doi.org/10.1016/j.phrs.2008.05.002

Rodríguez-Carpena, J. G., Morcuende, D., Andrade, M. J., Kylli, P., & Estevez, M. (2011). Avocado (*Persea americana* Mill.) phenolics, *in vitro* antioxidant and antimicrobial activities, and inhibition of lipid and protein oxidation in porcine patties. *Journal of Agricultural Food Chemistry, 59*, 5625-5635. http://dx.doi.org/10.1021/jf1048832

Rohman, A., Windarsih, A., Riyanto, S., Sudjadi, G., Ahmad, S. A. S., Rosman, A. S., & Yusof, F. M. (2016). Fourier transforms infrared spectroscopy combined with multivariate calibrations for the authentication of avocado oil. *International Journal of Food Properties, 19*, 680-687. http://dx.doi.org/10.1080/10942912.2015.1039029

Schaffer, B., Wolstenholme, B. N., & Whiley, A. W. (2012). *The Avocado: Botany, Production and Uses* (2nd ed.). Croydon, CD: CPI Group (UK) Ltd.

Villa-Rodríguez, J. A., Molina-Corral, F. J., Ayala-Zavala, J. F., Olivas, G. I., & Gonzalez-Aguilar, G. A. (2011). Effect of maturity stage on the content of fatty acids and antioxidant activity of 'Hass' avocado. *Food Research International, 44*, 1231-1237. http://dx.doi.org/10.1016/j.foodres.2010.11.012

Vinha, A. F., Moreira, J., & Barreira, S. V. P. (2013). Physicochemical parameters, phytochemical composition and antioxidant activity of the Algarvian avocado (*Persea americana* Mill.). *Journal of Agricultural Science, 5*, 100-109. http://dx.doi.org/10.5539/jas.v5n12p100

Zhang, L., Zhang, D. S., & Liu, K. D. (2015). Environmental analysis and countermeasures for industrial development of Hainan avocado. *Chinese Journal of Agricultural Resources and Regional Planning, 36*, 78-84.

23

Enhancing the Defensive Mechanism of Lead Affected Barley (*Hordeum vulgare* L.) Genotypes by Exogenously Applied Salicylic Acid

Taskeen Arshad[1], Nazimah Maqbool[1], Farrukh Javed[1], Abdul Wahid[1] & Muhammad Usman Arshad[2]

[1] Department of Botany, University of Agriculture, Faisalabad, Pakistan

[2] Department of Plant Breeding and Genetics, University of Agriculture, Faisalabad, Pakistan

Correspondence: Nazimah Maqbool, Department of Botany, University of Agriculture, Faisalabad 38040, Pakistan. E-mail: nazimahmaqbool@gmail.com

Abstract

Lead is a non-essential element reduced plant growth and development that may cause multifarious disturbance in physiological, biochemical and structural integrity of plants. SA is an efficient signal molecule induces systemic resistance responses that control local defense reactions in plants. To evaluate the effect of SA on photosynthetic activity of Pb affected plants therefore a pot experiment was conducted on barley genotype Juo-93 and Juo-87 in the Old Botanical Garden, University Of Agriculture, Faisalabad, Pakistan. Three treatment levels (0, 100 and 200 µM) of lead sulphate ($PbSO_4$) were applied in thrice replication with or without SA (0.5 mM) along half strength of Hoagland's solution till the termination of experiment. The experiment was arranged in a completely randomized design. During course of study growth and pigments modulations were recorded. The result indicated that growth parameters such as root and shoot length, leaf area, fresh weight of shoot and root, dry weight of shoot and root were reduced under lead toxicity. Pb stress damaged the photosynthetic pigments such as Chlorophyll a, chlorophyll b, total chlorophyll and carotenoids but chlorophyll a/b was increased under Pb stress whereas exogenous application of SA alleviated the negative effect of lead toxicity. Juo-93 showed more tolerance to Pb toxicity as compared to Juo-87.

Keywords: carotenoid, lead toxicity, barley, chlorophyll, salicylic acid

1. Introduction

Heavy metal stress is one of the major abiotic stresses that cause environmental pollution in recent decades (Gisbert et al., 2003; Castro et al., 2011). These metals unlike other organic pollutants are not degraded and converted into harmless compounds via biological processes. Elevated concentrations of both essential and non-essential heavy metals in the soil can lead to toxicity symptoms and growth inhibition in most plants (Hall, 2002; Li et al., 2010). Toxicity may result from the binding of metals to sulphydryl groups in proteins, leading to inhibition of activity or disruption of structure, or from displacement of an essential element (Capuana, 2011). Heavy metals effects chlorophyll content in plants by interfering with chlorophyll synthesis either through direct inhibition of an enzymatic step or by inducing deficiency of an essential nutrient (Meers et al., 2010). Lead is very toxic environmental pollutant which is widely distributed in the soil and contaminated water. The key enzyme for chlorophyll biosynthesis i.e. α-amino laevulinate dehydrogenase is strongly inhibited by Pb ions (D. D. K. Prasad & A. R. K. Prasad, 1987). Photosynthesis is especially affected by lead exposure (Bazzaz et al., 1975); chlorophyll and carotenoids contents, photosynthetic rate and CO_2 assimilation are strongly decreased (Eun et al., 2002).

Seed treatment or foliar application of chemicals like glycinebetaine, kinetin, salicylic acid (Gunes, 2007; Karlidag, 2009) may increase yield of different crops due to reduction in stress induced inhibition of plant growth (Khan et al., 2003; Elwana & El-Hamahmyb, 2009). Salicylic acid is an endogenous growth regulator of phenolic nature and acts as potential non-enzymatic antioxidant which participates in the regulation of many physiological processes in plants (Simaei et al., 2012; Horvath et al., 2007), such as stomatal closure, photosynthesis, ion uptake, inhibition of ethylene biosynthesis, transpiration and stress tolerance (Khan et al., 2003; Arfan et al., 2007). Chlorophyll and carotenoid contents of maize leaves were increased upon treatment with SA under lead stress (Bosch et al., 2007; Najafian et al., 2009). Barley (*Hordeum vulgare* L.) is a highly adaptable cereal grain and considered to be the most drought and salinity tolerant among cereals (Ceccarelli et al., 1987; Belaid & Morris, 1991).

The aim of this project is to explore the response of barley under Pb contamination and ameliorating potential of SA towards metal toxicity.

2. Materials and Method

Seeds of two barley genotypes Juo-93 and Juo-87 were obtained from Ayub Agriculture Research Institute (AARI), Faisalabad, Pakistan. Ten seeds of good vigor of both genotypes were graded and sown in sand filled (5 kg) plastic pots (30 × 20 cm size). The pots were kept in a net house under bright sunlight. The climatic conditions were 30±2 °C with 50% relative humidity at the time of experiment. Lead sulphate ($PbSO_4$) of 100 and 200 µM along with 0.5 mM SA was applied by mixing with Hoagland's solution till the termination of experiment. Control was without treatment of SA and Pb for both barley genotypes for comparison. After 58 days of germination, prior to uprooting of plants, the intact plants were measured for shoot length and leaf area. Leaves of Juo-93 and Juo-87 varieties for each treatment was sampled and stored in an ice bath, immediately brought in to the Lab and grinded in 80% acetone for the estimation of photosynthetic pigments following spectrophotometric method (Hitachi-U-2001, Japan) (Yoshida et al., 1976; Davies, 1976). Plants were uprooted from three replicates for the measurement of root length, shoot and root fresh weight. For dry weight determination, the shoot and root parts were properly washed with water, blotted dry, wrapped in labelled paper bags and kept in an oven at 70 °C for one week.

The three factors factorial experiment was arranged in a completely randomized way. The collected data was statistically analyzed using LSD Test and their means were compared applying DMRT comparison Test.

3. Results

Genotypes Juo-93 and Juo-87 showed significant ($P < 0.05$) differences for growth attributes under SA and Pb treatments. Juo-93 performed better than Juo-87 under high concentration of Pb. Salicylic acid (SA) application was found beneficial for the growth of both barley genotypes. Length as well as fresh and dry weights of shoot and root was enhanced by SA as compared to control. Juo-93 showed great shoot and root length, their fresh weights as compared to Juo-87 in control. Rooting medium application of SA enhanced length and fresh weights of both upper and lower parts of both genotypes. Pb level of 100 µM was less toxic for shoot length and fresh weights whereas 200 µM Pb was more damaging for root length and fresh weights. Although SA application in combination with Pb (100 µM and 200 µM) treated plants improved the growth of both genotypes. Similar response in shoot and root dry weights were observed in the presence of SA and Pb treatments as shown by other growth parameters (Figure 1).

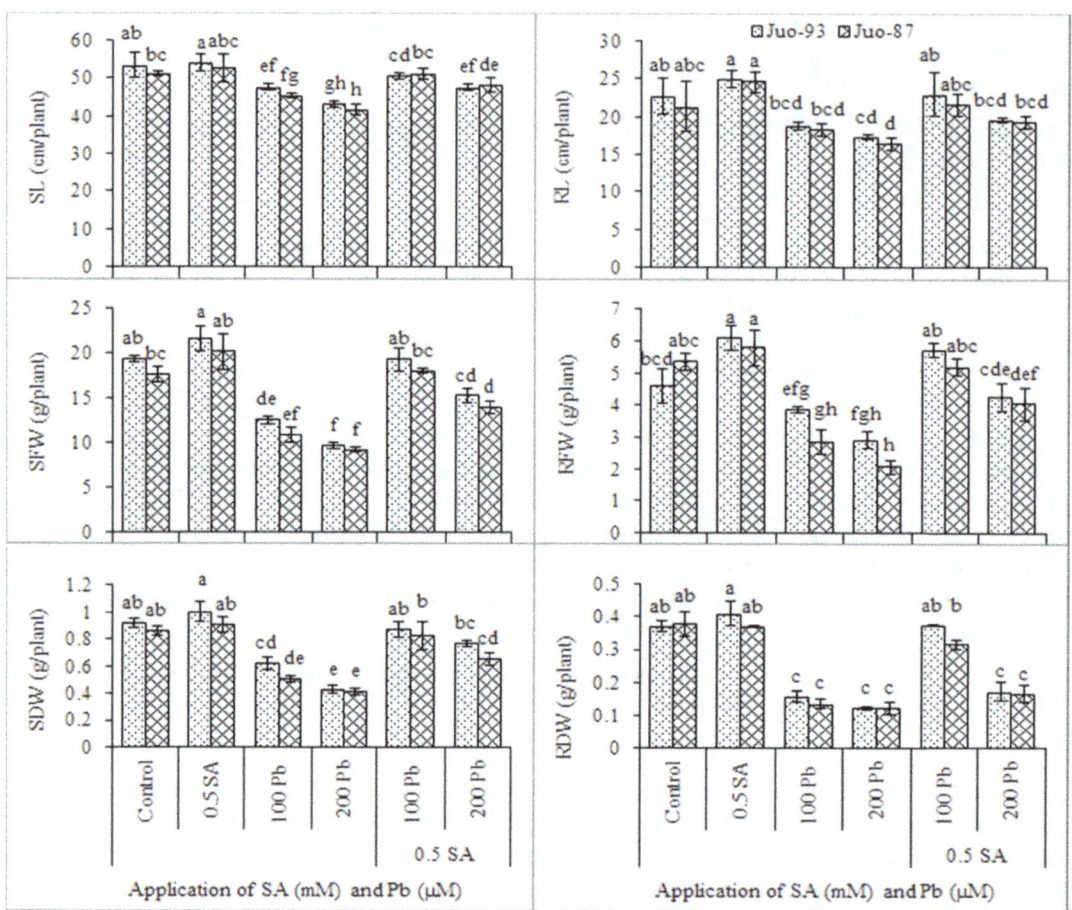

Figure 1. Graphical presentation of growth attributes of two barley genotypes (Juo-97 and Juo-87) treated with SA (0.5 mM) and Pb (100 and 200 µM)

Note. SL = shoot length, RL = root length, SFW, SDW = shoot fresh and dry weights, RFW, RDW = root fresh and dry weights. Histogram and Error bars represent Means±SD, while labels represent significance levels of treatments and genotypes (P < 0.05).

Juo-93 was found to be more efficient photosynthetically as compared to Juo-87 in control as well as treatment applications. SA broaden the leaf area in barley genotypes directly provide more surface area for photosynthetic activity. The chlorophyll and carotenoids contents increased in both by the increment of SA. Pb was proved to be toxic and directly affect the concentration of photosynthetic pigments in barley. Plants treated with 200 µM Pb had less chl-a, chl-b, chl-T in Juo-93 and Juo-87 as compared to 100 µM Pb. Application of SA in combination with both the treatments of Pb (100, 200 µM) enhanced the amount of chl-a, chl-b, chl-T and carotenoids, more in Juo-97 in comparison to Juo-87. Chl a/b ratio depicted that chl-a was more in both barley genotypes as compared to chl-b which shows that chl-b is more sensitive to Pb as compared chl-a (Figure 2).

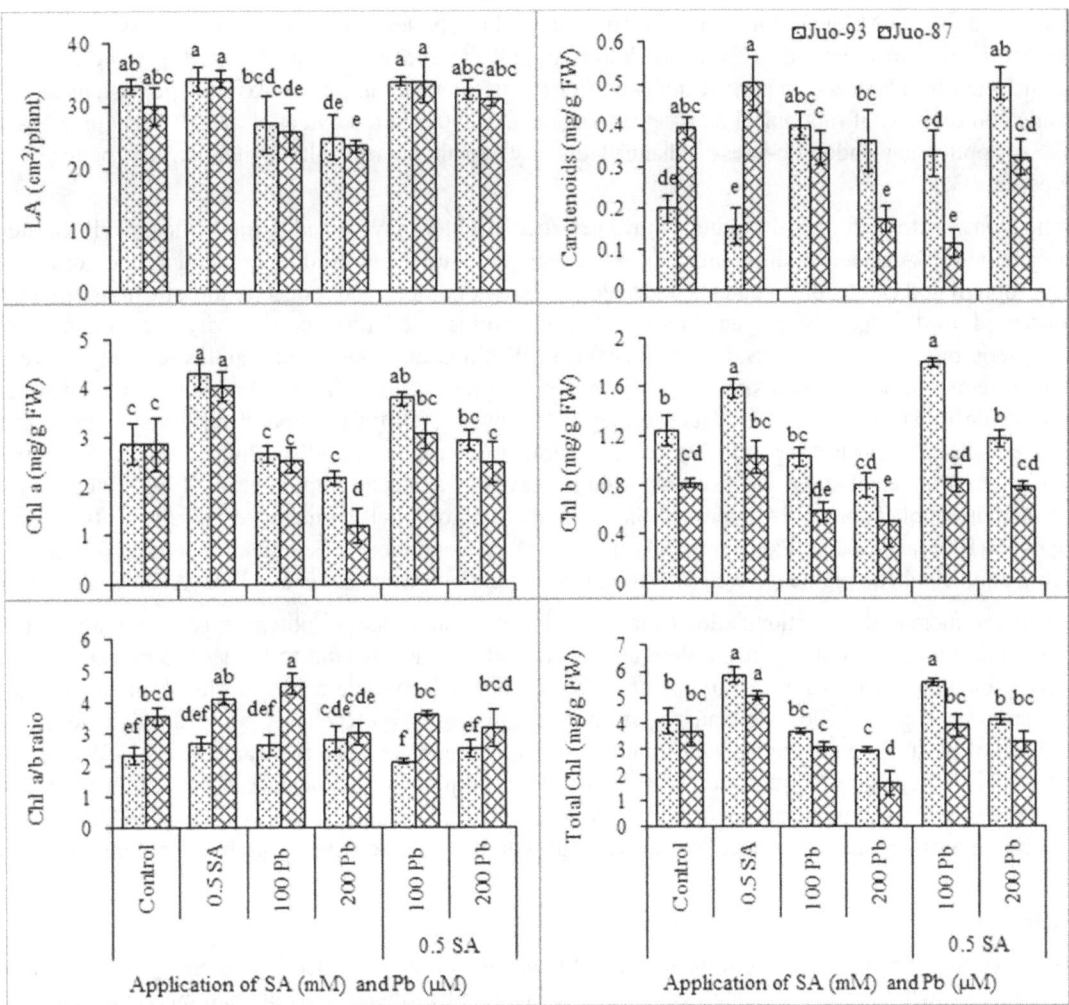

Figure 2. Graphical presentation of photosynthetic pigments of two barley genotypes (Juo-97 and Juo-87) treated with SA (0.5 mM) and Pb (100 and 200 µM)

Note. LA = leaf area, Chl a = Chlorophyll a, Chl b = Chlorophyll b. Histogram and Error bars represent Means±SD, while labels represent significance levels of treatments and genotypes (P < 0.05).

4. Discussion

Lead (Pb) concentration increases in cultivated soils due to irrigation with sewage effluents. The other major cause is the use of lead arsenate in pesticides (Paivoke, 2002). The exhausting of Pb from automobiles and industrial emissions pollutes atmosphere for plants as well as humans (Awofolu, 2004). Pb is non-essential heavy metal that sharply decreases crop productivity (Sengar et al., 2008). Pb is considered as protoplasmic poison which is taken up by plants from soil and compartmentalized in vacuolar sequestration by phytochelatins (Anjum et al., 2015; Sharma et al., 2016). The entry of Pb from leaves is dependent on leaf morphology and their ability to absorb Pb from the aerial sources (Sharma & Dubey, 2005) however, bulk of the Pb is taken up by plants from the soil through roots (Fahr et al., 2013). The translocation of Pb is highly restricted to upper parts of the plants through roots (Aziz et al., 2007; Tangahu et al., 2011). Calcareous or soil rich in phosphate precipitates Pb and make it non-toxic to plants (Li et al., 2013). Application of Pb decrease the shoot and root length of two barley genotypes as studied in Brassica (Pallavi & Rama, 2005; Ghani et al., 2016). Growth data result showed that 100 µM and 200 µM level of lead reduced the shoot and root length, their fresh and dry weight and leaf area per plant (Hussain et al., 2013). Reduction in the root biomass has also been reported in turnip and lettuce when treated with lead (Hassanein et al., 2013; Gupta & Sinha, 2007). Dry mass production of plant was dependant completely on light harvesting that was required for the generation of reducing powers (ATP and NADPH) during the photosynthesis light reaction and reducing powers use in dark reaction for photo-assimilate production (Taiz & Zeiger, 2010).

In present study, 200 μM level of lead was found toxic for both barley genotypes while applied level of SA (0.5 mM) in the rooting medium was effective in improving all the above mentioned growth parameters that were slightly affected by Pb-stress. Important indication of stress tolerance in plants was prolific system of roots by increasing cell division of root apical meristems (Yang et al., 2000; Shakirova et al., 2003; Vicente & Plasencia, 2011). SA application under Pb-stress enhanced the dry weight of root although relatively better in barley genotype Juo-93.

Pb application was toxic for photosynthetic attributes such as chlorophyll a and b, total chlorophyll, carotenoids, but application of SA enhanced the contents of these pigments. Juo-93 showed less reduction in chlorophyll a, b, total chlorophyll and carotenoids contents under Pb-stressed conditions and reduced ratio of Chlorophyll a/b was also observed in this genotype than Juo-87. Key determinant of final productivity was maintenance of photosynthetic pigments under stressful circumstances (Wahid et al., 2009). Carotenoids not only harvest light on photosystems but acts as antioxidant mainly protecting photosystems from the action of ROS generated in chloroplast (de Passcale et al., 2001; Huchzermeyer & Koyro, 2005). Chlorophyll ratio that is used as stress indicator enhanced with increase in heavy metal treatment (Monni et al., 2001). Improvement in chlorophylls ratio under environmental stress has been reported in leaves of spinach (Delfine et al., 1993). Under Pb stress rooting medium application of SA improved the contents of chlorophyll pigments, especially Chl-a, b and total chlorophylls (Khodary, 2004). Exogenous SA of 0.5 mM concentration is beneficial in improving growth and protecting photosynthetic pigments from salts toxicity as in barley (Metwally et al., 2003; Ananieva et al., 2004).

SA application increased the antioxidation capacity and protection of photosynthetic apparatus against Pb stress. Pb translocation through roots mainly follow apoplastic pathway. Roots tend to hinder the movement of Pb in apoplast by binding it with carboxyl groups of carbohydrates galacturonic acid and glucuronic acid in the cell wall (Sharma & Dubey, 2006). Pb transport from the soil to the root cells possibly through voltage gated Ca-channels of plasma membrane isolated from roots of wheat and corn plants (Marshall et al., 1994; Huang et al., 1994). This voltage gated Pb transport was blocked by nifidipine (a Ca-channel blocker) (Tomsig & Suszkiw, 1991). Exogenous SA might be involves in activation of nifedipine, blocking the Ca-channels for Pb translocation into the roots. SA resulted in improved growth of two barley genotypes by restricting the Pb at root levels.

5. Conclusion

In conclusion, SA (0.5 mM) is found to be beneficial in enhancing Pb tolerance in two barley genotypes. Juo-93 showed better growth than Juo-87 under lead and salicylic acid treatments. Anatomical, molecular and genomic studies of SA and Pb treated roots should be done in order to understand the tolerance mechanism and selected level of SA (0.5 mM) should be tested in field trails so that it can be recommended for metal toxic soils.

References

Ananieva, A. E., Christov, K. N., & Popova, L. P. (2004). Exogenous treatment with salicylic acid leads to increased antioxidant capacity of barley plants exposed to paraquat. *Journal of Plant Physiology, 161*, 319-328. https://doi.org/10.1078/0176-1617-01022

Anjum, N. A., Hasanuzzaman, M., Hossain, M. A., Thangavel, P., Roychoudhury, A., Gill, S. S., & Ahmad, I. (2015). Jacks of metal/metalloid chelation trade in plants-an overview. *Frontier Plant Sciences, 6*, 192. https://doi.org/10.3389/fpls.2015.00192

Arfan, M., Athar, H. R., & Ashraf, M. (2007). Does exogenous application of salicylic acid through the rooting medium modulate growth and photosynthetic capacity in two differently adapted spring wheat cultivars under salt stress? *Journal of Plant Physiology, 6*, 685-694. https://doi.org/10.1016/j.jplph.2006.05.010

Awofolu, O. R. (2004). Impact of automobile exhaust on levels of lead in a commercial food from bus terminal. *Journal of Applied Sciences and Environmental Management, 8*, 23-27.

Aziz, M. A., Ghafoor, A., Saifullah, H. R. A., & Sabir, M. (2007). Effect of glucose and acetic acid on Ni, Pb and Zn transformations in contaminated soil. *Pakistan Journal of Agricultural Sciences, 44*, 228-235.

Bazzaz, F. A., Carlson, R. W., & Rolfe, G. L. (1975). The inhibition of corn and sunflower photosynthesis by lead. *Physiologia Plantarum, 34*, 326-329. https://doi.org/10.1111/j.1399-3054.1975.tb03847.x

Belaid, A., & Morris, M. L. (1991). Wheat and Barley Production in Rainfed Marginal Environments of West Asia and North Africa. *Problems and prospects* (p. 91). CIMMYT Economics Working Paper.

Bosch, S. M., Penuelas, J., & Llusia, J. (2007). A deficiency in salicylic acid alters isoprenoid accumulation in water stressed transgenic Arabidopsis plants. *Plant Sciences, 172*, 756-762. https://doi.org/10.1016/j.plantsci.2006.12.005

Capuana, M. (2011). Heavy metals and woody plants-biotechnologies for phytoremediation. *Journal of Biogeosciences and Forestry, 4*, 7-15. https://doi.org/10.3832/ifor0555-004

Castro, R., Caetano, L., Ferreira, G., Padilha, P., Saeki, M., Zara, L., ... Castro, G. (2011). Banana peel applied to the solid phase extraction of copper and lead from river water. *Industrial and Engineering Chemistry and Reserach, 50*, 3446-3451. https://doi.org/10.1021/ie101499e

Ceccarelli, S., Grando, S., & Van Leur, J. (1987). Genetic diversity in barley landraces from Syria and Jordan. *Euphytica, 36*, 389-405. https://doi.org/10.1007/BF00041482

de Pascale, S., Maggio, A., Fogliano, V., Ambrosino, P., & Ritieni, A. (2001). Irrigation with saline water improves carotenoids content and antioxidant activity of tomato. *Horticultural and Science Biotechnology, 76*, 447-453. https://doi.org/10.1080/14620316.2001.11511392

Delfine, S., Alvino, A., Villiani, M. C., & Loreta, F. (1999). Restrictions to carbon dioxide conductance and photosynthesis in spinach leaves recovering from salt stress. *Plant Physiology, 119*, 1101-1106. https://doi.org/10.1104/pp.119.3.1101

Elwana, M. W. M., & El-Hamahmyb, M. A. M. (2009). Improved productivity and quality associated with salicylic acid application in greenhouse pepper. *Scientia Horticulturae, 122*, 521-526. https://doi.org/10.1016/j.scienta.2009.07.001

Eun, S. O., Youn, H. S., & Lee, Y. (2002). Lead disturbs microtubule organization in the root meristem of *Zea mays. Physiologia Plantarum, 110*, 357-365. https://doi.org/10.1111/j.1399-3054.2000.1100310.x

Fahr, M., Laplaze, L., Bendaou, N., Hocher, V., El Mzibri, M., Bogusz, D., & Smouni, A. (2013). Effect of lead on root growth. *Frontier Plant Sciences, 4*, 175. https://doi.org/10.3389/fpls.2013.00175

Gisbert, C., Ros, R., Deharo, A., Walker, D. J., Pilarbernal, M., Serrano, R., & Navarro-Avino, J. (2003). A plant genetically modified that accumulates Pb is especially promising for phytoremediation. *Biochemistry and Biophysics Reserach Communication, 303*, 440-445. https://doi.org/10.1016/S0006-291X(03)00349-8

Gunes, A., Inal, A., Alpaslan, M., Eraslan, F., Bagci, E. G., & Cicek, N. (2007). Salicylic acid induced changes on some physiological parameters symptomatic for oxidative stress and mineral nutrition in maize (*Zea mays* L.) grown under salinity. *Journal of Plant Physiology, 164*, 728-736. https://doi.org/10.1016/j.jplph.2005.12.009

Gupta, A. K., & Sinha, S. (2007). Assessment of different single extraction methods for the prediction of bioavailable metals from tannery waste contaminated soil to *Brassica juncea* L. Czern. (var. aibhav). *Journal of Hazardous Materials, 149*, 144-150. https://doi.org/10.1016/j.jhazmat.2007.03.062

Hall, J. L. (2002). Cellular mechanisms for heavy metal detoxification and tolerance. *Journal of Experimental Botany, 53*, 1-11. https://doi.org/10.1093/jexbot/53.366.1

Hassanein, R. A., Hashem, H. A., El-Deep, M. H., & Shouman, A. (2013). Soil contamination with heavy metals and its effect on growth, yield and physiological responses of vegetable crop plants (Turnip and Lettuce). *Journal of Stress Physiology and Biochemistry, 9*, 146-162.

Horvath, E., Szalai, G., & Janda, T. (2007). Induction of Abiotic Stress Tolerance by Salicylic Acid Signaling. *Journal of Plant Growth Regulation, 26*, 290-300. https://doi.org/10.1007/s00344-007-9017-4

Huang, J. W., Grunes, D. L., & Kochian, L. V. (1994). Voltage dependent Ca^{2+} influx into right-side-out plasmamembrane vesicles isolated from wheat roots: characteristics of a putative Ca^{2+} channel. *Proceeding of National Academy of Sciences of the United State of America, 91*, 3473-3477. https://doi.org/10.1073/pnas.91.8.3473

Huchzermeyer, B., & Koyro, H. W. (2005). Salt and drought stress effects on photosynthesis. In M. Pessarakli (Ed.), *Handbook of plant and crop stress* (2nd ed., pp. 751-778). Marcel Dekker Inc., New York, USA. https://doi.org/10.1201/9781420027877.ch39

Hussain, A., Abbas, N., Arshad, F., Akram, M., Khan, Z. I., Ahmad, K., Mirzael, F. (2013). Effects of diverse doses of Lead (Pb) on different growth attributes of *Zea mays* L. *Agriculture Science, 4*, 262-265.

Karlidag, H., Yildirim, E., & Turan, M. (2009). Salicylic acid ameliorates the adverse effect of salt stress on strawberry. *Scientia Agricola, 66*, 271-278. https://doi.org/10.1590/S0103-90162009000200006

Khan, W., Prithiviraj, B., & Smith, D. (2003). Photosynthetic response of corn and soybean to foliar application of salicylates. *Journal of Plant Physiology, 160*, 485-492. https://doi.org/10.1078/0176-1617-00865

Li, L., Xing, W., Scheckel, G., Xiang, G., Ji, H., & Li, H. (2013). Lead retention in a calcareous soil influenced by calcium and phosphate amendments. *Journal of Hazardous Materials, 15*, 262-250. https://doi.org/10.1016/j.jhazmat.2013.08.058

Li, Q., Cai, S., Mo, C., Chu, B., Peng, L., & Yang, F. (2010). Toxic effects of heavy metals and their accumulation in vegetables grown in a saline soil. *Ecotoxicology and Environmental Safety, 73*, 84-88. https://doi.org/10.1016/j.ecoenv.2009.09.002

Marashal, J., Corzo, A., Leigh, R. A., & Sanders, D. (1994). Membrane potential-dependent calcium transport in right-side-out plasma membrane vesicles from *Zea mays* L. roots. *The Plant Journal, 5*, 683-694. https://doi.org/10.1111/j.1365-313X.1994.00683.x

Meers, E., Van Slycken, S., Adriaensen, K., Ruttens, A., Vangronsveld, J., Laing, D., ... Tack, T. F. (2010). The use of bio-energy crops (*Zea mays*) for 'phytoattenuation' of heavy metals on moderately contaminated soils: a field experiment. *Chemosphere, 78*, 35-41. https://doi.org/10.1016/j.chemosphere.2009.08.015

Metwally, A., Finkmemeier, I., Georgi, M., & Dietz, K. J. (2003). Salicylic acid alleviates the Cd toxicity in barley seedlings. *Plant Physiology, 132*, 272-281. https://doi.org/10.1104/pp.102.018457

Monni, S., Uhlig, C., Junttila, O., Hansen, E., & Hynynen, J. (2001). Chemical composition and ecophysiological responses of *Empetrum nigrum* to above ground element application. *Environmental Pollution, 112*, 417-426. https://doi.org/10.1016/S0269-7491(00)00139-1

Najafian, S., Khoshkhui, M., Tavallali, V., & Saharkhiz, M. J. (2009). Effect of salicylic acid and salinity in thyme (*Thymus vulgaris* L.): Investigation on changes in gas exchange, water relations, and membrane stabilization and biomass accumulation. *Australian Journal of Basic and Applied Sciences, 3*, 2620-2626.

Paivoke, A. E. A. (2002). Soil lead alters phytase activity and mineral nutrient balance of *Pisum sativum*. *Environmental and Experimental Botany, 48*, 61-73. https://doi.org/10.1016/S0098-8472(02)00011-4

Prasad, D. D. K., & Prasad, A. R. K. (1987). Altered α-amino luvelinic acid metabolism by Pb and Hg in germinating seedling of Bajra (*Pennisetum typhoidenum*). *Journal of Plant Physiology, 127*, 241-249. https://doi.org/10.1016/S0176-1617(87)80143-8

Sengar, R. S., Gautam, M., Sengar, R. S., Garg, S. K., Sengar, K., & Chaudary, R. (2008). Lead stress effect on physiobiochemical activities of higher plants. *Rev. Environ. Contam. Toxicol., 196*, 73-93. https://doi.org/10.1007/978-0-387-78444-1_3

Shakirova, F. M., Sakhabutdinova, A. R., Bezrukova, M. V., Fatkhutdinova, R. A., & Fatkhutdinova, D. R. (2003). Changes in the hormonal status of wheat seedlings induced by salicylic acid and salinity. *Plant Science, 164*, 317-322. https://doi.org/10.1016/S0168-9452(02)00415-6

Sharma, P., & Dubey, R. S. (2005). Lead toxicity in plants. *Brazilian Journal of Plant Physiology, 17*, 35-52. https://doi.org/10.1590/S1677-04202005000100004

Sharma, S. S., Dietz, K.-J., & Mimura, T. (2016). Vacuolar compartmentalization as indispensable component of heavy metal detoxification in plants. *Plant Cell and Environment, 39*, 1112-1126. https://doi.org/10.1111/pce.12706

Simaei, M., Khavari-Nejad, R. A., & Bernard, F. (2012). Exogenous application of salicylic acid and nitric oxide on the ionic contents and enzymatic activities in NaCl-stressed soybean plants. *American Journal of Plant Sciences, 3*, 1495-1503. https://doi.org/10.4236/ajps.2012.310180

Taiz, L., & Zeiger, E. (2010). *Plant physiology* (5th ed.). Sinauer Associates Inc. Publishers, Sunderland.

Tangahu, B. V., Abdullah, S. R. S., Basri, H., Idris, M., Anuar, N., & Mukhlisin, M. (2011). A review on heavy metals (As, Pb, and Hg) uptake by plants through phytoremediation. *International Journal of Chemical Engineering, 2011*, 31. https://doi.org/10.1155/2011/939161

Tomsig, J. L., & Suszkiw, J. B. (1991). Permeation of Pb through calcium channels: fura-2 measurements of voltage and dihydropyridine sensitive Pb entry in isolated bovine chromaffin cells. *Biochimica et Biophysica Acta, 1069*, 197-200. https://doi.org/10.1016/0005-2736(91)90124-Q

Vicente, M. R. S., & Plasencia, J. (2011). Salicylic acid beyond defense: Its role in plant growth and development. *Journal of Experimental Botany, 62*, 3321-3338. https://doi.org/10.1093/jxb/err031

Wahid, A., Arshad, M., & Farooq, M. (2009). Cadmium phytotoxicity: Responses, mechanisms and mitigation strategies. In E. Lichtfouse (Ed.), *Advances in sustainable-book series* (Vol. 1, pp. 371-403). Springer, New York. https://doi.org/10.1007/978-1-4020-9654-9_17

Yang, Y. Y., Jung, J. Y., Song, W. Y., Suh, H. S., & Lee, Y. (2000). Identification of rice varieties with high tolerance or sensitivity to lead and characterization of the mechanism of tolerance. *Plant Physiology, 124*, 1019-1026. https://doi.org/10.1104/pp.124.3.1019

Permissions

List of Contributors

Bardales-Lozano Ricardo Manuel and Abanto-Rodriguez Carlos
Ret Bionorte (Multi-institutional Programme of Amazon), Brazil

Edvan Alves Chagas, Oscar Smiderle and Antonio Carlos Centeno Cordeiro
Empresa Brasileira de Pesquisa Agropecuária, Embrapa, Brazil

Pollyana Cardoso Chagas, Adamor Barbosa Mota Filho and Olisson Mesquita Souza
Universidade Federal de Roraima (UFRR), Brazil

Ana Veruska Cruz da Silva and Ana da Silva Ledo
Embrapa Coastal Tablelands, Aracaju, Sergipe, Brazil

Julie Anne Espíndola Amorim
State University of São Paulo 'Júlio de Mesquita Filho', Jaboticabal, São Paulo, Brazil

Marília Freitas de Vasconcelos Melo
State University of São Paulo 'Júlio de Mesquita Filho', Botucatu, São Paulo, Brazil

Allivia Rouse Carregosa Rabbani
Federal Institute of Bahia, Porto Seguro, Bahia, Brazil

Gislaine Aparecida Denardi Biasolo, Daniel Antonio Kucmanski, Sabrina Pinto Salamoni, João Elisandra Minotto and Cesar Milton Baratto
Nucleus of Biotechnology, University of the West of Santa Catarina, UNOESC, Videira, SC, Brazil

Peterson Pereira Gardin
Nucleus of Biotechnology, University of the West of Santa Catarina, UNOESC, Videira, SC, Brazil
Santa Catarina Agency for Agricultural Research and Rural Extension (Epagri), Videira, Brazl

Xicun Dong, Wenjian Li, Ruiyuan Liu and Wenting Gu
Department of Radiobiology, Institute of Modern Physics, Chinese Academy of Sciences, Lanzhou, China

Xia Yan
Key Laboratory of Inland River Ecohydrology, Cold and Arid Regions Environmental and Engineering Research Institute, Chinese Academy of Sciences, Lanzhou, China
Key Laboratory of Stress Physiology and Ecology in Cold and Arid Regions of Gansu Province, Cold and Arid Regions Environmental and Engineering Research Institute, Chinese Academy of Sciences, Lanzhou, China

Casinga Clérisse
International Institute of Tropical Agriculture, Bukavu, Sud-Kivu, Democratic Republic of Congo
Faculté des Sciences Agronomiques et Environnement, Université Evangélique en Afrique, Bukavu, Sud-Kivu, Democratic Republic of Congo

Haminosi Ghislain
Faculté des Sciences Agronomiques et Environnement, Université Evangélique en Afrique, Bukavu, Sud-Kivu, Democratic Republic of Congo

Cirimwami Legrand
Faculté des Sciences, Université de Kisangani, Kisangani, Province de la Tshopo, Democratic Republic of Congo

Pangirayi Tongoona and John Derera
African Centre for Crop Improvement, School of Agricultural, Earth and Environmental Sciences, College of Agriculture, Engineering and Science, University of KwaZulu Natal, Scottsville, Pietermaritzburg, Republic of South Africa

Lawrence Owere
Buginyanya Zonal Agricultural Research and Development Institute, Mbale, Uganda

Nelson Wanyera
National Semi-Arid Resources Research Institute, Serere, Private Bag, Soroti, Uganda

Adilson Nunes da Silva and Durval Dourado-Neto
Department of Crop Science, ESALQ/University of São Paulo, Piracicaba, SP, Brazil

Evandro Luiz Schoninger and Paulo Cesar Ocheuze Trivelin
Stable Isotopes Laboratory, CENA/University of São Paulo, Piracicaba, SP, Brazil

Victor Meriguetti Pinto and Klaus Reichardt
Soil Physics Laboratory, CENA/University of São Paulo, Piracicaba, SP, Brazil

Majid Shahi-Bajestani
Research and Innovation Center of Etka Organization, Tehran, Iran

Kheyzaran Dolatabadi
Department of Horticulture, Ferdowsi University of Mashhad, Mashhad, Iran

Paras Nath and Hemalatha Palanivel
College of Agriculture, Fisheries and Forestry, Fiji National University, Koronivia, Fiji

A. K. Panday and Akhilesh Kumar
Department of Entomology and Agricultural Zoology, Institute of Agricultural Sciences, Banaras Hindu University, Varanasi, India

Nasir S. A. Malik, Alberto Nuñez and Lindsay C. McKeever
USDA-ARS, ERRC, Wyndmoor, PA, USA

Lijuan Dang, Haizhen Zhao, Chong Zhang and Zhaoxin Lu
College of Food Science and Technology, Nanjing Agricultural University, Nanjing, Jiangsu, P.R. China

Meizhong Hu
College of Food Science and Technology, Nanjing Agricultural University, Nanjing, Jiangsu, P.R. China
Tongren Polytechnic College, Tongren Guizhou, P.R. China

Jiansheng Yu
Tongren Polytechnic College, Tongren Guizhou, P.R. China

Yingjian Lu
Department of Nutrition and Food Science, University of Maryland, College Park, Maryland, USA

Molly O. Akello, Felister Nzuve, Florence Olubayo and James Muthomi
Department of Plant Science and Crop Protection, University of Nairobi, Nairobi, Kenya

Godwin Macharia
Kenya Agricultural and Livestock Research Organization, Njoro, Kenya

Thomas Odong, Fred Kabi and Patrick Rubaihayo
Department of Agricultural Production, School of Agricultural Sciences, College of Agriculture and Environmental Sciences, Makerere University, Kampala, Uganda

Sally Chikuta
Department of Agricultural Production, School of Agricultural Sciences, College of Agriculture and Environmental Sciences, Makerere University, Kampala, Uganda
Department of Agriculture, Ministry of Agriculture, Chibombo District, Zambia

G. Lubadde
National Semi-Arid Resources Research Institute, Soroti, Uganda

P. Tongoona, J. Derera and J. Sibiya
University of KwaZulu Natal (UKZN), Pietermaritzburg Campus, Scottsville, South Africa

Paras Nath and Hemalatha Palanivel
College of Agriculture, Fisheries and Forestry, Fiji National University, Koronivia, Fiji

A. K. Panday and Akhilesh Kumar
Department of Entomology and Agricultural Zoology, BHU, Varanasi, India

A. B. Rai
Indian Institute of Vegetable Research, Varanasi, India

Csilla Deák, Veronika Anna Nagy, Réka Oszlányi and István Papp
Department of Plant Physiology and Plant Biochemistry, Faculty of Horticultural Science, Szent István University, Budapest, Hungary

Katalin Jäger and Beáta Barnabás
Agricultural Institute, Centre for Agricultural Research, Hungarian Academy of Sciences, Martonvásár Hungary

Cassio G. Freire
University Alto Vale do Rio do Peixe, St. Victor Baptista Adami, Center, Caçador, SC, Brazil

João P. P. Gardin and César M. Baratto
University of West of Santa Catarina, Unoesc, Videira, SC, Brazil

Renato L. Vieira
Agricultural Research and Rural Extension Company of Santa Catarina, Epagri, Experimental Station of Caçador, SC, Brazil

Juliane Mendes Lemos Blainski, Argus Cesar da Rocha Neto, Caroline Luiz, Márcio José Rossi and Robson Marcelo Di Piero
Center of Agricultural Sciences, Federal University of Santa Catarina, Florianópolis, Santa Catarina, Brazil

Wang Yue, Gao Qingrong and Zhang Weidong
State Key Laboratory of Crop Sciences, Shandong Agricultural University, Taian City, Shandong Province, China
Genetic and Breeding Department, Agronomy College, Shandong Agricultural University, Taian City Shandong Province, China

Sun Fulai, Zhang Yanxia and Wang Nan
Bureau of Agriculture, Binzhou City, Shandong Province, China

Nariman S. Ahmad, Shadia H. S. Kareem, Kamil M. Mustafa and Dastan A. Ahmad
Crop Science Department, Faculty of Agricultural Sciences, University of Sulaimani, Kurdistan, Iraq

Mirfat Ahmad Hasan Salahuddin and Erny Sabrina Mohd Noor
Agrobiodiversity and Environment Research Centre, MARDI Headquarters, Selangor, Malaysia

Zaulia Othman and Joanna Cho Lee Ying
Horticultural Research Centre, MARDI Headquarters, Selangor, Malaysia

Salma Idris
Seed and Genebank Centre, MARDI Headquarters, Selangor, Malaysia

Yu Ge, Zhaoxi Zhou and Weihong Ma
Haikou Experimental Station, Chinese Academy of Tropical Agricultural Sciences, Hainan, China

Xiongyuan Si
Biotechnology Center, Anhui Agricultural University, Anhui, China

Jianqiu Cao
College of Horticulture and Landscape Architecture, Hainan University, Hainan, China

Wenlin Wang
Guangxi South Subtropical Agricultural Science Research Institute, Guangxi, China

Taskeen Arshad, Nazimah Maqbool, Farrukh Javed and Abdul Wahid
Department of Botany, University of Agriculture, Faisalabad, Pakistan

Muhammad Usman Arshad
Department of Plant Breeding and Genetics, University of Agriculture, Faisalabad, Pakistan

Index

www.ingramcontent.com/pod-product-compliance
Lightning Source LLC
Chambersburg PA
CBHW080412190526
45161CB00003B/216